EXS 97

Plant Systems Biology

Edited by Sacha Baginsky and Alisdair R. Fernie

Birkhäuser Verlag
Basel • Boston • Berlin

Editors

Sacha Baginsky
Institute of Plant Sciences
Swiss Federal Institute of Technology
ETH Zentrum, LFW E
8092 Zürich
Switzerland

Alisdair R. Fernie
MPI for Molecular Plant Physiology
Am Mühlenberg 1
14476 Golm
Germany

Library of Congress Control Number: 2006937911

Bibliographic information published by Die Deutsche Bibliothek
Die Deutsche Bibliothek lists this publication in the Deutsche Nationalbibliografie; detailed bibliographic data is available in the Internet at <http://dnb.ddb.de>.

ISBN 13: 978-3-7643-7261-3 Birkhäuser Verlag, Basel – Boston – Berlin

The publisher and editor can give no guarantee for the information on drug dosage and administration contained in this publication. The respective user must check its accuracy by consulting other sources of reference in each individual case.
The use of registered names, trademarks etc. in this publication, even if not identified as such, does not imply that they are exempt from the relevant protective laws and regulations or free for general use.
This work is subject to copyright. All rights are reserved, whether the whole or part of the material is concerned, specifically the rights of translation, reprinting, re-use of illustrations, recitation, broadcasting, reproduction on microfilms or in other ways, and storage in data banks. For any kind of use, permission of the copyright owner must be obtained.

© 2007 Birkhäuser Verlag, P.O. Box 133, CH-4010 Basel, Switzerland
Part of Springer Science+Business Media
Printed on acid-free paper produced from chlorine-free pulp. TCF ∞
Cover illustration: see page 151. With friendly permission of Sven Schuchardt
Typesetting: Fotosatz-Service Köhler GmbH, Würzburg

Printed in Germany
ISBN 10: 3-7643-7261-3
ISBN 13: 978-3-7643-7261-3

e-ISBN 10: 3-7643-7439-X
e-ISBN 13: 978-3-7643-7439-6

9 8 7 6 5 4 3 2 1

www.birkhauser.ch

Preface

Given that the opening chapter by Bruggeman et al. will provide an introduction to systems biology, it is not our intention in this preface to cover this; rather we will give an overview of the contents of this book and outline our reasoning for compiling it in the way that we have. This book is intended to give a comprehensive overview of the research field, which given its diversity, should have appeal to graduate students wanting to broaden their knowledge as well as to specialists of any of the genomic sub-disciplines. The overall structure of our book is inspired by the different consequences of gene expression, ranging from DNA, via RNA to proteins and metabolites, before the last chapters dealing with computational considerations concerning data standardization, storage, distribution and finally integration.

Given the origins of systems biology, the opening chapter deals with theoretical and mathematical approaches toward understanding the cellular hierarchy of biological systems with the chapters that follow dealing either with the acquisition of multi-factorial datasets or with their subsequent bioinformatical and biological interpretation. First among these, the chapter by Causse and Rothan, explains the collection or generation and identification of genetic variance suitable for systems biology. Herein both reverse (genotype to phenotype) and forward (phenotype to genotype) genetic strategies are discussed as methods of studying the effect of allelic variation as a method of perturbing biological systems with particular focus on quantitative genetics approaches and on the technological advances that will likely facilitate systems biology in plants. The third chapter by Foyer et al. utilizes the signaling functions of ascorbate to present a case study for experimental and interpretational analysis of global transcription profiling. This chapter thus provides three important functions firstly providing an important example of the use of environmental perturbation as a method to study plant systems, secondly presenting important considerations that need to be borne in mind both in experimental planning and equally importantly in data analysis of microarray experimentation and finally illustrating how biological information can be extracted from such studies. As an alternative experimental strategy, collection and evaluation of experimental data across a time course to allow an analysis of the kinetic response to a given perturbation. In a complementary chapter to Foyer et al., Hennig and Köhler explore this approach using case studies involving the analysis of the function of the transcription factors PHERES and LEAFY. The approach they introduce is the complementation of mutants by reintroduction of an unmutated copy of the gene in question under the control of an inducible promoter. Hennig and Köhler lay a special emphasis on discussing experimental design strategies to accept, or reject a hypothesis generated from the high-throughput data. The final chapter concerned with transcriptional regulation that by Sundaresan, describes advances in the understanding of RNA interference presenting methods for their identification via computational analysis as well as discussing strategies to experimentally verify their function. RNA interference introduces an additional layer of regulation into a cellular system and may have an impact on how we understand RNA stability and posttranscriptional regulation in a complex biological "system".

Jumping to the next level of the cellular hierarchy, the subsequent two chapters deal with the analysis and characterization of proteins – those molecules that determine the metabolic and regulatory capacities of cells. Their high-throughput analysis has become possible by two parallel scientific achievements: the acquisition of genome information and the development of soft peptide ionization techniques for mass spectrometric applications. Brunner et al.'s chapter provides a thorough overview of different methods for the quantification of proteins, e.g. by comparing gel- and mass-spectral based proteomics methods for the differential display of proteins in two different samples and for their accurate quantification. Schuchardt and Sickmann's chapter provides a thorough overview of state-of-the art mass spectrometry (MS) equipment that is currently available for systematic protein analyses. Because mass spectrometric methods differ considerably each method has specific strength and weaknesses that determine its applicability to special experimental strategies. Therefore, this chapter has a special emphasis on the discussion of MS equipment for a certain experimental design. It furthermore covers the analysis of posttranslational modifications using phosphorylation as an example and lastly touches upon emerging issues of data analysis in proteomics.

The chapters by Steinhauser/Kopka and Sumner et al. deal with experimental considerations for measuring primary and secondary metabolites, respectively. Steinhauser and Kopka provide an overview of the requirements for establishing a GC-MS based metabolite profiling platform covering the entire experimental time frame from conceptual design through sample extraction and analysis to data analysis. The chapter additionally addresses the issue of quality by defining the widely used terminologies of fingerprinting, profiling and target application. Sumner et al. focus on the larger and more chemically diverse secondary metabolites. In this chapter Sumner and co-authors discuss the current state of the art in identifying and quantifying secondary metabolites of plant origin, and highlight the difficulties in doing so, as well as discussing potential solutions for the future. While the two preceding chapters are concerned with analysis of steady-state levels of metabolites, Dieuaide-Noubhani et al.'s chapter deals with the considerably more complex task of dynamic analysis of metabolism using techniques of metabolite flux analysis. The chapter covers both theoretical and experimental aspects of flux determination and also reviews recent key papers that attempt to integrate both experimental data and bioinfomatic modeling in order to allow a more comprehensive understanding of plant metabolism.

Having covered protocols for data acquisition the final module of this book will focus on what to do with global data sets post-acquisition. The first chapter in this section that of Nikiforova and Willmitzer describes the utility of correlation network visualisation and analysis utilizing the authors own studies on plant responses to nutrient deprivation to illustrate the power of this tool when applied to post-genomic datasets. The serious problem of non-standard ontology and the current status in adapting to a common language in the naming of both genes and proteins is discussed in Ahrens et al.'s chapter. As part of this issue, the authors highlight strategies to make data available to a wide scientific community in order to promote data distribution for the benefit of research progress.

The final chapters are both concerned with the integration of data from several different multi-factorial experiments and using them to model a biological system such that its reaction on a perturbation can be precisely predicted. Both of these chapters, by Steinfath et al. and by Schöner et al. highlight potentials and challenges of current modeling strategies and comment on their ability to retrieve biologically meaningful data. These final two chapters provide the full circle to the opening chapter, in wrapping up more theoretical considerations about biological systems that involve mathematical models and novel computer algorithms. We sincerely hope that our book presents an informative basic overview of the emergent discipline of systems biology from both experimental and theoretic perspectives and we both hope you enjoy reading it – we certainly did!

Sacha Baginsky
Alisdair Fernie   October 2006

Contents

List of contributors XI

Preface

*Frank J. Bruggeman, Jorrit J. Hornberg, Fred C. Boogerd
and Hans V. Westerhoff*
Introduction to systems biology   1

Christophe Rothan and Mathilde Causse
Natural and artificially induced genetic variability in crop
and model plant species for plant systems biology 21

Christine H. Foyer, Guy Kiddle and Paul Verrier
Transcriptional profiling approaches to understanding how plants
regulate growth and defence: A case study illustrated by analysis
of the role of vitamin C 55

Lars Hennig and Claudia Köhler
Case studies for transcriptional profiling 87

Cameron Johnson and Venkatesan Sundaresan
Regulatory small RNAs in plants 99

Erich Brunner, Bertran Gerrits, Mike Scott and Bernd Roschitzki
Differential display and protein quantification 115

Sven Schuchardt and Albert Sickmann
Protein identification using mass spectrometry: A method overview . . 141

Dirk Steinhauser and Joachim Kopka
Methods, applications and concepts of metabolite profiling:
Primary metabolism 171

*Lloyd W. Sumner, David V. Huhman, Ewa Urbanczyk-Wochniak
and Zhentian Lei*
Methods, applications and concepts of metabolite profiling:
Secondary metabolism 195

*Martine Dieuaide-Noubhani, Ana-Paula Alonso, Dominique Rolin,
Wolfgang Eisenreich and Philippe Raymond*
Metabolic flux analysis: Recent advances in carbon metabolism in plants . . 213

Victoria J. Nikiforova and Lothar Willmitzer
Network visualization and network analysis 245

*Christian H. Ahrens, Ulrich Wagner, Hubert K. Rehrauer, Can Türker
and Ralph Schlapbach*
Current challenges and approaches for the synergistic use of systems
biology data in the scientific community 277

*Matthias Steinfath, Dirk Repsilber, Matthias Scholz, Dirk Walther
and Joachim Selbig*
Integrated data analysis for genome-wide research 309

*Daniel Schöner, Simon Barkow, Stefan Bleuler, Anja Wille,
Philip Zimmermann, Peter Bühlmann, Wilhelm Gruissem and Eckart Zitzler*
Network analysis of systems elements 331

Index 353

List of contributors

Ana-Paula Alonso, Department of Plant Biology, Michigan State University, 166 Plant Biology Building, East Lansing, MI 48824, USA

Christian H. Ahrens, Functional Genomics Center Zürich, Winterthurerstrasse 190, Y32H66, 8057 Zürich, Switzerland; e-mail: christian.ahrens@fgcz.ethz.ch

Simon Barkow, Computer Engineering and Networks Laboratory, Swiss Federal Institute of Technology (ETH), Gloriastrasse 35, 8092 Zürich, Switzerland

Stefan Bleuler, Computer Engineering and Networks Laboratory, Swiss Federal Institute of Technology (ETH), Gloriastrasse 35, 8092 Zürich, Switzerland

Fred C. Boogerd, Molecular Cell Physiology, Institute for Molecular Cell Biology, BioCentrum Amsterdam, Faculty of Earth and Life Sciences, Vrije Universiteit, De Boelelaan 1085, 1081 HV Amsterdam, The Netherlands

Peter Bühlmann, Seminar for Statistics, Swiss Federal Institute of Technology (ETH), Leonhardstrasse 27, 8092 Zürich, Switzerland

Frank J. Bruggeman, Molecular Cell Physiology, Institute for Molecular Cell Biology, BioCentrum Amsterdam, Faculty of Earth and Life Sciences, Vrije Universiteit, De Boelelaan 1085, 1081 HV Amsterdam, The Netherlands; and Systems Biology Group, Manchester Centre for Integrative Systems Biology, Manchester Interdisciplinary Biocentre, School of Chemical Engineering and Analytical Science, University of Manchester, 131 Princess Street, Manchester M1 7ND, UK; e-mail: frank.bruggeman@falw.vu.nl

Erich Brunner, Institute of Molecular Biology, University of Zürich, Winterthurerstr. 190, 8057 Zürich, Switzerland; e-mail: erich.brunner@molbio.unizh.ch

Mathilde Causse, INRA-UR 1052, Unité de Génétique et Amélioration des Fruits et Légumes, BP 94, F-84143 Montfavet cedex, France; e-mail: Mathilde.Causse@avignon.inra.fr

Martine Dieuaide-Noubhani, INRA Université Bordeaux 2, UMR 619 "Biologie du Fruit", IBVI, BP 81, 33883 Villenave d'Ornon Cedex, France

Wolfgang Eisenreich, Lehrstuhl für Organische Chemie und Biochemie, Technische Universität München, Lichtenbergstraße 4, 85747 Garching, Germany

Christine H. Foyer, Crop Performance and Improvement Division, Rothamsted Research, Harpenden, Hertfordshire, AL5 2JQ, UK; e-mail: christine.foyer@bbsrc.ac.uk

Bertran Gerrits, Functional Genomics Center Zürich, Winterthurerstr. 190, 8057 Zürich, Switzerland

Wilhelm Gruissem, Plant Biotechnology, Institute of Plant Sciences, Swiss Federal Institute of Technology (ETH), Rämistrasse 2, 8092 Zürich, Switzerland

Lars Hennig, Swiss Federal Institute of Technology (ETH) Zürich, Plant Biotechnology, ETH Zentrum, LFW E47, Universitätstr. 2, 8092 Zürich, Switzerland; e-mail: lhennig@ethz.ch

Jorrit J. Hornberg, Molecular Cell Physiology, Institute for Molecular Cell Biology, BioCentrum Amsterdam, Faculty of Earth and Life Sciences, Vrije Universiteit, De Boelelaan 1085, 1081 HV Amsterdam, The Netherlands

David V. Huhman, Plant Biology Division, The Samuel Roberts Noble Foundation 2510 Sam Noble Parkway, Ardmore, OK 73401, USA

Guy Kiddle, Crop Performance and Improvement Division, Rothamsted Research, Harpenden, Hertfordshire, AL5 2JQ, UK

Claudia Köhler, Swiss Federal Institute of Technology (ETH) Zürich, Plant Developmental Biology, ETH Zentrum, LFW E 53.2, Universitätstr. 2, 8092 Zürich, Switzerland

Joachim Kopka, Max Planck Institute of Molecular Plant Physiology, Am Muehlenberg 1, 14476 Potsdam-Golm, Germany; e-mail: kopka@mpimp-golm.mpg.de

Zhentian Lei, Plant Biology Division, The Samuel Roberts Noble Foundation 2510 Sam Noble Parkway, Ardmore, OK 73401, USA

Victoria J. Nikiforova, Max-Planck-Institut für Molekulare Pflanzenphysiologie, Am Mühlenberg 1, 14476 Potsdam-Golm, Germany; e-mail: nikiforova@mpimp-golm.mpg.de

Philippe Raymond, INRA Université Bordeaux 2, UMR 619 "Biologie du Fruit", IBVI, BP 81, 33883 Villenave d'Ornon Cedex, France ; e-mail: raymond@bordeaux.inra.fr

Hubert K. Rehrauer, Functional Genomics Center Zürich, Winterthurerstrasse 190, Y32H66, 8057 Zürich, Switzerland

Dirk Repsilber, Institute for Biology and Biochemistry, University Potsdam c/o MPI-MP, Am Mühlenberg 1, 14476 Potsdam-Golm, Germany; e-mail: repsilber@mpimp-golm.mpg.de

Dominique Rolin, INRA Université Bordeaux 2, UMR 619 "Biologie du Fruit", IBVI, BP 81, 33883 Villenave d'Ornon Cedex, France

Bernd Roschitzki, Functional Genomics Center Zürich, Winterthurerstr. 190, 8057 Zürich, Switzerland

Christophe Rothan, INRA-UMR 619 Biologie des Fruits, IBVI-INRA Bordeaux, BP 81, 71 Av. Edouard Bourlaux, 33883 Villenave d'Ornon cedex, France; e-mail: rothan@bordeaux.inra.fr

Ralph Schlapbach, Functional Genomics Center Zürich, Winterthurerstrasse 190, Y32H66, 8057 Zürich, Switzerland

Matthias Scholz, Max Planck Institute of Molecular Plant Physiology, Am Mühlenberg 1, 14476 Potsdam-Golm, Germany, current address: ZIK-Center for functional Genomics, University of Greifswald, F.-L.-Jahn-Str. 15, 17487 Greifswald, Germany; e-mail: matthias.scholz@uni-greifswald.de

Daniel Schöner, Plant Biotechnology, Institute of Plant Sciences, Swiss Federal Institute of Technology, Rämistrasse 2, 8092 Zürich, Switzerland

List of contributors

Sven Schuchardt, Fraunhofer Institute of Toxicology and Experimental Medicine, Drug Research and Medical Biotechnology, Nikolai-Fuchs-Strasse 1, 30625 Hannover, Germany; e-mail: sven.schuchardt@item.fraunhofer.de

Mike Scott, Functional Genomics Center Zürich, Winterthurerstr. 190, 8057 Zürich, Switzerland

Joachim Selbig, Institute for Biology and Biochemistry, University Potsdam and Max Planck Institute of Molecular Plant Physiology, Am Mühlenberg 1, 14476 Potsdam-Golm, Germany; e-mail: Selbig@mpimp-golm.mpg.de

Albert Sickmann, Rudolf-Virchow-Center, DFG-Research Center for Experimental Biomedicine, University of Würzburg, Versbacherstr. 9, 97078, Würzburg, Germany; e-mail: Albert.Sickmann@virchow.uni-wuerzburg.de

Matthias Steinfath, Institute for Biology and Biochemistry, University Potsdam c/o MPI-MP, Am Mühlenberg 1, 14476 Potsdam-Golm, Germany; e-mail: steinfath@mpimp-golm.mpg.de

Dirk Steinhauser, Max Planck Institute of Molecular Plant Physiology, Am Muehlenberg 1, 14476 Potsdam-Golm, Germany

Lloyd W. Sumner, Plant Biology Division, The Samuel Roberts Noble Foundation 2510 Sam Noble Parkway, Ardmore, OK 73401, USA; e-mail: lwsumner@noble.org

Venkatesan Sundaresan, Plant Biology and Plant Sciences University of California, Street?? Davis, CA 95616, USA; e-mail: sundar@ucdavis.edu

Can Türker, Functional Genomics Center Zürich, Winterthurerstrasse 190, Y32H66, 8057 Zürich, Switzerland

Ewa Urbanczyk-Wochniak, Plant Biology Division, The Samuel Roberts Noble Foundation, 2510 Sam Noble Parkway, Ardmore, OK 73401, USA

Paul Verrier, Biomathematics and Bioinformatics Division, Rothamsted Research, Harpenden, Hertfordshire, AL5 2JQ, UK

Ulrich Wagner, Functional Genomics Center Zürich, Winterthurerstrasse 190, Y32H66, 8057 Zürich, Switzerland

Dirk Walther, Max Planck Institute of Molecular Plant Physiology, Am Mühlenberg 1, 14476 Potsdam-Golm, Germany; e-mail: walther@mpimp-golm.mpg.de

Hans V. Westerhoff, Molecular Cell Physiology, Institute for Molecular Cell Biology, BioCentrum Amsterdam, Faculty of Earth and Life Sciences, Vrije Universiteit, De Boelelaan 1085, 1081 HV Amsterdam, The Netherlands; and Systems Biology Group, Manchester Centre for Integrative Systems Biology, Manchester Interdisciplinary Biocentre, School of Chemical Engineering and Analytical Science, University of Manchester, 131 Princess Street, Manchester M1 7ND, UK

Anja Wille, Seminar for Statistics, Swiss Federal Institute of Technology (UETH), Leonhardstrasse 27, 8092 Zürich, Switzerland

Lothar Willmitzer, Max-Planck-Institut für Molekulare Pflanzenphysiologie, Am Mühlenberg 1, 14476 Potsdam-Golm, Germany

Eckart Zitzler, Computer Engineering and Networks Laboratory, Swiss Federal Institute of Technology (ETH), Gloriastrasse 35, 8092 Zürich, Switzerland; e-mail: zitzler@tik.ee.ethz.ch

Plant Systems Biology
Edited by Sacha Baginsky and Alisdair R. Fernie
© 2007 Birkhäuser Verlag/Switzerland

Introduction to systems biology

Frank J. Bruggeman[1,2], Jorrit J. Hornberg[1], Fred C. Boogerd[1] and Hans V. Westerhoff[1,2]

[1] Molecular Cell Physiology, Institute for Molecular Cell Biology, BioCentrum Amsterdam, Faculty of Earth and Life Sciences, Vrije Universiteit, De Boelelaan 1085, NL-1081 HV, Amsterdam, The Netherlands
[2] Systems Biology Group, Manchester Centre for Integrative Systems Biology, Manchester Interdisciplinary Biocentre, School of Chemical Engineering and Analytical Science, University of Manchester, 131 Princess Street, Manchester M1 7ND, UK

Abstract

The developments in the molecular biosciences have made possible a shift to combined molecular and system-level approaches to biological research under the name of *Systems Biology*. It integrates many types of molecular knowledge, which can best be achieved by the synergistic use of models and experimental data. Many different types of modeling approaches are useful depending on the amount and quality of the molecular data available and the purpose of the model. Analysis of such models and the structure of molecular networks have led to the discovery of principles of cell functioning overarching single species. Two main approaches of systems biology can be distinguished. *Top-down* systems biology is a method to characterize cells using system-wide data originating from the Omics in combination with modeling. Those models are often phenomenological but serve to discover new insights into the molecular network under study. *Bottom-up* systems biology does not start with data but with a detailed model of a molecular network on the basis of its molecular properties. In this approach, molecular networks can be quantitatively studied leading to predictive models that can be applied in drug design and optimization of product formation in bioengineering. In this chapter we introduce analysis of molecular network by use of models, the two approaches to systems biology, and we shall discuss a number of examples of recent successes in systems biology.

From a molecular to a systems perspective in biology

In the last century many of the molecular details of living organisms have been deciphered. The identification of molecular constituents was greatly speeded up by genome sequencing. Many of the processes occurring in cells have been characterized. For simple organisms, such as *Escherichia coli* or yeast, large parts of the metabolic network structure, the operon structure and their transcriptional regulators are now known [1–3].

This knowledge allows for combined molecular and system-level studies applying a synergistic approach involving modeling, theory, and experiment under the name of *Systems Biology*. Dynamics of entire cells cannot yet be modeled with detailed kinetic models but we anticipate that this may happen within a decade or two. Detailed stoichiometric models of entire organisms have already been studied [1, 4–6]. Those cannot deal with the dynamics of cells for they do not contain any kinetic data; they focus on distributions of steady-state flux or study network organization. However, the dynamics of a number of subsystems of cells have already been modeled in great detail (e.g., [7–12]). Such models describe the molecular mechanisms operative in cells. They contain all the molecular knowledge available of the systems under study; they are near replica of the real system. We term such models *silicon-cell models*. They allow for a 'completeness' test of our knowledge (e.g., [7, 9, 10]). This form of scientific rigidity is unprecedented in biology. In addition, those models allow for analysis of the system *in silico* in ways not (yet) achievable in the laboratory (e.g., [13, 14]). More importantly, they may allow for rational strategies of drug design in medicine and optimization of product formation in bioengineering (e.g., [11, 15, 16]). Also more qualitative models are of importance in systems biological approaches to illustrate principles (re-) occurring in molecular networks [17, 18]. Such models may be model reductions of complicated silicon-cell models to facilitate explanation of phenomena by focusing on the core mechanism responsible for some phenomenon of interest. In other cases, such models may be approximations of the real system to describe phenomena too complicated to grasp without usage of mathematical modeling [14, 18, 19].

Systems biology aims to provide a firm link between the molecular disciplines in biology, such as genetics, molecular biology, biochemistry, enzymology, and biophysics, and the disciplines within biology that study entire organisms, i.e., cell biology and physiology [20, 21]. It does so by quantitatively characterizing the molecular mechanisms in organisms on a molecular and system level. Such combined molecular and system-level studies are therefore a sort of unification; they 'unify' the molecular characterization of organisms with their physiological – behavioral or functional – characterization. That is, they indicate how the properties of organisms are brought about by the properties of their molecular constitution and organization and how the system can be altered molecularly to have it behave as desired.

Many associate this kind of strategy with reduction, i.e., that properties of organisms are reduced to properties of molecules; that properties of organisms are *just* properties of molecules. We disagree with such kinds of statements [22]. Rather, the type of reduction achieved here is that of mechanistic explanation [23, 24]. Properties of organisms that are unique to organisms – not found on the level of single molecules or simpler systems thereof – are explained in terms of the molecular mechanisms that manifest those properties. Accordingly, organisms display emergent behaviors not displayed by any of their molecules in isolation, such as adaptation, growth, robustness, and natural selection [22, 25]. Those emergent system properties do depend on the properties of the molecular constituents *but* even more

Introduction to systems biology

so on how they interact in the organism to function in mechanisms. Without the latter knowledge the emergent properties are not understood.

From a nested-level-of-organization point of view, systems biology is an *interlevel* approach to biology rather than an *intralevel* approach, which is more characteristic of molecular biology and genetics [22]. Comparing to physics, systems biology shares more similarities with statistical thermodynamics than with macroscopic thermodynamics, which is more a mirror image of physiology or molecular biology. Contrast the temperature of a system of particles, perceived in statistical thermodynamics as the average kinetic energy of the particles, which is an intrinsically interlevel concept, with the interpretation of the ideal gas law (pV=nRT) in macroscopic thermodynamics that merely expresses a relation among system properties and is therefore intralevel. Interlevel approaches are not so common in science [26] but are central to studies of complex systems [23, 27].

Organismal properties are not properties of molecules but of networks of molecules

A characterization of a (resting) bag of billiard balls leads to a list of many properties. None of them depend on how the billiard balls are organized within the bag. Many of them are retrievable by superposition of the properties of isolated individual billiard balls. Actually, according to any reasonable sense of organization, the billiard balls in the bag cannot be considered organized relative to each other. Even if all blue ones are on top it does not matter, for many of the characterizing properties of a bag of billiard balls do not depend on the color of the balls. This example, simple as it may be, indicates a number of interesting points. For instance, not all systems have properties that depend on the organization of their constituents. One could then argue that this is obviously so since the billiard balls are all the same; therefore one cannot speak of organization in this case. But changing their color does not have an effect, indicating that only some properties of parts matter for the systems characterization in terms of its organization – or in terms of its mechanisms.

Obviously, cells are *not* comparable to a bag of billiards balls in any meaningful biological sense. Cells *do* display behaviors that depend on their molecular organization. They consist of molecules of different types that occur in different abundances depending on conditions and history. Those molecules engage in interactions of high specificity; not all molecules interact and if some of them do interact then often by varying degree. The interactions and their effects are not retrievable from the isolated molecules without considering cells as molecular networks; that is, without integrating all the molecular properties, for instance by using mathematical models [22, 25]. This does not mean that all properties of cells depend on their molecular organization. For instance, their mass, total energy and the number of molecular constituents do not.

Let's consider a simple molecular network to make the dominant role of molecular organization in determining the properties of cells more transparent. Along the way, we shall introduce a number of general characteristics of cells perceived as molecular networks. The network we consider consists of enzyme 1 and 2. Enzyme 1 produces X out of S whereas enzyme 2 has X as a substrate and produces P:

$$S \xleftrightarrow{\text{enzyme 1}} X \xleftrightarrow{\text{enzyme 2}} P$$

We shall describe it in terms of a kinetic model (e.g., [28]); a type of modeling used often in systems biology; for examples see JWS online at www.jjj.bio.vu.nl [29, 30]. The system properties of interest are the concentration of X and the flux J through the pathway at steady state. Steady state is defined as the state where X remains constant while a net flux runs through the pathway. In contrast, an equilibrium state is defined as a net flux of zero while X is constant. Both enzymes have many different properties but only their kinetic properties matter for X and J at steady state; that is, their 3D-structure, gene sequence, or weight do not matter.

In terms of kinetic properties, the rate with which enzyme 1 produces X and enzyme 2 consumes X is described by the following reversible Michaelis-Menten rate equations [31]:

$$v_1 = \frac{V_{MAX,1} \cdot S/K_{1,S} \cdot (1 - X/(S \cdot K_{eq,1}))}{1 + S/K_{1,S} + X/K_{1,X}} \tag{1a}$$

$$v_2 = \frac{V_{MAX,2} \cdot X/K_{2,X} \cdot (1 - P/(X \cdot K_{eq,2}))}{1 + X/K_{2,X} + P/K_{2,P}} \tag{1b}$$

The maximal rates of the enzymes are denoted by $V_{MAX,1}$ and $V_{MAX,2}$, respectively. The affinity of the two enzymes for their substrates and products are given by Michaelis-Menten constants: $K_{1,S}$, $K_{1,X}$, $K_{2,X}$, and $K_{2,P}$. $K_{1,S}$ indicates that in the absence of X, the first enzyme operates at half-maximal rate if $S = K_{1,S}$ whereas if $S \gg K_{1,S}$ the rate of the first enzyme is maximal. Both reactions are inhibited by their products: by a thermodynamic term, involving an equilibrium constant, $K_{eq,1}$ for enzyme 1 or $K_{eq,2}$ for enzyme 2, and by a kinetic term involving a Michaelis-Menten constant. The equilibrium constants are determined by the standard free energies of the substrates and products of a reaction and do not depend on the properties of an enzyme (e.g., [32]).

The rate of change in the concentration of X is described by an ordinary differential equation:

$$\frac{dX}{dt} = v_1 - v_2 \tag{2}$$

The concentration of X increases, i.e., $dX/dt > 0$, if $v_1 > v_2$ and *vice versa*. This is a kinetic model of the simple network we are studying. To determine the dynamics of the concentration of X as function of time, given some initial concentration of X, a computer is most helpful. This type of kinetic modeling approach, using experimentally determined kinetic parameters and network structure, has proven very promising. Many of such type of models can be found on the JWS online website (at www.jjj.bio.vu.nl) [29, 30].

In thermodynamic equilibrium ($v_1 = v_2 = 0$), one finds that: $X = S \cdot K_{eq,1} = P / K_{eq,2}$. Apparently, the kinetic properties of the enzyme do not matter! This is a general result for systems in thermodynamic equilibrium irrespective of the complexity of the network [33]. This changes in a steady state. To attain a steady state, the concentrations of S and P should remain fixed (set by the experimentalist) and their ratio (P/S) should not be chosen equal to the product of the equilibrium constants of the two reactions. In the steady state, $v_1 = v_2 \neq 0$ and the concentration of X, i.e., $\bar{X}$, is a solution from the algebraic equation $v_1 - v_2 = 0$. We will not give the analytical solution here as it is given by a rather complicated equation that depends on all the kinetic properties. Graphically, the steady-state concentration of X and the flux J can be found by determining the intersection of the rate functions v_1 and v_2 as function of X for a given set of kinetic parameters. It is not hard to imagine that all kinetic parameters now effect $\bar{X}$ and J, for the shape of the rate curves of enzyme 1 and enzyme 2, and therefore their intersection, depends on them. The steady-state flux J now equals $v_1(\bar{X})$.

For illustrative purposes, let us consider a biologically unrealistic form of rate equations for enzyme 1 and 2; that is, mass-action kinetics:

$$v_1 = k_1^+ S - k_1^- X, \quad v_2 = k_2^+ X - k_2^- P \qquad (3)$$

The 'k' coefficients are referred to as elementary rate constants. The steady-state concentration of X now equals:

$$\bar{X} = \frac{k_1^+ S + k_2^- P}{k_1^- + k_2^+} \qquad (4)$$

Already in this simple example, with unrealistic kinetics and over-simplified network structure, we find that all the kinetic parameters of the reactions and a characterization of the environment, the fixed concentrations of S and P, determine the steady state concentration of X. The mathematical function describing the dependency of the steady state concentration of X on those parameters, i.e., Eq. 4, is also dependent on the network structure. This illustrates that only by integration of all those pieces of information, i.e., characterization of the environment, properties of reactions, and network structure, the steady-state system properties can be retrieved. Examples of such studies can be found on the online modeling website JWS online (www.jjj.bio.vu.nl).

To investigate whether all molecular properties of the network are equally important we return to the description of the system having biologically relevant kinetics. Suppose we want to determine whether enzyme 1 and 2 are as important for controlling the steady-state concentration of X by investigating the fractional change in $\bar{X}$ upon a fractional in the enzyme amount of enzyme 1 and 2 by changing their

V_{MAX}'s. This we accomplish for enzyme 1 by taking the total fractional derivative of the steady-state condition for X, i.e., $v_1(\bar{X}, V_{MAX,1}) - v_2(\bar{X}) = 0$:

$$\frac{\partial \ln v_1}{\partial \ln \bar{X}} \frac{d \ln \bar{X}}{d \ln V_{MAX,1}} + \frac{\partial \ln v_1}{\partial \ln V_{MAX,1}} - \frac{\partial \ln v_2}{\partial \ln \bar{X}} \frac{d \ln \bar{X}}{d \ln V_{MAX,1}} = 0 \qquad (5)$$

In terms of metabolic control analysis (MCA) [32, 34–36], those differentials are identified as control coefficients ('C' with proper subscript and superscript) and elasticity coefficients ('ε' with proper subscript and superscript):

$$C_1^X = \frac{d \ln \bar{X}}{d \ln V_{MAX,1}}, \quad \varepsilon_X^{v_1} = \frac{\partial \ln v_1}{\partial \ln \bar{X}}, \quad \varepsilon_X^{v_2} = \frac{\partial \ln v_2}{\partial \ln \bar{X}} \qquad (6)$$

This gives an expression for the dependence of the concentration control coefficient of the first enzyme on the steady-state concentration of X in terms of elasticity coefficients (note that: $\partial \ln v_1 / \partial \ln V_{MAX,1} = 1$):

$$C_1^X = \frac{-1}{\varepsilon_X^{v_1} - \varepsilon_X^{v_2}} \qquad (7)$$

Typically, the elasticity coefficient of the first enzyme for X shall be negative: X inhibits the rate of its producing enzyme. It activates the rate of the second enzyme. This leads to a positive control coefficient for enzyme 1, which can be intuitively understood: a higher activity of the first enzyme should lead to a higher concentration of X to allow for a higher rate of enzyme 2. For the second enzyme, we obtain (after the same operation as in Eq. 6 with respect to $V_{MAX,2}$):

$$C_2^X = -C_1^X \qquad (8)$$

Interestingly, the sum of the concentration control coefficients equals zero! This can be understood by considering that, if in steady state, $v_1(\bar{X}) - v_2(\bar{X}) = 0$, both rates are changed by the same factor α, the value of $\bar{X}$ shall remain unchanged. The steady-state flux will change with factor α, however; illustrating that the flux control coefficients of the two enzymes obey the following law:

$$C_1^J + C_2^J = 1 \qquad (9)$$

The flux control coefficient of enzyme 1, i.e., C_1^J, is defined as:

$$C_1^J = \frac{d \ln J}{d \ln V_{MAX,1}} = \frac{d \ln v_1}{d \ln V_{MAX,1}} = \varepsilon_X^1 C_1^X + 1 \Rightarrow$$
$$C_1^J = \frac{-\varepsilon_X^2}{\varepsilon_X^1 - \varepsilon_X^2} \qquad (10)$$

Introduction to systems biology

Interestingly, it has been proven mathematically that those two summation theorems (Eq. 8 and 9) hold irrespectively of the complexity of the network (having r reactions) and for all concentrations and fluxes [34, 35, 37]:

$$\sum_{i=1}^{r} C_i^X = 0, \quad \sum_{i=1}^{r} C_i^J = 1 \tag{11}$$

This can be understood by the same kind of reasoning as was given above. Networks with a level-structure or cascade-structure have additional summation theorems [38, 39].

Within the network studied so far two other theorems exist. They are referred to as connectivity theorems and relate control coefficients and elasticity coefficients:

$$C_1^X \varepsilon_X^1 + C_2^X \varepsilon_X^2 = -1, \quad C_1^J \varepsilon_X^1 + C_2^J \varepsilon_X^2 = 0 \tag{12}$$

Those relationships can be easily verified using Eq. 7, 8, 9 and 10. Those two equations can be easily understood by considering one of the assumptions of MCA: it assumes that the steady state is (asymptotically) stable with respect to fluctuations [32]. This stability means that the time-averaged concentration X in steady state, despite of thermally fluctuating reaction rates, equals $\bar{X}$ (and that the time-averaged flux equals J) with a variance depending on the distance from thermodynamic equilibrium and the non-linearity of the system at steady state [32, 40, 41]. The connectivity theorems express exactly this stability property for they indicate the outcome of the dissipating response of the system to restore any change in $\bar{X}$ and J upon a perturbation in $\bar{X}$ induced by thermally fluctuating reaction rates. In contrast to the summation theorems, the connectivity theorems do depend on the structure of the network [37, 42–44]. Together the summation and connectivity theorems allow one to derive control coefficients in terms of elasticity coefficients [42].

This section illustrated that many of the interesting properties of cells studied in cell biology and physiology are related to the properties of the molecules, the environment, and the network structure in a complicated nonlinear fashion. The exact dependency only becomes evident by integrating all those properties using models. This we illustrated using metabolic control analysis. Models then may indicate the existence of general relationships reminiscent of laws in physics [45].

Two approaches to systems biology: top-down and bottom-up

Two approaches to systems biology can be distinguished. *Top-down systems biology* starts with data, often generated by system-wide methods, and analyses this data using network models of various types and degrees of detail to discover molecular mechanisms, modules, and patterns of functional behavior (e.g., [4, 46–50]). Typically, the data analyzed originate from metabolomics, flux analysis, proteomics, transcriptomics, or combinations thereof. The following chapters will provide detailed information of how such data are acquired. This approach relies more on induction than *bottom-up system biology*. Top-down systems biology extracts information from the data rather than deducing it from pre-existing knowledge. In bottom-up systems biology experimentation is done not on the entire system level but on smaller subsystems and typically small quantitative heterogeneous datasets are used, containing steady-state and transient metabolite and flux data. The experiments are done on the basis of detailed models of the system to both validate and improve the model or to investigate hypotheses inspired by model analysis. The models used are typically silicon-cell models (e.g., [7–12, 51, 52]). Top-down systems biology is an interesting approach for determination of the network structure and the identification of the molecular mechanisms operative within cells that have not yet been fully characterized [53]. This approach may lead to a more complete picture of the molecular network inside cells. In later stages, top-down systems biological studies may develop into bottom-up approaches as soon as the network has been more carefully characterized. Bottom-up systems biology builds on pre-existing molecular data and allows for analysis of their systemic consequences for the cell [20].

Examples of systems biology research[1]

One aspect of systems biology is the analysis of the structure of the molecular networks and its consequences for the cell. In much the same way as genome sequencing has lead to the emergence of the theoretical analysis of genomes (bioinformatics), has the availability of the entire metabolic, signaling, and gene networks of cells led to the development of theoretical analyses of networks [6, 54]. Many interesting properties of molecular networks haven been discovered [54–56]. Most noticeably are small world organization [57, 58], modularity [59, 60], motifs [61–63], flux balance analysis, extreme pathway and elementary mode analysis [6, 64–67]. All these methods analyze large-scale molecular networks and induce general information regarding their structure and functional consequences. This is one exciting branch of systems biology that is anticipated to develop further and discover many new insights into the molecular organization of cells. Reviews on this aspect of systems biology can be found elsewhere [6, 54].

Another aspect of systems biology is the construction of kinetic models of molecular network functioning as was introduced briefly in the previous section [12, 17, 20]. The history of kinetic model construction and analysis is already long. The first models of metabolism were created in the 1960s and 1970s [68, 69]. Those models suffered mostly from a lack of sufficient system data. The introduction of desktop computers, the development of theory for the analysis of dynamics of nonlinear systems (e.g., [70]), and the development of non-equilibrium thermodynamics (e.g., [71, 72]) lead to the analysis of simplified models – core models – illustrating complex dynamics of molecular networks [19, 73–76]. As understanding progressed, those core models were interchanged for detailed models describing com-

---

[1] The models mentioned in this section can all be investigated online at the JWS online website (www.jjj.bio.vu.nl)

plex dynamics, e.g., compare core models of glycolysis [74, 75] with detailed models [77, 78]. The more detailed models are of interest in bioengineering as they may facilitate rational approaches to optimization of product formation [10, 11, 51, 79].

Hoefnagel et al. [11] developed a kinetic model of pyruvate metabolism in *Lactococcus lactis* to optimize the production rate of acetoin by this organism. All the rate equations of enzymes, as they were characterized in the literature, were incorporated in a kinetic model. They showed that two enzymes (lactate dehydrogenase (LDH) and NADH oxidase (NOX)), previously not identified as important for acetoin production, had most control on the acetoin production flux. By deleting LDH and overexpressing NOX in experiment they were able to redirect carbon flux to acetoin; 49% of pyruvate consumption flux in the mutant *versus* ~0% in the wild type. This result was of importance for industry.

Glycolysis is a catabolic pathway (Fig. 1A) that is present in all kinds of cells. Teusink et al. [80, 81] constructed a kinetic model of yeast glycolysis that was quite helpful in solving the puzzle of an unexpected phenotype of a particular mutant strain and at the same time lead to a surprising new insight about glycolysis. *Saccharomyces cerevisiae* strains with a lesion in the *TPS1* gene, which encodes trehalose-6-phosphate (Tre-6-P) synthase, cannot grow with glucose as the sole carbon and free energy source. Although this enzyme appeared to have little relevance to glycolysis – it was considered to function in the formation of storage carbohydrates and the acquisition of stress tolerance – it turned out to be crucial for growth on glucose. Using the detailed kinetic model of *S. cerevisiae* glycolysis it was shown that the turbo design of the glycolytic pathway (Fig. 1B), apart from being useful in allowing for rapid growth, also represents an inherent risk. A yeast cell investing ATP in the first part of glycolysis and producing a surplus of ATP in the downstream (lower) part of glycolysis runs the risk of an uncontrolled glycolytic flux. In the model, this resulted in the accumulation of hexose monophosphate and fructose-1,6-bisphosphate to levels that are considered toxic when established in the real yeast cell. The formation of trehalose-6-phosphate prevented glycolysis from going awry by inhibiting hexokinase (Fig. 2A), the first ATP-consuming step of glycolysis and thereby restricting the flux of glucose into glycolysis [80]. The importance of the trehalose branch of glycolysis for growth on glucose could only be discovered through the systems biological approach of combining experimental data with kinetic modeling as outlined above. Detailed models can also be used to calculate the outcome of experiments that are not yet achievable, too laborious or too costly to perform as a pilot experiment. Glycolysis in *Trypanosoma brucei* takes place in a special organel, the glycosome, except for the steps by which 3-phosphoglycerate is converted into pyruvate. In contrast to the situation described above for *S. cerevisiae*, the first step catalyzed by hexokinase is not at all regulated in trypanosomes. The glycosome is surrounded by a membrane (Fig. 2B). Bakker et al. [13] were able to calculate the effect of the removal of the glycosomal membrane in *T. brucei*. At the time, this experiment could not be performed experimentally. However, they could remove the membrane in a detailed kinetic model that was validated earlier [7]. The removal of the membrane was of interest because the biological advantage

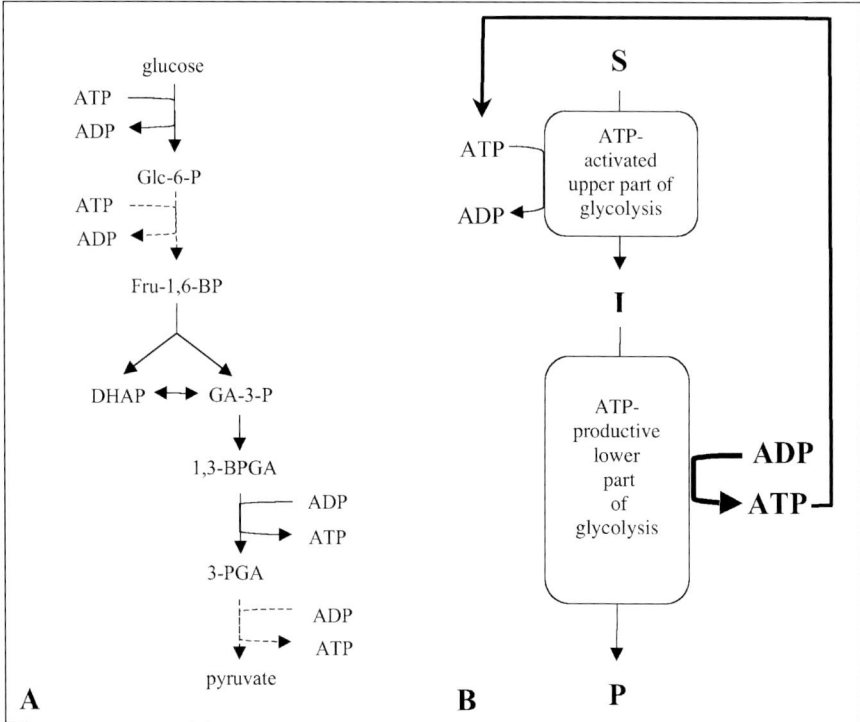

Figure 1. The dangerous turbo design of glycolysis. (A) A simplified scheme of glycolysis. Solid lines represent reactions catayzed by a single enzyme; dashed lines represent multiple sequential reactions. Glc-6P, glucose 6-phosphate; Fru-1,6-BP, fructose 1,6 bisphosphate; DHAP, dihydroxyaceton phosphate; GA-3-P, glyceraldehyde 3-phosphate; 1,3-BPGA, 1,3-bis-phosphoglycerate; 3-PGA, 3-phosphoglycerate. (B) The turbo design of glycolysis. Generalized scheme for glycolysis in which the upper part from substrate S to intermediate I combines the ATP-consuming reactions and the lower part from I to product P combines the ATP-producing reactions. The surplus of ATP produced in the lower part is depicted in bold capitals and the boosting effect on the upper part is indicated by thick lines.

of the glycosome was hypothesized by others to enable this organism to have an extremely high glycolytic flux. Bakker et al. [13] showed that yeast – which does not have glycosomes – can have fluxes as high as *T. brucei*. In addition, they showed that the removal of the glycosomal membrane did not cause a physiologically significant change in the glycolytic flux. Rather, the removal of the glycosome caused accumulation of glucose-6-phosphate and fructose-1,6-bisphosphate up to 100 mM. This would certainly represent a pathological situation for *T. brucei* involving phosphate depletion and possibly osmotic swelling. As it turned out, the glycosomal membrane makes sure that the upper part of glycolysis is not accelerated by the ATP produced by the lower part of glycolysis, because the surplus ATP producing step in the lower part of glycolysis (by pyruvate kinase) actually resides outside of the glycosome. Thus the glycosome is another implementation of a protective device

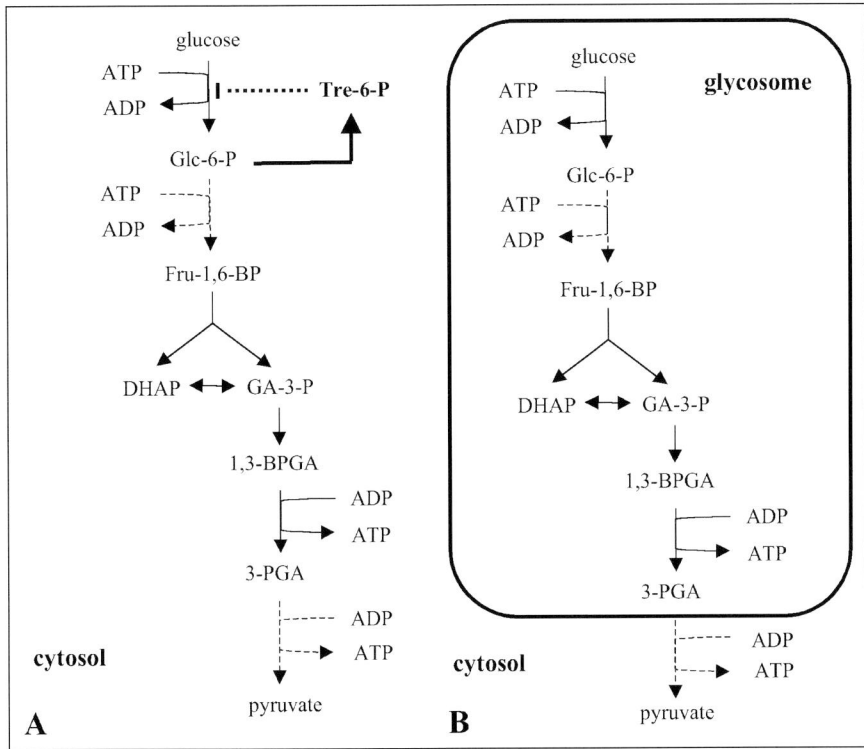

Figure 2. Two different solutions to the turbo design problem. (A) The trehalose branch in *S. cerevisiae*. The scheme is the same as the one shown in Figure 1A, except for the addition of the trehalose shunt in bold. Tre-6-P, trehalose 6-phosphate. The inhibition of hexokinase by Tre-6-P is indicated by a thick dashed line. (B) The glycosome in trypanosomes. Again, the scheme is the same as the one shown in Figure 1A, except for the addition of the glycosomal membrane in bold. The conversion of 3-PGA to pyruvate takes place outside of the glycosome.

against the potentially dangerous 'turbo' design of glycolysis. These two examples of models of glycolysis demonstrate the power of (bottom-up systems biological) kinetic models; when precise and detailed knowledge of the kinetics of the molecular components is available, so-called computer experimentation can be carried out which serves as an adequate substitute for true experimentation.

Regulation of metabolic flux is governed by many different mechanisms. They may function at the level of metabolism, transcription, translation, or at the level of degradation of mRNA or protein. At the level of metabolism, contributions to the regulation of enzymatic conversion rates are made by substrates and products, by effectors through allosteric feedback or feedforward loops, or by covalent modification. Recently a quantitative mathematical tool has been developed in our laboratory, referred to as hierarchical regulation analysis, that allows for the quantitative determination of the importance of all those mechanisms that contribute to the regulation of flux, given experimental data [82–84].

The regulation of the ammonium-assimilation flux by *Escherichia coli* is governed by a complicated mechanism involving multiple covalent modifications, feedback, substrate/product effects, gene expression and targeted protein degradation [85, 86]. This system has for a long time been a paradigm of flux regulation by way of covalent modification. We have recently integrated all molecular data of this network into a detailed kinetic model describing the short-term metabolic regulation of ammonium assimilation [12]. We confirmed many of the hypotheses postulated in the literature on how this system should function. We identified that covalent modification of glutamine synthetase is the most important determinant of the ammonium assimilation flux upon sudden changes in ammonium availability using hierarchical regulation analysis. Removal of the covalent modification of glutamine synthetase caused accumulation of glutamine and severe impairment of growth as was shown experimentally by others [87]. It was confirmed that indeed gene expression of glutamine synthetase alone can lead to regulation of ammonium assimilation; the ammonium assimilation flux was not sensitive to changes made in the level of any of the other enzymes. Finally, we predicted that one advantage of all this complexity is to allow *E. coli* to keep its ammonium assimilation flux constant despite of changes in the ammonium concentration and to change from an energetically unfavorable mode of ammonium uptake to a more favorable alternative as the ammonium level is increased.

The analysis and construction of models incorporating signal transduction networks at a high level of molecular detail has recently been pioneered because of their high potential in drug design [8, 15, 52, 88–90]. We have investigated one of the largest and most complete model of a signal transduction network for its control properties [90]. We determined the control coefficients of all the processes in the network on three characteristics of the transient activation profile of extracellular signal regulated kinase (ERK), which is a member of the mitogen activating protein kinase (MAPK) family. The model contained 148 reactions and 103 variable concentrations and it is an enlarged version of the model published by Schoeberl et al. [89]. To our surprise, we found that less than 10% of the reactions had a large control on ERK activation. We identified RAF as a candidate oncogene and indeed it was found frequently mutated in tumors. To cope with the enormous size of signal transduction network some systems biologists are presently developing theoretical methods for model reduction [91–93]. Such strategies may greatly facilitate understanding, analysis, and experimental design.

In model-driven experimentation, usage of simplified models that illuminate principles of system functioning and guide experimentation (experimental design) are extremely helpful. This approach is nicely illustrated by a series of papers by the group of Ferrell and co-workers [94–97] and Alon and co-workers [98–102]. In Pomerening et al. [97], Ferrell and co-workers investigate the core oscillator driving the cell cycle in *Xenopus laevis*. They study the entry into mitosis and the subsequent return to interphase by following the dynamics of the formation and degradation of the complex cdc2-cyclinB. The interphase-mitosis transition (mitosis: M-phase) is accompanied by synthesis and accumulation of cyclin-B and the subsequent formation of cdc2-cyclinB complex. The degradation of this complex is mediated by

APC-catalyzed degradation of cyclin-B and signals the exit of the M-phase and reentry into interphase. In addition, two net positive feedbacks play a role: via Myt1-Wee1 and cdc25. It was shown experimentally [103] that in the absence of the degradation of cyclin-B by APC the resulting network is bistable. In the presence of cyclin-B degradation, the network displays the oscillations characteristic for the cell cycle; more specifically, it functioned as a relaxation oscillator. Using a semi-detailed model (based on [18, 103]), the authors modeled the network in the absence and the presence of the degradation of cyclin-B and found bistability and oscillations, respectively. Then they investigated the effects of the two net positive feedbacks by inhibiting them. This caused the core oscillator to engage in damped oscillations rather than prolonged oscillations indicating the essentiality of the positive feedback for proper functioning of the cell cycle. The model they used was only quasi-detailed at best but still it had sufficient detail and reflection of reality facilitating model-driven experimentation. In our studies on MAPK signaling, we took a similar approach [45]. We used a simple core model of the MAPK pathway to investigate the difference between inhibition of phosphatases and kinases on the activation profile of ERK. We found that the core model could qualitatively predict the experimental data. It showed that phosphatases tend to control both the amplitude and duration of signaling whereas kinases tend to control only the amplitude. Those results were backed up by theory leading to new theorems in control analysis for signal transduction [45]. Another successful application of the use of simple models to drive experimentation is found in the work by Alon and co-workers [98–102]. They are characterizing the functional properties of motifs, small intracellular networks that occur more frequently in biological networks than in networks of similar size with a random structure. So far they focused mostly on gene circuitry and their activation by transcription factors. The reasoning behind the search and characterization for motifs is that if they occur significantly more frequently in biological networks their design is predicted to have a functional relevance for the cell. They have been successful in showing the functional significance of a number of these motifs. Synthetic biology takes the opposite approach. It tries to design new networks using simple models and implement those in cells to facilitate their analysis, as biosensors, and to endow them with new properties. One successful approach of synthetic biology has been the analysis of noise [104–111]. Noise occurs naturally in all physical systems. In cells noise, perceived as fluctuating copy numbers of molecules in cells, occurs because of fluctuating reaction rates due to local thermal fluctuations [40]. The magnitude of the fluctuations relative to the average copy number determines their influence and importance on intracellular dynamics. The effects of noise are most pronounced when the copy number of molecules are small, < 50 molecules/cell, but may become high even in systems with high average copy numbers, ~1,000s molecules/cell, if the system is sufficiently nonlinear [41, 112].

Conclusion

Systems biology is a rational continuation of successful experimental biology initiated by the molecular biosciences. It represents a combined molecular and systems approach to decipher how molecules jointly bring about cell behavior by cooperating in mechanisms. Those mechanisms can be studied individually (or in a small number) in bottom-up approaches of systems biology using either detailed models or core models. Top-down approaches of systems biology hope to identify such mechanisms and characterize them more roughly first before bottom-up approaches can home in on them in more detail. When the two approaches are combined a rational approach to discovery and characterization of molecular mechanisms, and therefore of cells, results that supplements pure molecular approaches.

References

1. Reed JL, Vo TD, Schilling CH, Palsson BO (2003) An expanded genome-scale model of *Escherichia coli* K-12 (iJR904 GSM/GPR). *Genome Biol* 4: R54
2. Keseler IM, Collado-Vides J, Gama-Castro S, Ingraham J, Paley S, Paulsen IT, Peralta-Gil M, Karp PD (2005) EcoCyc: a comprehensive database resource for *Escherichia coli*. *Nucleic Acids Res* 33: D334–337
3. Salgado H, Gama-Castro S, Peralta-Gil M, Diaz-Peredo E, Sanchez-Solano F, Santos-Zavaleta A, Martinez-Flores I, Jimenez-Jacinto V, Bonavides-Martinez C, Segura-Salazar J et al. (2006) RegulonDB (version 5.0): *Escherichia coli* K-12 transcriptional regulatory network, operon organization, and growth conditions. *Nucleic Acids Res* 34: D394–397
4. Stelling J, Klamt S, Bettenbrock K, Schuster S, Gilles ED (2002) Metabolic network structure determines key aspects of functionality and regulation. *Nature* 420: 190–193
5. Forster J, Famili I, Fu P, Palsson BO, Nielsen J (2003) Genome-scale reconstruction of the *Saccharomyces cerevisiae* metabolic network. *Genome Res* 13: 244–253
6. Price ND, Reed JL, Palsson BO (2004) Genome-scale models of microbial cells: evaluating the consequences of constraints. *Nat Rev Microbiol* 2: 886–897
7. Bakker BM, Michels PAM, Opperdoes FR, Westerhoff HV (1997) Glycolysis in bloodstream from *Trypanosoma brucei* can be understood in terms of the kinetics of the glycolytic enzymes. *J Biol Chem* 272: 3207–3215
8. Kholodenko BN, Demin OV, Moehren G, Hoek JB (1999) Quantification of short term signaling by the epidermal growth factor receptor. *J Biol Chem* 274: 30169–30181
9. Rohwer JM, Meadow ND, Roseman S, Westerhoff HV, Postma PW (2000) Understanding glucose transport by the bacterial phosphoenolpyruvate:glycose phosphotransferase system on the basis of kinetic measurements *in vitro*. *J Biol Chem* 275: 34909–34921
10. Teusink B, Passarge J, Reijenga CA, Esgalhado E, van der Weijden CC, Schepper M, Walsh MC, Bakker BM, van Dam K, Westerhoff HV et al. (2000) Can yeast glycolysis be understood in terms of *in vitro* kinetics of the constituent enzymes? Testing biochemistry. *Eur J Biochem* 267: 5313–5329
11. Hoefnagel MH, Starrenburg MJ, Martens DE, Hugenholtz J, Kleerebezem M, Van S, II, Bongers R, Westerhoff HV, Snoep JL (2002) Metabolic engineering of lactic acid bacteria, the combined approach: kinetic modelling, metabolic control and experimental analysis. *Microbiol* 148: 1003–1013

12. Bruggeman FJ, Boogerd FC, Westerhoff HV (2005) The multifarious short-term regulation of ammonium assimilation of *Escherichia coli*: dissection using an *in silico* replica. *Febs J* 272: 1965–1985
13. Bakker BM, Mensonides FI, Teusink B, van Hoek P, Michels PA, Westerhoff HV (2000) Compartmentation protects trypanosomes from the dangerous design of glycolysis. *Proc Natl Acad Sci USA* 97: 2087–2092
14. Bruggeman FJ, Hornberg JJ, Bakker BM, Westerhoff HV (2005) Introduction to computational models of biochemical reaction networks. In: A Kriete, R Eils (eds): *Computational Systems Biology*, Elsevier
15. Cascante M, Boros LG, Comin-Anduix B, de Atauri P, Centelles JJ, Lee PW (2002) Metabolic control analysis in drug discovery and disease. *Nat Biotechnol* 20: 243–249
16. Michels PAM, Bakker BM, Opperdoes FR, Westerhoff HV (In press) On the mathematical modelling of metabolic pathways and its use in the identification of the most suitable drug target. In: H Vial, A Fairlamb, R Ridley (eds): *Tropical disease guidelines and issues: discoveries and drug development,* WHO, Geneva.
17. Tyson JJ, Chen K, Novak B (2001) Network dynamics and cell physiology. *Nat Rev Mol Cell Biol* 2: 908–916
18. Tyson JJ, Chen KC, Novak B (2003) Sniffers, buzzers, toggles and blinkers: dynamics of regulatory and signaling pathways in the cell. *Curr Opin Cell Biol* 15: 221–231
19. Selkov EE, Reich JG (1981) *Energy metabolism of the cell*. Academic Press, London
20. Westerhoff HV, Palsson BO (2004) The evolution of molecular biology into systems biology. *Nat Biotechnol* 22: 1249–1252
21. Alberghina L, Westerhoff HV (eds) (2005) *Systems biology: definitions and perspectives (topics in current genetics),* Springer-Verlag Berlin, Heidelberg GmbH
22. Bruggeman FJ, Westerhoff HV, Boogerd FC (2002) BioComplexity: a pluralist research strategy is necessary for a mechanistic explanation of the "live" state. *Philosophical Psychology* 15: 411–440
23. Kauffman SA (1971) Articulation of parts explanations in biology. In: RC Buck, RS Cohen (eds): *Boston studies in the philosophy of science.* Kluver Academic Publishers, 257–272
24. Machamer P, Darden L, Craver CF (2000) *Thinking about mechanisms. Philosophy of Science* 67: 1–25
25. Boogerd FC, Bruggeman FJ, Richardson R, Stephan S (2005) Emergence and its place in nature: A case study of biochemical networks. *Synthese* 145: 131–164
26. Darden L, Maull N (1977) Interfield theories. *Philosophy of Sci* 44: 43–64
27. Auyang SY (1998) *Foundation of complex-system theories: in economics, evolutionary biology, and statistical physics*. Cambridge University Press, Cambridge
28. Tyson JJ, Novak B, Odell GM, Chen K, Thron CD (1996) Chemical kinetic theory: Understanding cell cycle regulation. *Trends Biochem Sci* 21: 89–96
29. Olivier BG, Snoep JL (2004) Web-based kinetic modelling using JWS Online. *Bioinformatics* 20: 2143–2144
30. Snoep JL, Bruggeman F, Olivier BG, Westerhoff HV (2005) Towards building the silicon cell: A modular approach. *Biosystems* 83: 207–216
31. Cornish-Bowden A (1995) *Fundamentals of enzyme kinetics*. Portland Press, London
32. Westerhoff HV, Van Dam K (1987) *Thermodynamics and control of biological free-energy transduction*. Elsevier Science Publishers BV (Biomedical Division), Amsterdam
33. Alberty RA (2002) Thermodynamics of systems of biochemical reactions. *J Theor Biol* 215: 491–501
34. Kacser H, Burns JA (1973) The control of flux. *Symp Soc Exp Biol* 27: 65–104

35. Heinrich R, Rapoport TA (1974) A linear steady-state treatment of enzymatic chains. General properties, control and effector strength. *Eur J Biochem* 42: 89–95
36. Fell DA (1997) Understanding the control of metabolism, First Edition. Portland Press, London and Miami
37. Westerhoff HV, Chen YD (1984) How do enzyme activities control metabolite concentrations? An additional theorem in the theory of metabolic control. *Eur J Biochem* 142: 425–430
38. Kahn D, Westerhoff HV (1991) Control theory of regulatory cascades. *J Theor Biol* 153: 255–285
39. Hofmeyr JH, Westerhoff HV (2001) Building the cellular puzzle: control in multi-level reaction networks. *J Theor Biol* 208: 261–285
40. Van Kampen NG (1992) Stochastic processes in chemistry and physics. North-Holland, Amsterdam
41. Elf J, Ehrenberg M (2003) Fast evaluation of fluctuations in biochemical networks with the linear noise approximation. *Genome Res* 13: 2475–2484
42. Reder C (1988) Metabolic control theory: a structural approach. *J Theor Biol* 135: 175–201
43. Kholodenko BN, Westerhoff HV, Puigjaner J, Cascante M (1995) Control in channeled pathways – a matrix-method calculating the enzyme control coefficients. *Biophys Chem* 53: 247–258
44. Westerhoff HV, Kell DB (1996) What bio technologists knew all along? *J Theor Biol* 182: 411–420
45. Hornberg JJ, Bruggeman FJ, Binder B, Geest CR, de Vaate AJ, Lankelma J, Heinrich R, Westerhoff HV (2005b) Principles behind the multifarious control of signal transduction. ERK phosphorylation and kinase/phosphatase control. *Febs J* 272: 244–258
46. Eisen MB, Spellman PT, Brown PO, Botstein D (1998) Cluster analysis and display of genome-wide expression patterns. *Proc Natl Acad Sci USA* 95: 14863–14868
47. Spellman PT, Sherlock G, Zhang MQ, Iyer VR, Anders K, Eisen MB, Brown PO, Botstein D, Futcher B (1998) Comprehensive identification of cell cycle-regulated genes of the yeast *Saccharomyces cerevisiae* by microarray hybridization. *Mol Biol Cell* 9: 3273–3297
48. Ideker T, Thorsson V, Ranish JA, Christmas R, Buhler J, Eng JK, Bumgarner R, Goodlett DR, Aebersold R, Hood L (2001) Integrated genomic and proteomic analyses of a systematically perturbed metabolic network. *Science* 292: 929–934
49. Daran-Lapujade P, Jansen ML, Daran JM, van Gulik W, de Winde JH, Pronk JT (2004) Role of transcriptional regulation in controlling fluxes in central carbon metabolism of *Saccharomyces cerevisiae*. A chemostat culture study. *J Biol Chem* 279: 9125–9138
50. Ihmels JH, Bergmann S (2004) Challenges and prospects in the analysis of large-scale gene expression data. *Brief Bioinform* 5: 313–327
51. Chassagnole C, Noisommit-Rizzi N, Schmid JW, Mauch K, Reuss M (2002) Dynamic modeling of the central carbon metabolism of *Escherichia coli*. *Biotechnol Bioeng* 79: 53–73
52. Lee E, Salic A, Kruger R, Heinrich R, Kirschner MW (2003) The roles of APC and Axin derived from experimental and theoretical analysis of the Wnt pathway. *PLoS Biol* 1: E10
53. Ideker T, Galitski T, Hood L (2001) A new approach to decoding life: systems biology. *Annu Rev Genomics Hum Genet* 2: 343–372
54. Barabasi AL, Oltvai ZN (2004) Network biology: understanding the cell's functional organization. *Nat Rev Genet* 5: 101–113

55. Albert R, Barabasi AL (2002) Statistical mechanics of complex networks. *Revs Mod Physics* 74: 47–97
56. Newman MEJ (2003) The structure and function of complex networks. *SIAM Rev* 45: 167–256
57. Fell DA, Wagner A (2000) The small world of metabolism. *Nat Biotechnol* 18: 1121–1122
58. Jeong H, Tombor B, Albert R, Oltvai ZN, Barabasi AL (2000) The large-scale organization of metabolic networks. *Nature* 407: 651–654
59. Ravasz E, Somera AL, Mongru DA, Oltvai ZN, Barabasi AL (2002) Hierarchical organization of modularity in metabolic networks. *Science* 297: 1551–1555
60. Tanay A, Sharan R, Kupiec M, Shamir R (2004) Revealing modularity and organization in the yeast molecular network by integrated analysis of highly heterogeneous genome-wide data. *Proc Natl Acad Sci USA* 101: 2981–2986
61. Milo R, Shen-Orr S, Itzkovitz S, Kashtan N, Chklovskii D, Alon U (2002) Network motifs: simple building blocks of complex networks. *Science* 298: 824–827
62. Shen-Orr SS, Milo R, Mangan S, Alon U (2002) Network motifs in the transcriptional regulation network of *Escherichia coli*. *Nat Genet* 31: 64–68
63. Yeger-Lotem E, Sattath S, Kashtan N, Itzkovitz S, Milo R, Pinter RY, Alon U, Margalit H (2004) Network motifs in integrated cellular networks of transcription-regulation and protein–protein interaction. *Proc Natl Acad Sci USA* 101: 5934–5939
64. Schuster S, Dandekar T, Fell DA (1999) Detection of elementary flux modes in biochemical networks: a promising tool for pathway analysis and metabolic engineering. *Trends Biotechnol* 17: 53–60
65. Schilling CH, Letscher D, Palsson BO (2000) Theory for the systemic definition of metabolic pathways and their use in interpreting metabolic function from a pathway-oriented perspective. *J Theor Biol* 203: 229–248
66. Covert MW, Schilling CH, Palsson B (2001) Regulation of gene expression in flux balance models of metabolism. *J Theor Biol* 213: 73–88
67. Papin JA, Stelling J, Price ND, Klamt S, Schuster S, Palsson BO (2004) Comparison of network-based pathway analysis methods. *Trends Biotechnol* 22: 400–405
68. Garfinkel D, Hess B (1964) Metabolic control mechanisms. Vii.A Detailed computer model of the glycolytic pathway in ascites cells. *J Biol Chem* 239: 971–983
69. Rapoport TA, Heinrich R, Jacobasc G, Rapoport S (1974) Linear steady-state treatment of enzymatic chains – mathematical-model of glycolysis of human erythrocytes. *Eur J Biochem* 42: 107–120
70. Guckenheimer J, Holms P (1983) *Nonlinear oscillations, dynamical systems, and bifurcations of vector fields*. Springer-Verlag, New York
71. Nicolis G, Prigogine I (1977) *Self-organization in nonequilibrium systems: from dissipative structures to order through fluctuations*. John Wiley & Sons, New York
72. Nicolis G, Prigogine I (1989) *Exploring complexity: An introduction*. WH Freeman & Co. San Francisco
73. Lefever R, Nicolis G (1971) Chemical instabilities and sustained oscillations. *J Theor Biol* 30: 267–284
74. Goldbeter A, Lefever R (1972) Dissipative structures for an allosteric model – application to glycolytic oscillations. *Biophysical J* 12: 1302
75. Selkov E (1975) Stabilization of energy charge, generation of oscillations and multiple steady states in energy metabolism as a result of purely stoichiometric regulation. *Eur J Biochem* 59: 151–157
76. Goldbeter A (1997) *Biochemical oscillations and cellular rhythms: the molecular bases of periodic and chaotic behaviour*. Cambridge University Press, Cambridge

77. Hynne R, Dano S, Sorensen PG (2001) Full-scale model of glycolysis in *Saccharomyces cerevisiae*. *Biophys Chem* 94: 121–163
78. Reijenga KA, van Megen YM, Kooi BW, Bakker BM, Snoep JL, van Verseveld HW, Westerhoff HV (2005) Yeast glycolytic oscillations that are not controlled by a single oscillophore: a new definition of oscillophore strength. *J Theor Biol* 232: 385–398
79. Kremling A, Bettenbrock K, Laube B, Jahreis K, Lengeler JW, Gilles ED (2001) The organization of metabolic reaction networks. III. Application for diauxic growth on glucose and lactose. *Metab Eng* 3: 362–379
80. Teusink B, Walsh MC, van Dam K, Westerhoff HV (1998) The danger of metabolic pathways with turbo design. *Trends Biochem Sci* 23: 162–169
81. Teusink B, Passarge J, Reijenga CA, Esgalhado E, Van der Weijden CC, Schepper M, Walsh MC, Bakker BM, Van Dam K, Westerhoff HV et al. (2000) Can yeast glycolysis be understood in terms of *in vitro* kinetics of the constituent enzymes? Testing biochemistry. *Eur J Biochem* 267: 5313–5329
82. ter Kuile BH, Westerhoff HV (2001) Transcriptome meets metabolome: hierarchical and metabolic regulation of the glycolytic pathway. *FEBS Lett* 500: 169–171
83. Even S, Lindley ND, Cocaign-Bousquet M (2003) Transcriptional, translational and metabolic regulation of glycolysis in *Lactococcus lactis* subsp. *cremoris* MG 1363 grown in continuous acidic cultures. *Microbiol* 149: 1935–1944
84. Rossell S, van der Weijden CC, Kruckeberg AL, Bakker BM, Westerhoff HV (2005) Hierarchical and metabolic regulation of glucose influx in starved *Saccharomyces cerevisiae*. *FEMS Yeast Res* 5: 611–619
85. Rhee SG, Chock PB, Stadtman ER (1989) Regulation of *Escherichia coli* glutamine synthetase. *Adv Enzymol Relat Areas Mol Biol* 62: 37–92
86. Ninfa AJ, Jiang P, Atkinson MR, Peliska JA (2000) Integration of antagonistic signals in the regulation of nitrogen assimilation in *Escherichia coli*. *Curr Top Cell Regul* 36: 31–75
87. Kustu S, Hirschman J, Burton D, Jelesko J, Meeks JC (1984) Covalent modification of bacterial glutamine synthetase: physiological significance. *Mol Gen Genet* 197: 309–317
88. Hoffmann A, Levchenko A, Scott ML, Baltimore D (2002) The IkappaB-NF-kappaB signaling module: temporal control and selective gene activation. *Science* 298: 1241–1245
89. Schoeberl B, Eichler-Jonsson C, Gilles ED, Muller G (2002) Computational modeling of the dynamics of the MAP kinase cascade activated by surface and internalized EGF receptors. *Nat Biotechnol* 20: 370–375
90. Hornberg JJ, Binder B, Bruggeman FJ, Schoeberl B, Heinrich R, Westerhoff HV (2005) Control of MAPK signalling: from complexity to what really matters. *Oncogene* 24: 5533–5542
91. Kruger R, Heinrich R (2004) Model reduction and analysis of robustness for the Wnt/beta-catenin signal transduction pathway. *Genome Inform Ser Workshop Genome Inform* 15: 138–148
92. Borisov NM, Markevich NI, Hoek JB, Kholodenko BN (2005) Signaling through receptors and scaffolds: independent interactions reduce combinatorial complexity. *Biophys J* 89: 951–966
93. Conzelmann H, Saez-Rodriguez J, Sauter T, Kholodenko BN, Gilles ED (2006) A domain-oriented approach to the reduction of combinatorial complexity in signal transduction networks. *BMC Bioinformatics* 7: 34
94. Ferrell JE Jr, Machleder EM (1998) The biochemical basis of an all-or-none cell fate switch in *Xenopus* oocytes. *Science* 280: 895–898

95. Bagowski CP, Ferrell JE Jr (2001) Bistability in the JNK cascade. *Curr Biol* 11: 1176–1182
96. Brandman O, Ferrell JE Jr, Li R, Meyer T (2005) Interlinked fast and slow positive feedback loops drive reliable cell decisions. *Science* 310: 496–498
97. Pomerening JR, Kim SY, Ferrell JE Jr (2005) Systems-level dissection of the cell-cycle oscillator: bypassing positive feedback produces damped oscillations. *Cell* 122: 565–578
98. Rosenfeld N, Elowitz MB, Alon U (2002) Negative autoregulation speeds the response times of transcription networks. *J Mol Biol* 323: 785–793
99. Mangan S, Alon U (2003) Structure and function of the feed-forward loop network motif. *Proc Natl Acad Sci USA* 100: 11980–11985
100. Mangan S, Zaslaver A, Alon U (2003) The coherent feedforward loop serves as a sign-sensitive delay element in transcription networks. *J Mol Biol* 334: 197–204
101. Dekel E, Mangan S, Alon U (2005) Environmental selection of the feed-forward loop circuit in gene-regulation networks. *Phys Biol* 2: 81–88
102. Mangan S, Itzkovitz S, Zaslaver A, Alon U (2006) The incoherent feed-forward loop accelerates the response-time of the gal system of *Escherichia coli*. *J Mol Biol* 356: 1073–1081
103. Pomerening JR, Sontag ED, Ferrell JE Jr (2003) Building a cell cycle oscillator: hysteresis and bistability in the activation of Cdc2. *Nat Cell Biol* 5: 346–351
104. Elowitz MB, Levine AJ, Siggia ED, Swain PS (2002) Stochastic gene expression in a single cell. *Science* 297: 1183–1186
105. Ozbudak EM, Thattai M, Kurtser I, Grossman AD, van Oudenaarden A (2002) Regulation of noise in the expression of a single gene. *Nat Genet* 31: 69–73
106. Swain PS, Elowitz MB, Siggia ED (2002) Intrinsic and extrinsic contributions to stochasticity in gene expression. *Proc Natl Acad Sci USA* 99: 12795–12800
107. Paulsson J (2004) Summing up the noise in gene networks. *Nature* 427: 415–418
108. Thattai M, van Oudenaarden A (2004) Stochastic gene expression in fluctuating environments. *Genetics* 167: 523–530
109. Golding I, Paulsson J, Zawilski SM, Cox EC (2005) Real-time kinetics of gene activity in individual bacteria. *Cell* 123: 1025–1036
110. Pedraza JM, van Oudenaarden A (2005) Noise propagation in gene networks. *Science* 307: 1965–1969
111. Rosenfeld N, Young JW, Alon U, Swain PS, Elowitz MB (2005) Gene regulation at the single-cell level. *Science* 307: 1962–1965
112. Elf J, Paulsson J, Berg OG, Ehrenberg M (2003) Near-critical phenomena in intracellular metabolite pools. *Biophys J* 84: 154–170

Natural and artificially induced genetic variability in crop and model plant species for plant systems biology

Christophe Rothan[1] and Mathilde Causse[2]

[1] INRA-UMR 619 Biologie des Fruits, IBVI-INRA Bordeaux, BP 81, 71 Av. Edouard Bourlaux, 33883 Villenave d'Ornon cedex, France
[2] INRA-UR 1052, Unité de Génétique et Amélioration des Fruits et Légumes, BP 94, 84143 Montfavet cedex, France

Abstract

The sequencing of plant genomes which was completed a few years ago for *Arabidopsis thaliana* and *Oryza sativa* is currently underway for numerous crop plants of commercial value such as maize, poplar, tomato grape or tobacco. In addition, hundreds of thousands of expressed sequence tags (ESTs) are publicly available that may well represent 40–60% of the genes present in plant genomes. Despite its importance for life sciences, genome information is only an initial step towards understanding gene function (functional genomics) and deciphering the complex relationships between individual genes in the framework of gene networks. In this chapter we introduce and discuss means of generating and identifying genetic diversity, i.e., means to genetically perturb a biological system and to subsequently analyse the systems response, e.g., the changes in plant morphology and chemical composition. Generating and identifying genetic diversity is in its own right a highly powerful resource of information and is established as an invaluable tool for systems biology.

Introduction

In the plant genomic era, huge amounts of sequence data have been obtained, mostly for model plants but also for an ever increasing number of non model plant species. Genome sequencing, which was completed a few years ago for *Arabidopsis* and rice, is currently underway for numerous crop plants of high commercial value such as maize, poplar, tomato, grape or tobacco. In addition, hundreds of thousands of EST sequences are publicly available for many plant species (e.g., at TIGR, http://www.tigr.org/tdb/tgi/plant.shtml) and may represent between 40 and 60% of the genes present in plant genomes. However, the identification of very large sets of gene sequences in any plant species is only an initial step towards (i) understanding gene function in the plant (functional genomics) and (ii) deciphering and representing the complex relationships between gene sequence and protein expression variation, corresponding pathways and networks, and changes in plant morphology and chemical composition (plant systems biology).

The recent development of high throughput methods for transcriptional profiling of genes using microarrays (Chapters by Foyer et al. and Hennig and Köhler) and for metabolite profiling using various separation and analytical techniques (metabolome) (Chapters by Steinhauser and Kopka, and Sumner et al.), as well as the current progress in large scale protein analysis (proteomics, Chapters by Brunner et al. and Schuchardt and Sickmann) and morphological phenotyping of plants, has revolutionised the way we now envisage plant systems biology. By studying plants to find out where and when, and under what conditions, whole sets of genes and proteins are expressed, and by analysing the correlations with corresponding changes in plant phenotype (development, morphology and chemical composition), we are now able to infer the putative functions of genes and to deduce the possible relationships between pathways, regulatory networks and phenotypes.

Linking phenotype to genotype: Strategies

Basically, two strategies, usually named forward and reverse genetics, will help bridge the gap between genotypic variations and associated phenotypic changes. Both are based on the use of natural or artificially induced allelic gene variation to gain insights into the relationship between genes, their function and their influence on phenotypic traits. The forward (traditional) genetic approach aims at discovering the gene(s) responsible for variations of known single Mendelian traits or of quantitative traits (Quantitative Trait Loci or QTL) previously identified through phenotypic screening of natural populations. In contrast, the main objective of reverse genetics is to unravel the physiological role of a target gene and to establish its effect on the plant phenotype.

Forward genetic approaches
Forward genetic approaches have been hampered until recently in many crop plants by the lack of detailed genetic maps, genomic resources (BACs, bacterial artificial chromosome) and genomic sequences. Due to the remarkable development of genetic marker technology over the last 15 years, genetic linkage maps are now available for most crop species, allowing the comparative mapping of crop species and model plants, the location of loci controlling Mendelian traits or QTL on linkage groups and finally the isolation by map-based cloning of the gene responsible for the phenotype. Today, the availability and use of high throughput and precise analytical tools for metabolic profiling (Chapters by Steinhauser and Kopka, and Sumner et al.) has considerably increased the number of compounds that can be identified and quantified in plants. This will enable the decomposition of previously identified complex quantitative traits into multiple single quantitative traits, potentially unravelling loci controlling whole metabolic pathways. The use of transcriptome or proteome profiling and genome sequence information will provide new candidate genes for characterising the sequences responsible for natural genetic variation.

Reverse genetic approaches
Genome and EST sequencing, and large scale analyses of transcript, protein and metabolite profiles, can give rise to a large number of candidate genes whose function needs to be evaluated in the context of the plant. Very efficient reverse genetic tools, mostly based on insertional mutagenesis and targeted silencing of specific genes by RNAi-based technology (Chapter by Johnson and Sundaresan), have therefore been developed in model plants. However, a comparable strategy is clearly impossible for most crop plants, due to cost or technical limitations such as a large genome size or the unfeasibility of large scale genetic transformation. One might consider that the information gained from model plants can easily be transferred to plant species. Currently, recent advances in plant studies indicate that results obtained from a model plant are not always applicable to other plant species, not only because many crop plants have specialised organs not present in the model plants *Arabidopsis* and rice (e.g., tubers in potato, root in sugar beet or fruit in tomato) but also because a considerable fraction of the genes are probably unique to the different taxa or even to the particular species to which they belong [1]. In addition, for certain categories of genes, e.g., those involved in signalling pathways or in regulatory processes such as transcription factors or kinases, knockout mutations can be lethal for the plant, induce phenotypic variations only distantly related to the real function of the target gene or, in some cases, give weaker phenotypes than those observed with missense mutations that produce dominant-negative mutants [2]. In these circumstances, natural or artificially induced allelic variants appear as the most appropriate strategy.

Forward genetics: Gene and QTL characterisation

The possibility of saturating the genome with molecular markers has allowed Mendelian mutations and QTL to be systematically mapped. Since the early 1990s, hundreds of studies have been conducted to map Mendelian mutations and QTL in plants. Several genes have been cloned through map-based cloning [3–5], but only a few QTL have been cloned and characterised. QTL are not different in nature from loci responsible for discrete variations, but, rather than a 'mutant-wild-type' opposition, there are moderate differences (of effects) between 'wild-type' (or active) alleles, which are responsible for the variation of quantitative characters. One can believe that systems biology and high-throughput genomic approaches will lead to a rapid increase in the number of gene/QTL cloned and of our understanding of the genetic basis of natural variation.

Principles and methods of QTL mapping

QTL mapping is based on a systematic search for association between the genotype at marker loci and the average value of a trait. It requires:

- a segregating population derived from the cross of two individuals contrasted for the character of interest.
- that the genotype of marker loci distributed over the entire genome is determined for each individual of the population (and thus a saturated genetic map is constructed).
- the measurement of the value of the quantitative character for each individual of the population.
- the use of biometric methods to find marker loci whose genotype is correlated with the character, and estimation of the genetic parameters of the QTL detected.

Several biometric techniques to find QTL have been proposed, from the most simple, based on analysis of variance or Student's test, applied marker by marker, to those that take into account simultaneously two or more markers [6]. The QTL are characterised by three parameters (a, d, R^2). The additive effect a is equal to $(m_{22} - m_{11})/2$, where m_{22} and m_{11} are the mean values of homozygous genotypes *A1A1* and *A2A2*, respectively. The degree of dominance is the difference between the mean of the heterozygotes *A1A2*, and half the sum of the homozygotes: $d = m_{12} - (m_{11} + m_{22})/2$ (Fig. 1). Each segregating QTL contributes to a certain fraction of the total phenotypic variation, which is quantified by the R^2, which is the ratio of the sum of squares of the differences linked to the marker locus genotype to the sum of squares of the total differences. Epistasis (interaction between QTL) may also be searched for by screening for interaction between every pair of markers, but due to the number of tests, very stringent thresholds must be applied and thus only very highly significant interactions are detected, unless a specific design is used. The advantage of QTL detection on individual markers is its simplicity. Other more powerful methods have been developed that allow us to precisely position QTL in the interval between the markers and to estimate their effects at this position. The most widespread method for testing for the presence of a QTL in an interval between two markers is based on the calculation of a LOD score. At each position on a chromosome (with a step of 2 cM for example), the decimal logarithm of the probability ratio below is calculated:

$$\text{LOD} = \log_{10} \frac{V(a_1, d_1)}{V(a_0, d_0)}$$

where $V(a_1, d_1)$ is the value of the probability function for the hypothesis of QTL presence, in which the estimations of parameters are a_1 and d_1, and where $V(a_0, d_0)$ is the value of the probability function for the hypothesis of QTL absence, that is, when $a_0 = 0$ and $d_0 = 0$ [7]. A LOD of 2 thus signifies that the presence of a QTL at a given point is 100 times more probable than its absence; a LOD of 3 means 1,000 times more probable, etc. A curve of LOD can thus be traced as a function of the position on a linkage group. The maximum of the curve, if it goes beyond a certain

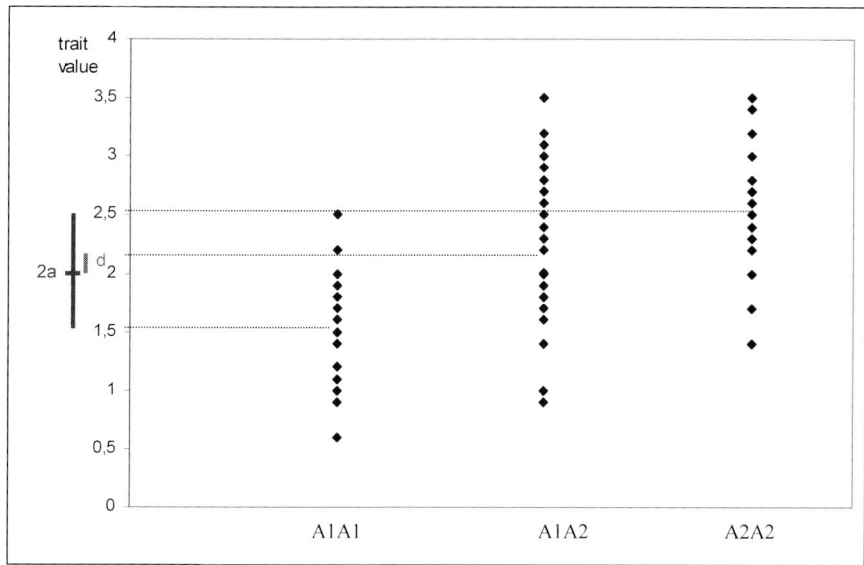

Figure 1. Genetic parameters related to a QTL. The plot shows average values of the three genotypic classes at the marker B (of Fig. 1) for the quantitative character studied. A significant difference between the means signifies that the effects of two alleles at the QTL are sufficiently different to have detectable consequences. The parameters *a* and *d* are then estimated. R^2 is related to the intraclass variance s^2 and to the sample size.

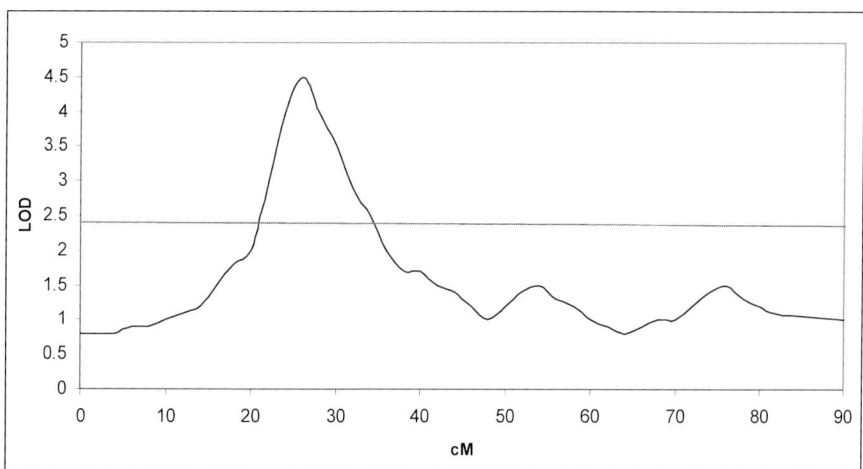

Figure 2. Example of Lod plot along a 90 cM chromosome. The most likely position of the QTL is shown with the confidence interval associated.

threshold, indicates the most probable position of the QTL (Fig. 2). The confidence interval of the QTL position is thus conventionally defined as the chromosomal fragment corresponding to a reduction in LOD of 1 unit in relation to the maximum LOD, which indicates that the probability ratio has fallen by a factor of 10. This method was first implemented in the Mapmaker/QTL software [8], which is coupled with the Mapmaker software for the construction of genetic maps. Several related methods have then been proposed including the composite interval mapping that takes the other QTL present in the genome, represented by markers that are close to them, as co-factors in the model. This reduces the residual variation induced by their segregation [9–10] and then substantially improves the precision of estimation of QTL effects and positions. These methods are implemented in several software. Access to most of these software is free and the addresses of sources can be found in databases including http://www.stat.wisc.edu/~yandell/qtl/software.

Factors influencing QTL detection

Although the principle of QTL detection is relatively simple, several parameters influence the results and must be taken into account to optimise the experimental setup. For a given sample size, the efficiency of QTL detection depends partly on the additive effect of QTL (a very small difference of effects between alleles will not be found significant) and partly on the variance within the genotypic classes. This variance depends on environmental effects (the environmental control of variations increases the efficiency of the test) on other segregating QTL in the genome, on the presence of epistasis and on the distance between markers and QTL (this is particularly important if the density of markers is low). Because of the large number of analyses carried out, low values of α must be chosen. For interval mapping, a global risk of $\alpha = 0.05$ for the entire genome imposes a fairly high LOD threshold per interval, which depends on the density of markers and the genetic length of the genome [7]. Thresholds are now usually estimated following permutation tests, based on a random resampling of data [11].

Efficiency of QTL detection and precision of QTL location depends more on population size than on marker density [12]. Once a mean marker density of 20 or 25 cM is attained, any supplementary means must be invested in analysing additional individuals rather than in increasing the number of markers. A QTL with a strong effect will be detected with a high probability whatever the population size, but for detection of a QTL with moderate effect (R^2 about 5%), it is necessary to use a larger number of individuals. It must also be noted that it is better to increase the number of genotypes in the population rather than the number of replications per genotype.

The populations in which QTL mapping is most efficient are those derived from crosses between two homozygous lines, such as F2, recombinant inbred lines (RIL), doubled haploid (DH) and backcross (BC). F2 are the only populations allowing the dominance effect to be estimated, while a mixture of *a* and *d* is estimated with BC. Highly recombinant inbred lines (HRIL) obtained after several cycles of intercrossing individuals were proposed to increase the precision of marker ordering and subsequently also to increase the precision of QTL mapping [13]. When no homozygous parental lines are available (in allogamous species and species with a long generation time, such as trees), QTL detection is complicated because the parents may differ by more than two alleles, and because the phase (coupling or repulsion) of the marker-QTL linkage may change from one family to another. Various populations may nevertheless be used, from F1, BC or populations using information from two generations in families of full siblings [14]. Knowledge of the grandparent genotypes at marker loci can improve detection by allowing phases of associations between adjacent markers to be identified [15].

Tanksley and Nelson [16] proposed to search for QTL in populations of advanced backcross (BC2, BC3, BC4). Although the power of QTL detection is reduced, this strategy is interesting when screening positive alleles from a wild species, as it will allow the identification of mostly additive effects and will reduce linkage with unfavourable alleles and thus simultaneously advance the production of commercially desirable lines.

The efficiency of detecting a particular QTL in a segregating population is low because other QTL are segregating and major QTL mask minor ones. For this reason, Eshed and Zamir [17] proposed the use of introgression lines in which each line possesses a unique segment from a wild progenitor introgressed in the same genetic background. The whole genome has been covered with 75 lines and has created a sort of 'genome bank' of a wild species in the genome of a cultivated tomato. These lines can then be compared with the parental cultivated line to search for QTL carried by the introgressed fragments. The detection is more efficient than in a classical progeny because of the fixation of the rest of the genome. Greater test efficiency and a significant economy in terms of time and effort can also be achieved by molecular genotyping exclusively individuals showing the extreme values of the character studied (through selective genotyping) [18]. Nevertheless this approach is only useful for detecting QTL with major effects and can be applied only if one character is studied.

What have we learnt from QTL studies?

Ever since the mapping of QTL became possible, several studies have showed that even with populations of moderate size (sometimes less than 100 individuals), some QTL are almost always found, for all types of characters and plants [19–20]. Data compiled from maize and tomato, where many QTL have been mapped, indicate that the effects of QTL measured by their R^2 are distributed according to a marked L curve, with a few QTL having a strong or very strong effect, and most QTL having a weak or very weak effect. With populations of normal size (60 to 400 individuals), R^2 are usually overestimated [21] and depending on the characters, one to ten QTL are usually detected with an average of 4 QTL detected per study [22]. These numbers constitute a minimum estimate of the number of segregating QTL in the populations studied for several reasons: (i) Some QTL have an effect below the detection threshold, (ii) some chromosomal segments may contain several linked QTL when only one is apparent and (iii) if two QTL of comparable effect are closely linked, but in repulsion phase, i.e., if the positive alleles at the two loci do not come from the same parent, no QTL will be detected, until fine mapping is attempted [23]. Moreover, the monomorphic QTL in a given population cannot be detected. For species and traits where a large number of studies have been performed with several progenies, it is frequent to compile more than 30 QTL [24, 25]. Using meta-analysis, Chardon and colleagues [26] summarised 22 studies and identified at least 62 QTL controlling flowering time in maize.

Transgressive QTL are frequently discovered. Even when highly contrasted individuals have been chosen as parents of a population, it is not rare to find a QTL showing an effect opposite to that expected from the value of the parents. Results from advanced backcross experiments in tomato showed for example unexpected positive transgressions from wild relatives, for various fruit traits [27].

When comparative mapping data are available, some QTL of a given character are frequently found at homologous positions on the genomes of species that are more or less related. This is the case for grain weight in several legume species [28–30], for domestication traits in cereals [31, 32] and for fruit-related traits in Solanaceae species [33].

Epistasis between QTL is rarely detected with classical populations [34], but this is mostly due to statistical limits of the populations studied. A way of increasing the reliability of epistasis analysis is to eliminate the 'background noise' due to other QTL by using near isogenic lines (differing only by a chromosome fragment) for a particular QTL as parents of the populations studied [35]. On the other hand, it is not because a QTL does not show epistatic interactions with other QTL taken individually that its effect is independent of the genetic background. For instance, the effects of two maize domestication QTL are much weaker when they are segregating in a 'teosinte' genetic background than in an F2 maize x teosinte background [36]. Similarly, significant QTL by genetic background interaction was shown in tomato by transferring the same QTL regions into three different lines [37].

QTL mapping is particularly interesting in attempting to analyse the determinism of complex characters, by focusing on components of these characters [38–40]. QTL mapping thus provides access to the genetic basis of correlations between characters. When characters are correlated, at least some of their QTL will be common (or at least genetically linked). In the case of apparent co-location of QTL controlling different characters, there is no direct method to highlight the existence of a single QTL with a pleiotropic effect or of two linked QTL. Korol and colleagues [41] proposed a statistical test to use the information of correlated traits to locate QTL simultaneously controlling several traits. They showed that this approach increased the power of QTL detection when compared to a trait by trait search. Nevertheless the best way to distinguish pleiotropy from linkage is through fine mapping experiments. Many fine mapping experiments have separated QTL that were initially thought to control two related traits [42–44].

The environment may have a significant impact on the effect of QTL: a QTL detected in one environment may no longer be detected in another, or its effect may vary. This has been frequently observed, even though the environmental influence differs according to the characters and the range of environments studied. Certain QTL are detected in all or almost all the environments tested, while others are specific to a single environment. Several statistical methods for the estimation of QTL x environment interactions have been proposed [45–48]. Certain studies look directly at QTL involved in the response to environmental changes such as soil nitrogen [49] or drought [50]. Ecophysiological modelling may also be used to identify the biological processes underlying QTL and to distinguish loci affected by the environment [51–53].

Characterisation of QTL: Still a difficult task

Today, in plants, several Mendelian mutations have been characterised by positional cloning in plants, but still very few QTL have been definitively characterised at the molecular level ([54, 55], Tab. 1). Direct cloning of a QTL is more difficult than cloning a major gene because the QTL only partially influences character variation and its effect can only be appreciated by statistical methods. For this reason, the resources required are more considerable and the first QTL cloned by map-based cloning correspond to QTL with strong effects that are independent of the environment. Figure 3 illustrates the general strategy used to characterise a QTL. If nothing is known about the physiological and molecular determinism of the character, positional cloning is the most straightforward method to characterise a QTL. If on the other hand some genes involved in the expression of the character are known, it is possible to test whether the polymorphism of one of them (the 'candidate' gene) could explain the variation of the character. In both cases it is necessary to reduce the interval around the QTL through fine mapping.

The population sizes conventionally used do not allow for precise location of QTL with moderate effects (confidence intervals usually range from 10–30 cM). Such segments may comprise several hundreds of genes, so any attempt at characterising or positional cloning of QTL is impracticable. To fine map a QTL it is necessary to compare several near-isogenic lines differing only for a region containing the QTL that has to be located precisely. The QTL can be located more precisely by comparing these new lines to the initial recurrent line [42]. Such lines can be derived through backcrosses or using residual heterozygosity of RILs [56]. The QTL is 'mendelised' when it is the only source of variation for the trait. Introgression lines constitute another point of departure for fine mapping and cloning a QTL. By deriving an F_2 population from a cross between an introgression line and a cultivated line, then self-fertilising the individuals carrying a recombination in the fragment of interest, fixed lines for different subgroups of the initial fragment can be created [57].

Positional cloning can only really be considered when the QTL is precisely located in an interval much smaller than one centimorgan, in which case large insert libraries (YAC or BAC) can be screened. Ideally the distance between marker and QTL should be around the size of a BAC clone. This is obtained by studying a population of several thousands plants [58] and obtaining polymorphic markers closely linked to the QTL. To confirm that the isolated gene corresponds to the QTL

Table 1. Summary of the QTL cloned in plants. The gene function is indicated. When a candidate gene was proposed, it is indicated if it was early (E) or late (L) in the cloning process. Adapted from Salvi and Tuberosa [55]

Species	Trait	QTL	Function	Method	Candidate Gene*	QTN	Functional proof	Ref
Arabidopsis	Flowering time	ED1	CRY2 Cryptochrome	Pos. cloning	Yes (L)	a.a. substitution	Transformation	[69]
	Flowering time	FLW	Transcription Factor	Pos. cloning	Yes (E)	Gene deletion	Transformation	[76]
	Insect resistance	GS-elong	MAM synthase	Pos. cloning	Yes (E)	Nucleotide and indels	No	[78]
	Root morphology	BRX	Transcription Factor	Pos. cloning	No	Premature stop codon	Transformation	[73]
Rice	Heading time	Hd1	Transcription Factor	Pos. cloning	Yes (L)	Unidentified	Transformation	[79]
	Heading time	Hd3a	Unknown	Pos. cloning	Yes (L)	Unidentified	Transformation	[80]
	Heading time	Hd6	Protein kinase	Pos. cloning	No	Premature stop codon	Transformation	[72]
	Heading time	Ehd1	B-type response regulator	Pos. cloning	No	a.a. substitution	Transformation	[70]
	Salt tolerance	SKC1	HKT-type transporter	Pos. cloning	Yes (L)	a.a. substitution	Complementation	[71]
Tomato	Fruit sugar content	Brix9-2-5	Invertase	Pos. cloning	Yes (L)	a.a. substitution	Complementation	[59, 60]
	Fruit shape	Ovate	Unknown	Pos. cloning	No	Premature stop codon	Transformation	[74]
	Fruit size	fw2.2	Unknown	Pos. cloning	No	Unknown	Transformation	[61, 75]

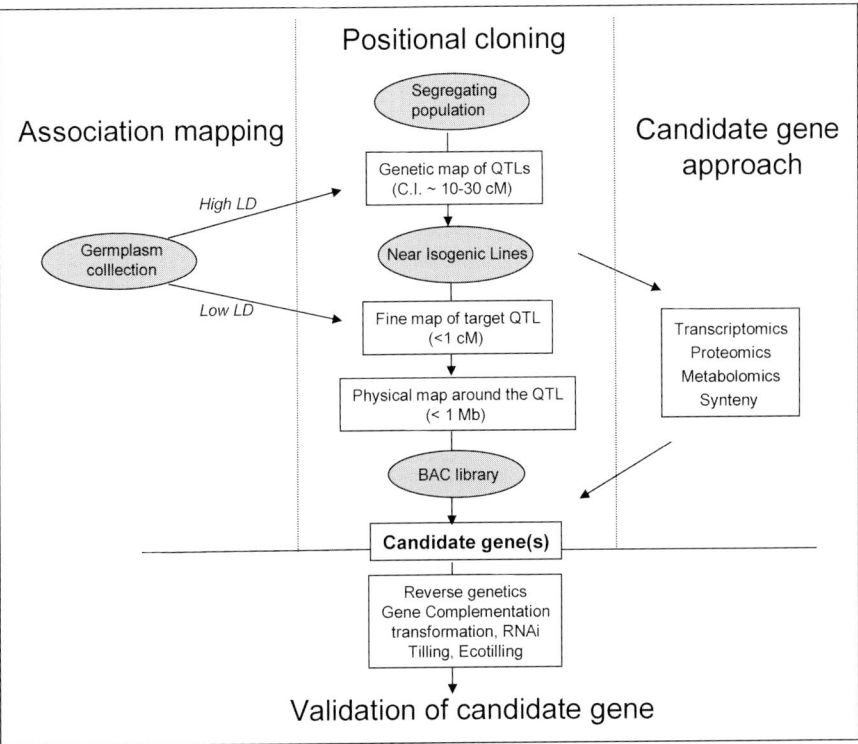

Figure 3. General strategy used to characterise a QTL.

of interest, the ideal situation is to obtain a recombinant within the candidate gene that leads to different values of the trait. For example the cloning of a QTL controlling the variation in sugar content of tomato fruit followed fine mapping [59] and benefited from the existence of recombinations within the gene to localise the QTL in a region of 484bp covering the sequence of a cell-wall invertase. The functional polymorphism was then delimited to an amino acid near the catalytic site which affects enzyme kinetics and fruit sink strength [60]. Transformation with contrasted alleles may allow us to definitively prove that the candidate gene is the QTL. A fruit weight QTL in tomato responsible for about 30% of the variation of this character has been isolated using the classical strategy of high resolution mapping by screening 3472 F2 plants, identifying 53 recombinants (between two markers 4.2 cM apart) and screening a YAC library. From a YAC likely to contain the required gene, a cosmid library was screened and three clones used to transform a tomato variety. The cosmid leading to differences in fruit size after transformation was sequenced and the two sequences corresponding to ORFs were used in a second round of transformation. This allowed the definitive identification of the clone corresponding to the QTL [61]. Certain problems may arise from validation by transformation, as generally we aim to modify the value of a trait by introducing a favourable allele, no easy task when the effect of the environment, the genetic background, and the transformation (dose effects, gene silencing) may interfere. Constructions to overexpress the gene can be used but carry a risk of seeing artefactual positive effects on the trait.

For certain quantitative characters, the physiology of the plant indicates what the functions in question might be. For others, mutants with phenotypes resembling extreme variations of the character are available. If the corresponding genes are available, whether they are responsible for the QTL of the character studied depends on whether they are polymorphic and whether this polymorphism has repercussions on the variation of the character considered [40, 62, 63].

The confirmation of the role of a candidate gene in the variation of a character is not direct and must proceed via:

- fine mapping of the QTL; testing for co-segregation of the candidate gene and the QTL with thousands of plants may allow the rejection of several candidate genes
- the search for correlations between polymorphisms of the candidate gene and variation of the character in populations in which linkage disequilibrium is minimal (in such populations, only a cause-effect relationship ensures the durability of the correlation throughout the generations). This association mapping approach has already been useful to characterise several QTLs [64–68]
- analysis of the variation at biochemical and metabolic levels. A necessary but not sufficient condition for a gene coding for an enzyme to be a QTL is that the activity of the enzyme must be variable. This has allowed elucidation of the origin of variation at the Lin5 QTL [60]
- molecular analysis of alleles to find the molecular basis of variation; the identification of the polymorphism responsible for the QTL is not straightforward, as it can be either a nucleotide substitution (or indel) causing an amino acid modification [59, 60, 69–71], a stop codon [72, 73], a gene deletion [74] or a mutation in a regulatory sequence that may be very distant from the gene [75–77]. The exact nature substitutions or indels are detected [78]
- transformation, even though this poses specific problems in the case of QTL [77–80]
- complementation of a known mutation corresponding to the same gene [59, 60, 71, 81].

How can systems biology help QTL characterisation?

Functional genomics facilitates gene or QTL cloning at different levels. Due to high throughput technologies, the number of ESTs sequenced and mapped is rapidly increasing for many species, providing new candidate genes [82]. Apart to the access to all the ORFs carried by a genome fragment, this will provide a non limited number of molecular markers useful for map-based cloning. In *Arabidopsis*, the access to the whole genome sequenced has considerably reduced the time to positionally clone a gene [4]. Although the number of genomes fully sequenced is still limited, their number is rapidly growing, now covering a range of botanical families. Synteny with model species should then assist in identifying molecular markers and candidate genes in related crop species [83]. Even distantly related species exhibit microsynteny (see for example tomato and *Arabidopsis* genomes [84]), thus markers and candidate genes can be transferable across species.

Microarray-based techniques may be helpful for high throughput identification of polymorphisms (SNP or Indels) at thousands of loci simultaneously [85]. Screening for candidate genes is also much more efficient when utilising high throughput tools for genome expression studies. Transcriptional profiling between near isogenic lines may provide a list of differentially expressed genes. Those which map in the QTL region are strong good candidates [86]. Expression profiling may also be used on a mapping population considering the level of expression of a gene as a trait (the QTL are thus expression QTL, called eQTL). These analyses provide important information about the organisation of regulatory networks [87], as eQTL are either located in the region of the corresponding gene (*cis*-regulation) or in a distant region (*trans*-regulation). A review of the first eQTL mapping experiments shows that (i) major effect eQTL are often detected, (ii) up to one-third of eQTL are *cis*-acting, and (iii) eQTL hot spots that explain variation for multiple transcripts are frequent [88]. Correspondence between eQTL and morpho-physiological QTL can then be researched [89]. It almost goes without saying however that this approach is limited by the fact that all the QTL are governed by alterations in RNA amounts.

An alternative approach consists of identifying loci affecting the quantities of protein (*Protein Quantity Loci* or PQLs) or loci responsible for the charge or molecular mass of protein isoforms (*Position Shift Loci* or PSLs) as detected by two-dimensional gel electrophoresis [90]. When a PQL cosegregates with a PSL, the variation of protein quantity can be due to a polymorphism within the protein itself. On the other hand, if PSL and PQL are mapped to distinct regions of the genome, the variation in protein quantity can be due to a trans-acting regulatory factor/gene [91]. In maize, this approach has been useful in discovering genes involved in water-stress tolerance [92]. Proteomic approaches, by revealing polymorphisms within genes as well as differences in protein expression are therefore complementary to DNA marker and mapping approaches. Metabolomic profiling combined to genetic studies may also provide insight on the physiological bases of quantitative trait and give clues on the candidate genes to screen [93]. At last, all the tools available for reverse genetics, collections of mutants, TILLING (Targeting Induced Local Lesions IN Genomes), RNAi (presented below) may be used to validate a candidate.

To recapitulate, forward genetics approaches are thus powerful tools for deciphering natural genotypic variability. They have also been applied to artificially induced mutants in crop and model plant species. In *Arabidopsis* for example, this strategy is yielding remarkable results by allowing the isolation of unknown genes involved in the control of specific phenotypes [94].

Reverse genetics strategies in plants

Several genome-wide gene targeting techniques have been widely developed in plants. In the absence of efficient and routine methods for homologous recombination in plants, insertional mutatagenesis using transferred DNA (T-DNA) from *Agrobacterium* or transposable elements has been the method of choice for genome size reverse genetics approaches in the model plants *Arabidopsis* and rice. Several populations of tens of thousand of mutagenised plants have been created with the objective to reach near saturation of the collections (e.g., *Arabidopsis* genetic resources at http://www.arabidopsis.org/portals/mutants/worldwide.jsp). Knockout mutants in a given gene can be screened by PCR-search of *Arabidopsis* insertion collections or even by BLAST search of the insertion flanking sequences. Since the probability to hit the gene is lower for small genes than for large genes, loss-of-function mutants for the target gene are not always identified and very large numbers of mutagenised plants are needed to reach near saturation of the collection [95]. Nonetheless, insertion collections have proved to be powerful reverse genetics tools for studying gene function in the context of the plant (as reviewed in [94]). In much the same way, collections of activation tagging lines resulting in gain-of-function phenotypes have been created. Target genes are activated by random insertion in the genome of T-DNA or transposable elements carrying strong promoters [96]. More recently, downregulation of specific genes by using RNAi-based technology [97] has been scaled up to genome-wide level in *Arabidopsis* (e.g., the AGRIKOLA project, http://www.agrikola.org/objectives.html). Genome-scale RNAi approaches take advantage of the easiness of *Agrobacterium* transformation of *Arabidopsis* using the floral dipping technique and of the recent development of site-specific recombination-based cloning vectors allowing efficient and high throughput insertion of inverted repeats of a gene sequence in plant transformation vectors [97, 98]. Though silencing efficiency may vary according to the gene studied, which often results in the observation of a range of more or less severe phenotypic effects in the RNAi silenced plants, this approach is particularly useful when analysing large gene families or classes of genes. In addition to the detailed functional analysis of individual genes, it also allows the study of detectable phenotypes by targeting the regions conserved among several genes in a multigene family, which is very useful when loss-of-function phenotypes are difficult to observe due to the high functional redundancy of plant genes [99]. This strategy may alleviate the need for multiple knockout mutants in order to detect phenotypic changes linked with the mutations in target genes belonging to the same family.

However, these strategies are mostly used for *Arabidopsis* [94] and, to a lesser extent, for rice [100, 101]. Most crop plants still await the development of similar high throughput methods for functional genomics. Considering the case of tomato is instructive. Tomato is the model plant for fleshy fruit development and for *Solanaceae* (among others: potato, tobacco, pepper), and at the same time, a commercial crop of prime importance. Tomato genome size is 950 Mb, i.e., several fold larger than the 125 Mb of *Arabidopsis* but much smaller than the 2,700 Mb of pepper and the 17,000 Mb of wheat, for example. Transposon-based insertional mutagenesis using the non-autonomous mobile elements Activator(Ac)/Dissociation(Ds) from maize have been developed in tomato and shown to be very effective for creating knockout mutants and for promoter-trap studies [102–104]. Activation-tagging lines using T-DNA insertions have also been developed, yielding very interesting gain-of-function phenotypes (Mathews et al., 2003). However, given the genome size of tomato, near to 200,000 to 300,000 transposon-tagged lines are necessary to obtain 95% saturation of the genome, according to some estimates [106]. Since tomato genetic transformation is based on the low throughput *in vitro* somatic embryogenesis, this goal is still out of reach for most groups, including large consortiums, even when using the miniature tomato cultivar MicroTom suitable for high throughput reverse genetics approaches [102]. Insertional mutagenesis with T-DNA in tomato, which necessitates a plant transformation step to obtain each insertion line, would require even more efforts.

The two rate-limiting steps pointed out for tomato, i.e., large genome size and lack of high throughput transformation methods are common features to most crop plants. Ideally, mutagenesis methods for genome-wide reverse genetics should be applicable to any plant whatever the genome size, remain independent of the availability of high throughput transformation methods for that plant (if such method exists) and give a range of mutations prone to be detected by easy, robust, automated and cheap techniques. With the overwhelming increase in sequence data for model and most field-grown crop plants, such alternatives have been developed in recent years. These methods, based on the use of chemical or physical mutagenesis techniques and previously employed for decades for creating genetic variability, have been mostly exploited until recently in plant breeding programs and in forward genetics approaches aimed at identifying the genes behind the phenotypes.

Chemical mutagens and ionising radiations usually create high density of irreversible mutations ranging from point mutations to very large deletions, depending on the mutagenic agent used. As a consequence, saturated mutant collections can be obtained with only a few thousand mutagenised lines, which should be compared to the hundreds of thousand of lines necessary for reaching near saturation collections of insertional mutants [95]. Unknown mutations in target genes can be screened using low throughput classical methods, including DNA sequencing, which may eventually become the method of choice due to the large decrease in DNA sequencing prices over the last years. The recent development of PCR-based technologies allowing the detection of unknown mutations triggered the rapid development of mutant collections in crop and model plants and of high throughput mutation screening methods aimed at discovering the phenotypes behind the genes. An additional advantage of mutant plants in many countries, especially in some European countries opposed to GMO plants, is that they are not genetically modified organisms and, as such, not subjected to regulatory or public acceptance barriers. Mutant alleles can thus be used for crop improvement using traditional and marker assisted breeding programs.

The following section will describe two of the major reverse genetic techniques recently developed for functional genomics approaches in model and crop species: (i) fast neutron mutagenesis and detection [107] and (ii) TILLING (Targeting Induced Local Lesions IN Genomes) [108, 109].

Fast neutron mutagenesis and mutation detection

Fast neutron bombardment is a highly efficient mutagenic method that creates DNA deletions with size distribution ranging from a few bases to more than 30 kb. As a consequence, knockout mutants are obtained. Since the large deletions generated may encompass several genes, this general reverse-genetics strategy can be particularly useful in plant species where duplicated genes, which often show functional redundancy, are arranged in tandem repeats. Availability of tandem repeat knockouts may overcome the very difficult (or even impossible) task of obtaining double mutants. In addition, similar mutation frequencies are observed whatever the size of the genome of the plant [110], which renders this method very attractive for many crop species. One of its disadvantages is that the occurrence of large deletions may be problematic for subsequent genetic analyses. The construction of a deletion mutant collection is straightforward [102, 107, 111]. Basically, after conducting pilot studies aimed at determining the optimal dose necessary to achieve the rate of mutations desired (typically, half of the mutagenised M1 plants should be fertile enough; [112]), M0 seeds are mutagenised, giving M1 seeds which are sown. The M2 seeds are individually collected from the resulting M1 plants and a fraction of them are sown for collecting plant material for DNA extraction. The remaining M2 seeds can be sown for performing phenotypic and segregation analyses on the M2 families and/or stored until further use.

Screening the collection for mutations is a simple PCR-based technique (named Deleteagene for Delete-a-gene) described for rice and *Arabidopsis* [107, 112]. A region of the target gene is PCR-amplified from DNA samples collected from M2 plants using gene-specific primers. The primers and the length of the PCR extension time are carefully chosen so that deletions in target gene can be detectable by PCR in deletion mutants (typically, 1 kb deletions) but not in wild-type plants (wild-type DNA fragment with larger size is not amplified since extension time is too short). In addition, since PCR methods are highly sensitive, pools of up to 2,500 lines can be screened. Once a positive pool is detected, individual mutants can be detected using the same strategy by deconvolution of the pools and of the subpools, and further confirmed by DNA sequencing of the mutated target gene. Based on screenings performed in *Arabidopis*, about 50,000 mutagenised lines would be necessary to achieve an objective of deletion mutants in about 85% of the targeted loci. While possibly realistic in crop plants bearing dry fruits that are easy to collect (e.g., seeds), this objective is probably very difficult to achieve in some other species where seed harvesting is the limiting step, e.g., in the fleshy fruits such as tomato, melon or grape or in species with long reproductive cycles, e.g., the perennial trees. In tomato for example, the largest fast neutron mutagenesis collection includes several thousand M2 families in cv. M82 [102, 111] (http://zamir.sgn.cornell.edu/mutants/), which is already a huge task to produce. In addition, preliminary knowledge of genomic sequence is preferably needed for efficient PCR screening of deletion mutants thereby reducing the range of species for which this method can be used at the present time. For many crop species, forward genetics will probably remain the best adapted approach for using deletion mutant collections in the few coming years.

TILLING

TILLING is a general reverse-genetics strategy first described by McCallum et al. [108] who used this method for allele discovery for chromomethylase gene in *Arabidopsis* [113]. This method combines random chemical mutagenesis by EMS (ethylmethanesulfonate) with PCR-based methods for detecting unknown point mutations in regions of interest in target genes. Since the early description of the method, which was then performed by using heteroduplex analysis with dHPLC [108], the method has been refined and adapted to high throughput screening by using enzymatic mismatch cleavage with CEL1 endonuclease, a member of the S1 nuclease family [109, 114]. TILLING technology is quite simple, robust, cost-effective and thus affordable for many laboratories. In addition, it allows the identification of allelic series including knockout and missense mutations. For these reasons, this genome-wide reverse-genetics strategy has been applied very rapidly to a growing number of plants, including model plants and field-grown crops of diverse genome size and ploidy levels, and even to insects (*Drosophila* [115]). A number of TILLING efforts in plants have been reported for *Arabidopsis* [109, 116], *Lotus japonicus* [117], barley [118], maize [119] and wheat [120]. Recent reviews give excellent insights on the TILLING methods, from the production of the mutagenised population to the current technologies for mutation detection, and on the future prospects for TILLING [121–124]. In addition, a number of TILLING facilities have been created for various plants including facilities for *Arabidopsis* which already delivered >6,000 EMS-induced mutations in *Arabidopsis* and is also opened to other species [124] (ATP, http://tilling.fhcrc.org:9366/), maize at Purdue University (http://genome.purdue.edu/ maizetilling/), *Lotus* in Norwich (USA) (http://www.lotusjaponicus.org/tillingpages/ Homepage.htm), barley in Dundee (UK) (http://germinate.scri.sari.ac.uk/barley/mutants/), sugar beet in Kiel (Germany) (http://www.plantbreeding.uni-kiel.de/project_tilling.shtml), pea at INRA (Evry, France; http://www.evry.inra.fr/public/projects/tilling/tilling.html) and ecotilling at CanTILL (Vancouver, Canada) (http://www.botany.ubc.ca/can-till/).

Mutagenesis

EMS (ethylmethanesulfonate) is the mutagenic agent used for most of the plant TILLING projects cited above. As a result of EMS alkylation of guanine, more than 99% of mutations are G/C-to-A/T transitions, as experimentally shown by analysing (EMS)-induced mutations in *Arabidopsis* [116]. Other mutagens with genotoxic effects inducing point mutations, frameshifts or small insertion/deletions (InDel) are also likely to be applicable to a TILLING project using CEL1 endonuclease. Indeed, CEL1 technology allows the efficient detection of a broad range of mutations, i.e., the natural allelic variants found in different plant genotypes or ecotypes or the artificially-induced mutations in zebrafish induced by the *N*-ethylnitro-*N*-nitrosourea (ENU) mutagen [125]. With EMS, similar mutation frequencies are expected whatever the plant genome size [110], rendering this approach applicable to most crop species. However, considering the results from the diverse TILLING projects in different species, the mutation density detected by TILLING may actually range from 1 mutation/Mb in barley [118] and 1 mutation/500 kb in maize [119] to 1 mutation/40 kb in tetraploid wheat and even 1 mutation/25 kb in hexaploid wheat [120]. By comparison, mutation densities are 1 mutation/170 kb in *Arabidopsis* (ATP project [116]) and 1 mutation/125 kb in MicroTom tomato (our own unpublished results). Polyploidy may confer tolerance to EMS mutations, thus explaining the high density of mutations found in wheat [124].

EMS treatment is usually done by soaking the seeds (referred to as M0 seeds) in EMS solution for several hours (usually 12–16 h overnight); mutagenised seeds are then referred to as M1 seeds (Fig. 4). Pollen can also be mutagenised, as done in maize [119, 124]. At this step, a delicate balance has to be found between (i) the primary objective of mutagenesis for TILLING, which is to obtain saturated mutagenesis (i.e., the highest density of mutations possible in the plant genome) in order to analyse a reduced number of lines, and (ii) the amount of mutagenesis that a plant can withstand without overwhelming problems of seed lethality or plant lethality and sterility. In tomato, we obtained high density mutations using EMS doses giving 50–70% of seed lethality after EMS treatment (M1 seeds) and 40–50% of sterile plants in the M1 plants. Since the necessary EMS concentrations may vary considerably according to the species, the physiological state of the seeds and even from batch to batch, pilot studies with different EMS concentrations (from 0.2–1.5%) should be carried out before large scale mutagenesis. The M1 plants obtained by sowing the mutagenised seeds are chimeric and cannot be further used for mutation detection. Indeed, in the embryo, each cell is independently mutagenised. Only a few cells in the apical meristem (e.g., two to three cells in tomato, A. Levy, personal communication) will give rise to reproductive organs and thus to gametes. In contrast, mutations in other embryonic cells are not inherited by the next generation (somatic mutations) and will give rise to chimeric tissues in M1 plants (e.g., the variegated plants with dark green and light green or white sectors often observed in M1 plants).

The M2 seeds, obtained after selfing (or crossing when necessary) the M1 plants, are individually collected from each plant and stored. One or a few M2 plants are usually grown in order to provide plant material for DNA extraction (Fig. 4). Another strategy that we use in tomato, though it involves a time-consuming step, is to grow 12 individual plants per M2 family and to collect M3 seeds and tissue samples from these plants. In addition to enabling the multiplication of the seeds, this strategy allows the description of the plant phenotypes and the segregation analyses of visible mutations in the M2 families. These data are collected and further compiled in a phenotypic description database. The rationale is that once a mutation in a target gene is detected in an individual M2 family, the information on the phenotypic and segregation data can give a first hint on the severity of the mutation and the functional role in the plant of the target gene without having to wait for the observations made on M3 plants. This approach can be particularly useful when dealing with crop species that have a long developmental cycle and/or with specific plant tissues (e.g., fruits or seeds).

In addition to the artificially-induced mutants obtained by using various physical or chemical mutagens in species such as rice [126] or tomato [111], natural allelic

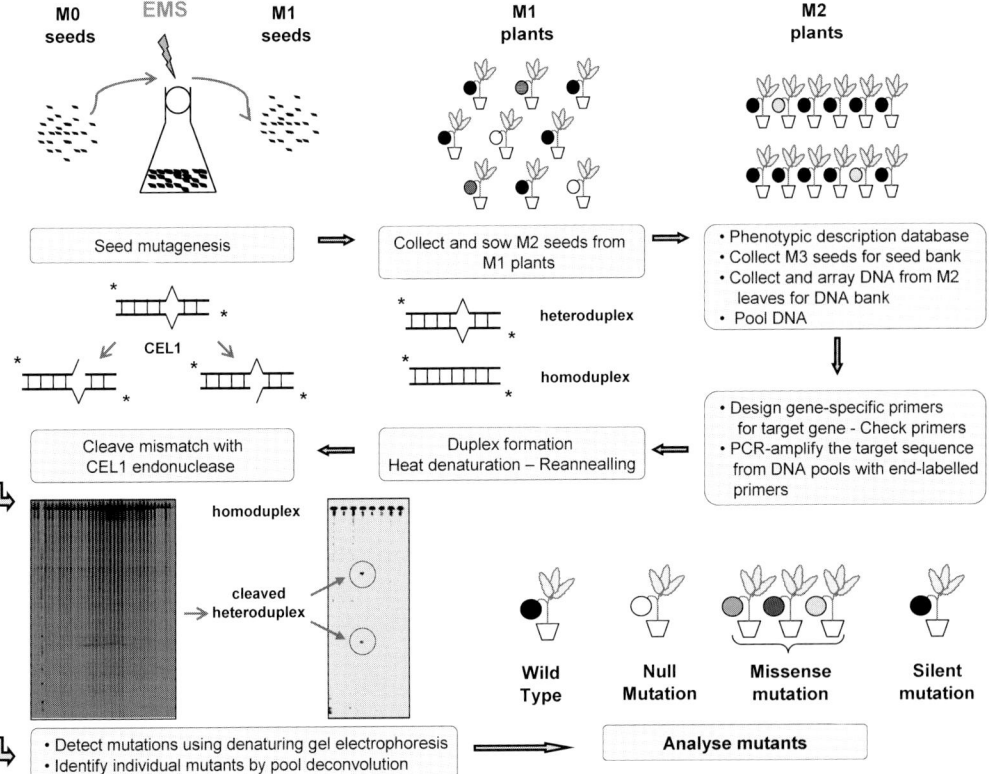

Figure 4. Schematic description of the TILLING procedure. Tomato TILLING strategy is shown. Seeds (M0) are mutagenised with ethylmethanesulfonate (EMS) giving M1 seeds, which are sown. M2 seeds from the resulting M1 plants are collected and sown. For each M2 family, 12 plants are grown and used for: (i) description of plant phenotype (data stored in a tomato mutant database); (ii), extraction of DNA from leaf tissue, later used for mutation detection; and (iii), collection of M3 seeds stored in a seed bank. For mutation detection, eightfold DNA pools are generated from M2 family DNA and gene-specific primers are designed to PCR-amplify the target gene from these pools. The resulting amplicon is heat denatured and reannealed, producing both homoduplexes and heteroduplexes (presence of a mismatch in the duplex). Heteroduplexes are cleaved at the 3' side of the mismatch by the CEL1 endonuclease and further detected by denaturing gel electrophoresis. Identification of the individual M2 family harbouring the mutation is done by deconvolution of the DNA pools using the same technology. Screening tomato mutant collection for a target gene (e.g., a gene involved in fruit colour) yields a series of mutant alleles. Some mutations (~5%) will create knockout mutants (null mutations, ~5%) or affect the biological function of the encoded protein (missense mutations, ~50%) while many mutations (~45%) will remain silent.

variants are already present in germplasm resources, which represent a large source of genetic variability for most crop and model species [57, 127]. Core collections may include related species, various accessions with high genetic diversity often collected near the centre of origin of the species, and cultivated lines and mutants obtained by breeders worldwide (e.g., the Tomato Genomic Resource Center at Davis: http://tgrc.ucdavis.edu/). In addition to the populations of artificially-induced mutants, these collections provide very useful resources for identifying natural alleles for a target gene using Ecotilling. This approach refers to the detection, using high throughput TILLING technology with CEL1 type endonuclease, of allelic variants in the species germplasm (e.g., ecotypes in *Arabidopsis*, hence the name of Ecotilling) [128]. This can be particularly useful in association genetics approaches, for example for the confirmation of the role of a candidate gene previously shown to be co-localised with a QTL.

Mutation detection

A recent review [124] describes in detail the current technologies for mutation and polymorphism detection while Yeung et al. [114] analyses and compares the diverse enzymatic mutation detection technologies available. Basically, three different technologies are used for high throughput mutation discovery in TILLING: (i), the denaturing high performance liquid chromatography (dHPLC), originally used in the first plant TILLING project described [108] and further improved since [118]. The dHPLC is a duplex DNA melting temperature-based system that allows the detection of duplex DNA fragments destabilised by mismatches using temperature-controlled hydrophobic columns. The system is automated and can be used for screening four family DNA pools. However, this technology displays best results with DNA fragment ranging from 300–600 bp and does not allow the precise location of the point mutation; (ii), the single-strand conformational polymorphism (SSCP), which detects conformational changes caused by point mutations and has been improved and automated for capillary DNA sequencers. However, it shows the same limitations as dHPLC, i.e., the limitation to pools of four DNA samples, the detection of fragments <500 bp, and the unknown location of the mismatch; and (iii) enzymatic mismatch cleavage using endonuclease enzymes, members of the S1 nuclease family, followed by electrophoresis separation of the cleaved fragments [109]. This technology has become the method of choice for high throughput TILLING [122].

Originally extracted from celery and later from other plant species, the CEL1 endonuclease is a mismatch cleavage enzyme showing very little sequence bias [114]. In addition, CEL1 has an exonuclease activity that cleaves the 5' end of DNA fragments, thus releasing the labelled end used for detecting DNA fragments (Fig. 4), which can decrease the sensitivity of the detection. Efficient CEL1 enzyme preparations can be purified from many plant sources [124]. In addition, enzymes performing similar functions have been cloned and are commercially available such as the Surveyor mutation detection kit (http://www.transgenomic.com/flash/surveyor/Surveyor.asp [129]) or the ENDO1 enzyme (http://www.evry.inra.fr/public/projects/tilling/tilling.html).

The technology used for high throughput TILLING with CEL1 is very simple (Fig. 4) and affordable in main research centres. First, a DNA fragment of 0.5–2 kb is amplified from DNA pools (usually eight-fold pools when detecting heterozygous mutations, i.e., 1 genome in 16) with differentially labelled primers. The design of the primer will depend on the previous knowledge of the protein (the most interesting region to target for functional analysis according to the user, e.g., the interacting domain in a transcription factor or the catalytic site in an enzyme), the probability of finding knockout or missense mutations in the region, which can be estimated using the CODDLE (Codons Optimised to detect Deleterious Lesions) software developed by the Seattle group (http://www.proweb.org/coddle) or, more simply for many crop plants, the availability of EST or genomic sequences. Amplification of the DNA fragment with unlabelled primers is usually done in a first round to check the primers, especially when amplifying DNA fragments with no previous knowledge of genomic sequence of the target gene, e.g., EST sequences. In order to reduce the costs of labelled primers specifically designed for a target gene, a two-step strategy can also be followed for amplifying labelled DNA fragments [115]. The labelling of the primers will depend on the electrophoresis equipment used: infrared-based sequencers such as LI-COR, which is commonly used for TILLING due to its robustness and sensitivity [109, 121, 124], or fluorescence-based sequencers such as ABI sequencers [114]. Once the labelled DNA fragment has been amplified, the amplicon is subjected to a high temperature-denaturation/low temperature-reannealing cycle, in order to allow the formation of DNA homoduplexes and heteroduplexes. By using CEL1 endonuclease, which cuts at the 3' side of the mismatch, the heteroduplexes are then cleaved while homoduplexes are left intact by the enzyme (Fig. 4). The cleaved end-labelled DNA fragments can be readily separated from non-cleaved DNA fragments by electrophoresis on denaturing gel. Furthermore, the use of differentially labelled primers allows the precise location on the gel of the two cleaved fragments and thus the detection of the region in the DNA sequence where mutation occurs. In addition to the use of Photoshop software for gel image analysis and band detection, newly developed free software called GelBuddy (www.proweb.org/gelbuddy/index.html) facilitates image analysis of TILLING gels [124].

Once a mutant is detected in a pool of families, the deconvolution of the pool and the detection of the mutated family or plant can be done using the same technology (PCR amplification of target gene, CEL1 cleavage and denaturing gel detection). The mutation in the target gene can thus be confirmed, usually by using DNA sequencing or alternative Single Nucleotide Polymorphism (SNP) detection technologies [124].

Linking mutation to phenotype

EMS induces point mutations, mostly G/C-to-A/T transitions. Single-base change in protein-coding genes may be classified as silent, missense or truncation. Silent mutations do not affect the protein. Missense mutations arise when single base change in a given codon induces changes in the amino acid encoded. Amino acid substitutions can be conservative (similar function is expected) or non conservative (e.g., the substitution of the neutral amino acid glycine by the basic amino acid arginine, which is expected to modify the function of the protein). The SIFT (Sorting Intolerant From Tolerant) program can be used to predict the damage to protein function caused by missense mutation (http://blocks.fhcrc.org/sift/SIFT.html). Truncations of the protein resulting in knockout mutants are expected from single-base changes converting an amino acid codon to a stop codon or from mutations in splice junctions. From the TILLING experimental results obtained in *Arabidopsis* [116] the proportion of nonsilent mutations that may affect the biological function of the protein and hence the phenotype of the plant, was estimated to be 55%, including 5% of truncations and 50% of missense mutations. Interestingly, there was a considerable bias in favour of heterozygotes for the detection of the most severe mutations (truncations), suggesting that corresponding knockouts mutations in homozygotes were lethal. These overall results highlight the potential of TILLING for discovering allelic series, including knockouts and hypomorphic mutations that are highly informative for functional studies of target genes.

Once a mutation is discovered in a target gene and the corresponding family identified, the effect on the plant resulting from a possible lesion on the protein must be screened phenotypically, usually on the M3 plants. At this point, a major issue is how to differentiate the mutation in the target gene detected by TILLING from the other background mutations in the plant introduced by EMS mutagenesis. Actually, the strategy will depend on the objective of TILLING, i.e., for mutation breeding purposes or for functional study of a target gene. For crop improvement, a number of cycles of backcrossing are necessary before agronomic use. In the highly mutagenised wheat for example, Slade et al. [120] estimated that four backcrosses should be sufficient to derive lines very similar to the parents but did not exclude the need for additional backcrosses. For functional studies, it is generally considered that the fastest method for demonstrating that the mutant phenotype results from a mutation in the target gene is to isolate additional mutant alleles [94].

The optimum number of mutated alleles necessary for functional studies of a gene of interest will mostly depend on the target gene studied. Based on the results obtained in *Arabidopsis* [116], an allelic series including one knockout mutation and ~10 missense mutations that can possibly affect the biological function of the protein should roughly comprise 20 mutated alleles. Depending on the species and the density of mutations in the collection of mutants, this objective usually involves the screening of 3,000–6,000 mutant lines. According to calculations made with *Arabidopsis* TILLING collections [121], the frequency of misattributing a phenotype observed in M3 plants in these collections to a mutation in the target gene can be estimated to ~0.05% when the parent M2 plant is heterozygous. When the M2 plant is homozygous, a backcross is necessary before selfing and analysing the plants. Another possibility is to cross two independent lines mutated in the same target gene. Background mutations are heterozygous in the resulting plants carrying the two non-complementing mutations that can therefore be considered as responsible for the phenotype observed.

Plant systems biology and reverse genetics approaches

During the last few years, tremendous efforts have been made in developing genome-size reverse genetics tools and genetic resources in model and crop plants for studying gene function in the context of the plant. At the same time, the development of high-throughput approaches for global analyses of transcripts, proteins and metabolites paved the way for a comprehensive description of complex networks involved in signal transduction cascades, in regulation and activity of primary or secondary metabolism pathways, and in many other aspects of plant development. These studies have major consequences on our present way of studying plants. First, they allow the discovery of new candidate genes putatively involved in the operation of plant functional networks [94]. Other candidate genes are being generated in both model and crop plants by the forward genetic approaches aimed at identifying the genes underlying the QTLs controlling traits of interest, as previously described. Second, beyond the mere functional study of a single gene, genomic-scale approaches now allow the study of plant biology from the systems level. Visualisation of metabolic pathways and cell functions is already facilitated in some model and crop plants by tools such as MAPMAN which uses transcriptome and metabolome data [130, 131], and models describing complex networks begin to be constructed in plants [132].

Plant mutants have already proved valuable tools for plant functional genomic studies, e.g., for the discovery of the function of new candidate genes and the analysis of their possible contribution to functional complexes or metabolic pathways [94, 133]. Given the very large collections of insertional mutants available in *Arabidopsis*, most of the studies have been focused on knockout mutants. Indeed, null mutants can be very helpful genetic tools for systems biology approaches, as demonstrated in yeast [134], for example. In this genome-scale study, knockout mutants with functions in central metabolism used in combination with computational analyses, flux data and phenotypic analyses gave access to the relative contribution of network redundancy and of alternative pathways to genetic network robustness in yeast. Although comparable studies are still difficult to carry out in plants, integrated analyses of plant primary and secondary metabolic networks using null mutants or overexpressing lines have been attempted [132, 133] and should progress with the availability of new mutant collections and analytical technologies.

In that context, the recent development of large-scale RNAi in *Arabidopsis* and, especially, of the TILLING and Ecotilling approaches in model and crop plants is very promising. The RNAi approach is already used in some model organisms such as *C. elegans* for inducing systematic perturbations of networks in order to study the functional relationships between the components of interacting complexes involved in a signalling pathway [135]. Systems biology approaches can also make use of TILLING and Ecotilling, which reveal allelic series corresponding to several independent point mutations or other small mutations in target genes. Point mutations are more prone than null mutants to cause a range of discrete variations close to those observed in natural populations, where most traits are controlled by Quantitative Trait Loci (QTLs). One advantage of the artificially-induced mutants for systems biology studies is that they share exactly the same genetic background and can thus be directly compared, while the lines containing the natural allelic variants usually differ by several tens or hundreds of genes, even in Nearly Isogenic Lines. Nonsilent point mutation usually results in protein lesion, the severity of which will cause a more or less profound effect on the biological function of the protein. Point mutations may also produce dominant-negative mutants, which are very useful tools for revealing functional interactions between the components of a complex or a signalling pathway [2], or even gain-of-function mutants such as the tomato LIN5 invertase variant with altered kinetic properties [60], originally cloned as a QTL controlling solid soluble solids content in tomato fruit [59]. The wide collection of mutants available for a gene of interest identified through TILLING should be particularly amenable for systems biology approaches since a range of quantitative effects, and not only of qualitative effects as in null mutants, can be obtained.

How to use these mutants? One of the most immediate applications in network analysis for mutants detected by TILLING is probably the study of the regulation of metabolic pathways. Although few TILLING results have been published to date, two of the target genes analysed were involved in sugar metabolism, either in starch synthesis [120] or in the synthesis of callose, a beta-1,3-glucan [136]. Metabolite profiling is a high throughput technology with limited cost per sample that allows the initial screening of the allelic mutants identified, even those showing no visual phenotype. Furthermore, since the establishment of network regulation needs large-scale studies involving as many different mutants in several target genes as possible [132, 134], metabolic profiling can be reduced in a first step to rapid metabolic fingerprinting of the mutants, as already experimented with mutants displaying a silent phenotype [137, 138]. In this approach, the most interesting mutants showing significant perturbations in metabolite profiles can be subsequently subjected to more detailed analyses, including transcriptome, proteome and metabolome profiling. The global set of data obtained can be further combined and analysed with the array of tools already available ([130, 139] and Chapters by Dieuaide-Noubhani et al., Nikiforova and Willmitzer, and Ahrens et al.), in order to validate the underlying hypotheses on the functional role of the target gene studied and/or to give a comprehensive view of the metabolic network [140]. One delicate step for fully understanding the changes in the metabolic network induced by the mutation in the target genes remains the analysis of the metabolic fluxes ([141] and Chapter by Dieuaide-Noubhani et al.), which can hardly be carried out in a high throughput manner in plants, and, therefore, will probably remain restricted to a limited number of mutants previously selected through global analyses.

Summary

The first experiments on gene and QTL mapping date from the late 1980s. Since that time, hundreds of mapping experiments have been performed, providing information on the genetic basis of individual traits or allowing complex traits to be dissected into their component parts. The number of Mendelian mutations characterised by a candidate gene approach or positional cloning has rapidly increased, but very few QTL have been characterised to date. Accumulated data from several species suggest a continuum between discrete variations (mutant genes) and continuous variations (QTL), and the identification of QTL will improve our understanding of the molecular and physiological basis to complex character variation. In this context, gene maps and large EST data sets will prove useful as sources of candidates. The access to a growing number of sequenced genomes, and to transcriptomic and proteomic approaches, should increase the efficiency of QTL characterisation. Furthermore ecophysiological modelling and metabolomic profiling will give clues to the physiological processes underlying QTL and the potential candidate genes. In this context, fine mapping of the QTL and validation of the candidate genes will become the most restrictive steps.

The development of large scale DNA sequencing facilities and of high throughput gene and protein expression and metabolite profiling technologies in model and crop plants has triggered the development of genome-wide reverse genetics tools aimed at identifying and characterising the function of candidate genes in the context of the plant. Insertional mutagenesis using T-DNA or transposons that creates knockout or activation-tagged mutants and, more recently, large scale gene targeting by RNAi have been the methods of choice for functional genomics in the model plants *Arabidopsis* and rice. However, most of the above mentioned tools are unavailable in crop plants due to limitations (low throughput genetic transformation technologies, size of the genome) inherent to the species. For these reasons, new technologies for detecting unknown mutations created by chemical mutagens or ionising radiations have emerged in the recent years. Among them, the TILLING (Targeting Induced Local Lesions In Genomes) technology, which is mostly based on the generation by a chemical mutagen (EMS) of high density point mutations evenly distributed in the genome and on the subsequent screening of the mutant collection by a PCR-based enzymatic assay, has become very popular and is currently applied to a wide variety of model and crop plants. Chemical mutagenesis used in the TILLING procedure generates a range of mutated alleles for a target gene, including knockouts and missense mutations, thereby affecting more or less severely the biological function of the corresponding protein and the phenotype of the plant. These allelic series should prove valuable tools for plant systems biology studies by enabling the comparative analysis of metabolic or other complex networks in plants showing genetic variability for a target gene with the help of genomics (transcriptome, proteome, metabolome) and data analysis/modelling tools.

References

1. Rensink WA, Lee Y, Liu J, Iobst S, Ouyang S, Buell RC (2005) Comparative analyses of six solanaceous transcriptomes reveal a high degree of sequence conservation and species-specific transcripts. *BMC Genomics* 6: 124
2. Diévart A, Clark SE (2003) Using mutant alleles to determine the structure and function of leucine-rich repeat receptor-like kinases. *Curr Opin Plant Biol* 6: 507–516
3. Tanksley SD, Ganal MW, Martin GB (1995) Chromosome landing – a paradigm for map-based gene cloning in plants with large genomes. *Trends Genet* 11: 63–68

4. Jander G, Norris SR, Rounsley SD, Bush DF, Levin IM, Last RL (2002) *Arabidopsis* map-based cloning in the post-genome era. *Plant Physiol* 129: 440–450
5. Xu YB, McCouch SR, Zhang QF (2005) How can we use genomics to improve cereals with rice as a reference genome? *Plant Mol Biol* 59: 7–26
6. Hackett CA (2002) Statistical methods for QTL mapping in cereals. *Plant Mol Biol* 48: 585–599
7. Lander ES, Botstein D (1989) Mapping Mendelian factors underlying quantitative traits using RFLP linkage maps. *Genetics* 121: 185–199
8. Lincoln SE, Daly MJ, Lander ES (1992) Constructing genetic maps with MAPMAKER/EXP version 3.0. A tutorial and reference manual http://linkage.rockeffeller.edu/soft/mapmaker
9. Jansen RC, Stam P (1994) High-resolution of quantitative traits into multiple loci via interval mapping. *Genetics* 136: 1447–1455
10. Zeng ZB (1994) Precision mapping of quantitative trait loci. *Genetics* 136: 1457–1468
11. Churchill GA, Doerge RW (1996) Empirical threshold values for quantitative trait mapping. *Genetics* 138: 963–971
12. Darvasi A, Soller M (1997) Simple method to calculate resolving power and confidence interval of QTL map location. *Behav Genet* 27: 125–132
13. Lee M, Sharopova N, Beavis WD, Grant D, Katt M, Blair D, Hallauer A (2002) Expanding the genetic map of maize with the intermated B73 x Mo17 (IBM) population. *Plant Mol Biol* 48: 453–461
14. Grattapaglia D, Bertolucci FL, Sederoff RR (1995) Genetic mapping of QTLs controlling vegetative propagation in *Eucalyptus grandis* and *E. urophylla* using a pseudo-testcross mapping strategy and RAPD markers. *Theor Appl Genet* 90: 933–947
15. Bradshaw HD, Stettler RF (1995) Molecular genetics of growth and development in *Populus*. IV. Mapping QTLs with large effects on growth, form and phenology traits in a forest tree. *Genetics* 139: 963–973
16. Tanksley SD, Nelson JC (1996) Advanced backcross QTL analysis: a method for the simultaneous discovery and transfer of valuable QTLs from unadapted germplasm into elite breeding lines. *Theor Appl Genet* 92: 191–203
17. Eshed Y, Zamir D (1995) An introgression line population of *Lycopersicon pennellii* in the cultivated tomato enables the identification and fine mapping of yield associated QTLs. *Genetics* 141: 1147–1162
18. Ronin Y, Korol A, Shtemberg M, Nevo E, Soller M (2003) High-resolution mapping of quantitative trait loci by selective recombinant genotyping. *Genetics* 164: 1657–1666
19. Lynch M, Walsh B (eds) (1998) Genetics and analysis of quantitative traits. Sinauer Associates
20. De Vienne D, Causse M (2002) Mapping and characterizing quantitative trait loci. In: D de Vienne (ed): *Molecular markers in Plant genetics and Biotechnology*. Sci Publisher Inc, 89–125
21. Xu SZ (2003) Theoretical basis of the Beavis effect. *Genetics* 165: 2259–2268
22. Kearsey MJ, Farquhar AGL (1998) QTL analysis in plants; where are we now? *Heredity* 80: 137–142
23. Kroymann J, Mitchell-Olds T (2005) Epistasis and balanced polymorphism influencing complex trait variation. *Nature* 435: 95–98
24. Grandillo S, Ku HM, Tanksley SD (1999) Identifying the loci responsible for natural variation in fruit size and shape in tomato. *Theor Appl Genet* 99: 978–987

25. Fulton TM, Bucheli P, Voirol E, López J, Pétiard V, Tanksley SD (2002) Quantitative trait loci (QTL) affecting sugars, organic acids and other biochemical properties possibly contributing to flavor, identified in four advanced backcross populations of tomato. *Euphytica* 127: 163–177
26. Chardon F, Virlon B, Moreau L, Falque M, Joets J, Decousset L, Murigneux A, Charcosset A (2004) Genetic architecture of flowering time in maize as inferred from quantitative trait loci meta-analysis and synteny conservation with the rice genome. *Genetics* 168: 2169–2185
27. Bernacchi D, Beck-Bunn T, Emmatty D, Eshed Y, Inai S, Lopez J, Petiard V, Sayama H, Uhlig J, Zamir D et al. (1998) Advanced back-cross QTL analysis of tomato. II. Evaluation of near-isogenic lines carrying single-donor introgressions for desirable wild QTL-alleles derived from *Lycopersicon hirsutum* and *L. pimpinellifolium*. *Theor Appl Genet* 97: 1191–1196
28. Fatokun CA, Menanciohautea DI, Danesh D, Young ND (1992) Evidence for orthologous seed weight genes in cowpea and mung bean based on rflp mapping. *Genetics* 132: 841–846
29. Timmerman-Vaughan GM, Mccallum JA, Frew TJ, Weeden NF, Russell AC (1996) Linkage mapping of quantitative trait loci controlling seed weight in pea (*Pisum sativum* L.). *Theor Appl Genet* 93: 431–439
30. Maughan PJ, Maroof MAS, Buss GR (1996) Molecular-marker analysis of seed-weight: genomic locations, gene action, and evidence for orthologous evolution among three legume species. *Theor Appl Genet* 93: 574–579
31. Paterson AH, Lin YR, Li Z, Schertz KF, Doebley JF, Pinson SR, Liu SC, Stansel JW, Irvine JE (1995) Convergent domestication of cereal crops by independent mutations at corresponding genetic loci. *Science* 269: 1714–1718
32. Devos M, Gale D (1997) Comparative genetics in the grasses. *Plant Mol Biol* 35: 3–15
33. Frary A, Doganlar S, Daunay MC, Tanksley S D (2003) QTL analysis of morphological traits in eggplant and implications for conservation of gene function during evolution of solanaceous species. *Theor Appl Genet* 107: 359–370
34. Tanksley SD (1993) Mapping Polygenes. *Annu Rev Genet* 27: 205–233
35. Eshed Y, Zamir D (1996) Less-than-additive epistatic interactions of quantitative trait loci in tomato. *Genetics* 143: 1807–1817
36. Doebley J, Stec A, Hubbard L (1997) The evolution of apical dominance in maize. *Nature* 386: 485–488
37. Chaib J, Lecomte L, Buret M, Causse M (2006) Stability over genetic backgrounds, generations and years of quantitative trait locus (QTLs) for organoleptic quality in tomato. *Theor Appl Genet* 112: 934–944
38. Prioul JL, Quarrie SA, Causse M, de Vienne D (1997) Dissecting complex physiological functions through the use of molecular quantitative genetics. *J Exp Bot* 48: 1151–1163
39. Lefebvre V, Palloix A (1996). Both epistatic and additive effects of QTLs are involved in polygenic induced resistance to disease: a case study, the interaction pepper – *Phytophthora capsici* Leonian. *Theor Appl Genet* 93: 503–511
40. Sergeeva LI, Keurentjes JJB, Bentsink L, Vonk J, van der Plas LHW, Koornneef M, Vreugdenhil D (2006) Vacuolar invertase regulates elongation of *Arabidopsis thaliana* roots as revealed by QTL and mutant analysis. *Proc Natl Acad Sci USA* 103: 2994–2999
41. Korol AB, Ronin YI, Itskovich AM, Peng J, Nevo E (2001) Enhanced efficiency of quantitative trait loci mapping analysis based on multivariate complexes of quantitative traits. *Genetics* 157: 1789–1803

42. Paterson AH, de Verna JW, Lanini B, Tanksley SD (1990) Fine mapping of quantitative trait loci using selected overlapping recombinant chromosomes, in an interspecies cross of tomato. *Genetics* 124: 735–742
43. Monforte AJ, Tanksley SD (2000) Fine mapping of a quantitative trait locus (QTL) from *Lycopersicon hirsutum* chromosome 1 affecting fruit characteristics and agronomic traits: breaking linkage among QTLs affecting different traits and dissection of heterosis for yield. *Theor Appl Genet* 100: 471–479
44. Lecomte L, Saliba-Colombani V, Gautier A, Gomez-Jimenez MC, Duffé P, Buret M, Causse M (2004) Fine mapping of QTLs of chromosome 2 affecting the fruit architecture and composition of tomato. *Mol Breeding* 13: 1–14
45. Zeng ZB (1993) Theoretical basis for separation of multiple linked gene effects in mapping quantitative trait loci. *Proc Natl Acad Sci USA* 90: 10972–10976
46. Jansen RC, van Ooijen JW, Stam P, Lister C, Dean C (1995) Genotype-by environment interaction in genetic mapping of multiple quantitative trait loci. *Theor Appl Genet* 91: 33–37
47. Romagosa I, Ullrich SE, Han F, Hayes PM (1996) Use of the additive main effects and multiplicative interaction model in QTL mapping for adaptation in barley. *Theor Appl Genet* 93: 30–37
48. Moreau L, Charcosset A, Gallais A (2004) Use of trial clustering to study QTL x environment effects for grain yield and related traits in maize. *Theor Appl Genet* 110: 92–105
49. Loudet O, Chaillou S, Merigout P, Talbotec J, Daniel-Vedele F (2003) Quantitative trait loci analysis of nitrogen use efficency in *Arabidopsis*. *Plant Physiol* 131 : 345–358
50. Juenger TE, Mckay JK, Hausmann N, Keurentjes JJB, Sen S, Stowe KA, Dawson TE, Simms EL, Richards JH (2005) Identification and characterization of QTL underlying whole-plant physiology in *Arabidopsis thaliana*: delta C-13, stomatal conductance and transpiration efficiency. *Plant Cell Environ* 28: 697–708
51. Reymond M, Muller B, Leonardi A, Charcosset A, Tardieu F (2003) Combining quantitative trait loci analysis and an ecophysiological model to analyze the genetic variability of the response of maize leaf growth to temperature and water deficit. *Plant Physiol* 131: 664–675
52. Quilot B, Kervella J, Genard M, Lescourret F (2005) Analysing the genetic control of peach fruit quality through an ecophysiological model combined with a QTL approach. *J Exp Bot* 56: 3083–3092
53. Yin XY, Struik PC, Kropff MJ (2004) Role of crop physiology in predicting gene-to-phenotype relationships. *Trends Plant Sci* 9: 426–432
54. Paran I, Zamir D (2003) Quantitative traits in plants: beyond the QTL. *Trends Genet* 19: 303–306
55. Salvi S, Tuberosa R (2005) To clone or not to clone plant QTLs: present and future challenges. *Trends Plant Sci* 10: 298–304
56. Tuinstra MR, Ejeta G, Goldbrough PB (1997) Heterogeneous inbred families (HIF) analysis: a method for developing near-isogenic lines that differ at quantitative trait loci. *Theor Appl Genet* 95: 1005–1011
57. Zamir D (2001) Improving plant breeding with exotic genetic libraries. *Nat Rev Genet* 2: 983–988
58. Durrett RT, Chen KY, Tanksley SD (2002) A simple formula useful for positional cloning. *Genetics* 160: 353–355
59. Fridman E, Pleban T, Zamir D (2000) A recombination hotspot delimits a wild-species quantitative trait locus for tomato sugar content to 484 bp within an invertase gene. *Proc Natl Acad Sci USA* 97: 4718–4723

60. Fridman E, Carrari F, Liu YS, Fernie AR, Zamir D (2004) Zooming in on a quantitative trait for tomato yield using interspecific introgressions. *Science* 305: 1786–1789
61. Frary A, Nesbitt TC, Grandillo S, Knaap E, Cong B, Liu J, Meller J, Elber R, Alpert KB, Tanksley SD (2000) fw2.2: a quantitative trait locus key to the evolution of tomato fruit size. *Science* 289: 85–88
62. Pflieger S, Lefebvre V, Causse M (2001) The candidate gene approach. A Review. *Molecular Breeding* 7: 275–291
63. Byrne PF, McMullen MD, Snooks ME, Musket TA, Theuri JM, Widstrom NW, Wiseman BR, Coe EH (1996) Quantitative trait loci and metabolic pathways: genetic control of the concentration of maysin, a corn earworm resistance factor, in maize silks. *Proc Natl Acad Sci USA* 93: 8820–8825
64. Thornsberry JM, Goodman MM, Doebley J, Kresovich S, Nielsen D, Buckler ES (2001) Dwarf8 polymorphisms associate with variation in flowering time. *Nat Genet* 28: 286–289
65. Olsen KM, Halldorsdottir SS, Stinchcombe JR, Weinig C, Schmitt J, Purugganan MD (2004) Linkage disequilibrium mapping of *Arabidopsis* CRY2 flowering time alleles. *Genetics* 167: 1361–1369
66. Osterberg MK, Shavorskaya O, Lascoux M, Lagercrantz U (2002) Naturally occurring indel variation in the *Brassica nigra* COL1 gene is associated with variation in flowering time. *Genetics* 161: 299–306
67. Gupta V, Mukhopadhyay A, Arumugam N, Sodhi YS, Pental D, Pradhan AK (2004) Molecular tagging of erucic acid trait in oilseed mustard (*Brassica juncea*) by QTL mapping and single nucleotide polymorphisms in FAE1 gene. *Theor Appl Genet* 108: 743–749
68. Guillet-Claude C, Birolleau-Touchard C, Manicacci D, Rogowsky PM, Rigau J, Murigneux A, Martinant JP, Barriere Y (2004) Nucleotide diversity of the ZmPox3 maize peroxidase gene: Relationships between a MITE insertion in exon 2 and variation in forage maize digestibility. *BMC Genetics* 5: Art. No. 19
69. El-Assal SED, Alonso-Blanco C, Peeters AJM, Raz V, Koornneef M (2001) A QTL for flowering time in *Arabidopsis* reveals a novel allele of CRY2. *Nat Genet* 29: 435–440
70. Doi K, Izawa T, Fuse T, Yamanouchi U, Kubo T, Shimatani Z, Yano M, Yoshimura A (2004) Ehd1, a B-type response regulator in rice, confers short-day promotion of flowering and controls FT-like gene expression independently of Hd1. *Genes Dev* 18: 926–936
71. Ren ZH, Gao JP, Li LG, Cai XL, Huang W, Chao DY, Zhu MZ, Wang ZY, Luan S, Lin HX (2005) A rice quantitative trait locus for salt tolerance encodes a sodium transporter. *Nat Genet* 37: 1141–1146
72. Takahashi Y, Shomura A, Sasaki T, Yano M (2001) Hd6, a rice quantitative trait locus involved in photoperiod sensitivity, encodes the alpha subunit of protein kinase CK2. *Proc Natl Acad Sci USA* 98: 7922–7927
73. Mouchel CF, Briggs GC, Hardtke CS (2004) Natural genetic variation in *Arabidopsis* identifies BREVIS RADIX, a novel regulator of cell proliferation and elongation in the root. *Genes Dev* 18: 700–714
74. Liu JP, Van Eck J, Cong B, Tanksley SD (2002) A new class of regulatory genes underlying the cause of pear-shaped tomato fruit. *Proc Natl Acad Sci USA* 99: 13302–13306
75. Cong B, Liu JP, Tanksley SD (2002) Natural alleles at a tomato fruit size quantitative trait locus differ by heterochronic regulatory mutations. *Proc Natl Acad Sci USA* 99: 13606–13611
76. Werner JD, Borevitz JO, Warthmann N, Trainer GT, Ecker JR, Chory J, Weigel D (2005) Quantitative trait locus mapping and DNA array hybridization identify an FLM deletion as a cause for natural flowering-time variation. *Proc Natl Acad Sci USA* 102: 2460–2465

77. Clark RM, Linton E, Messing J, Doebley JF (2004) Pattern of diversity in the genomic region near the maize domestication gene tb1. *Proc Natl Acad Sci USA* 101: 700–707
78. Kroymann J, Donnerhacke S, Schnabelrauch D, Mitchell-Olds T (2003) Evolutionary dynamics of an *Arabidopsis* insect resistance quantitative trait locus. *Proc Natl Acad Sci USA* 100: 14587–14592
79. Yano M, Katayose Y, Ashikari M, Yamanouchi U, Monna L, Fuse T, Baba T, Yamamoto K, Umehara Y, Nagamura Y et al. (2000) Hd1, a major photoperiod sensitivity quantitative trait locus in rice, is closely related to the *Arabidopsis* flowering time gene CONSTANS *Plant Cell* 12: 2473–2483
80. Kojima S, Takahashi Y, Kobayashi Y, Monna L, Sasaki T, Araki T, Yano M (2002) Hd3a, a rice ortholog of the *Arabidopsis* FT gene, promotes transition to flowering downstream of Hd1 under short-day conditions. *Plant Cell Physiol* 43: 1096–1105
81. Doebley J, Stec A, Hubbard L (1997) The evolution of apical dominance in maize. *Nature* 386: 485–488
82. Van der Hoeven R, Ronning C, Giovannoni J, Martin G, Tanksley S (2002) Deductions about the number, organization, and evolution of genes in the tomato genome based on analysis of a large expressed sequence tag collection and selective genomic sequencing. *Plant Cell* 14: 1441–1456
83. Schmidt R (2000) Synteny: recent advances and future prospects. *Current Opin Plant Biol* 3: 97–102
84. Fulton TM, Van der Hoeven R, Eannetta NT, Tanksley SD (2002) Identification, analysis, and utilization of conserved ortholog set markers for comparative genomics in higher plants. *Plant Cell* 14: 1457–1467
85. Borevitz JO, Liang D, Plouffe D, Chang HS, Zhu T, Weigel D, Berry CC, Winzeler E, Chory J (2003) Large-scale identification of single-feature polymorphisms in complex genomes. *Genome Res* 13: 513–523
86. Wayne ML, McIntyre LM (2002) Combining mapping and arraying: An approach to candidate gene identification. *Proc Natl Acad Sci USA* 99: 14903–14906
87. Schadt EE, Monks SA, Drake TA, Lusis AJ, Che N, Colinayo V, Ruff TG, Milligan SB, Lamb JR, Cavet G et al. (2003) Genetics of gene expression surveyed in maize, mouse and man. *Nature* 422: 297–302
88. Gibson G, Weir B (2005) The quantitative genetics of transcription. *Trends Genet* 21: 616–623
89. Guillaumie S, Charmet G, Linossier L, Torney V, Robert N, Ravel C (2004) Colocation between a gene encoding the bZip factor SPA and an eQTL for a high-molecular-weight glutenin subunit in wheat (*Triticum aestivum*). *Genome* 47: 705–713
90. Damerval C, Maurice A, Josse JM, de Vienne D (1994) Quantitative trait loci underlying gene product variation: a novel perspective for analyzing regulation of genome expression. *Genetics* 137: 289–301
91. de Vienne D, Leonardi A, Damerval C, Zivy M (1999) Genetics of proteome variation for QTL characterization: application to drought-stress responses in maize. *J Exp Bot* 50: 303–309
92. Consoli L, Lefevre A, Zivy M, de Vienne D, Damerval C (2002) QTL analysis of proteome and transcriptome variations for dissecting the genetic architecture of complex traits in maize. *Plant Mol Biol* 48: 575–581
93. Schauer N, Semel Y, Roessner U, Gur A, Balbo I, Carrari F, Pleban T, Perez-Melis A, Brudigam C, Kopka J et al. (2006) Genetics of metabolite in fruits of interspecific introgressions of tomato. *Nat Biotechnol* 24: 447–454

94. Ostergaard L, Yanofsky MF (2004) Establishing gene function by mutagenesis in *Arabidopsis thaliana*. *Plant J* 39: 682–696
95. Krysan PJ, Young JC, Sussman MR (1999) T-DNA as an insertional mutagen in *Arabidopsis*. *Plant Cell* 11: 2283–2290
96. Schneider A, Kirch T, Gigoshvili T, Mock H-P, Sonnewald U, Simon R, Flügge U-I, Werr W (2005) A transposon-based activation-tagging population in *Arabidopsis thaliana* (TAMARA) and its application in the identification of dominant developmental and metabolic mutations. *FEBS Lett* 579: 4622–4628
97. Waterhouse PM, Heliwell CA (2003) Exploring plant genomes by RNAi-induced gene silencing. *Nat Rev Genet* 4: 29–38
98. Karimi M, Inzé D, Depicker A (2002) GATEWAY vectors for *Agrobacterium*-mediated plant transformation. *Trends Plant Sci* 7: 193–195
99. Meinke DW, Meinke LK, Showalter TC, Schissel AM, Mueller LA, Tzafrir I (2003) A sequence-based map of *Arabidopsis* genes with mutant phenotypes. *Plant Physiology* 131: 409–418
100. An G, Jeong D-H, Jung K-H, Lee S (2005) Reverse genetics approaches for functional genomics of rice. *Plant Mol Biol* 59: 111–123
101. Hirochika H, Guiderdoni E, An G, Hsing Y-I, Eun MY, Han C-D, Upadhyaya N, Ramachandran R, Zhang Q, Pereira A et al. (2004) Rice mutant resources for gene discovery. *Plant Mol Biol* 54: 325–334
102. Meissner R, Jacobson Y, Melamed S, Levyatuv S, Shalev G, Ashri A, Elkind Y, Levy AA (1997) A new model system for tomato genetics. *Plant J* 12: 1465–1472
103. Meissner R, Chague V, Zhu Q, Emmanuel E, Elkind Y, Levy AA (2000) A high throughput system for transposon tagging and promoter trapping in tomato. *Plant J* 38: 861–872
104. Gidoni D, Fuss E, Burbidge A, Speckmann GJ, James S, Nijkamp D, Mett A, Feiler J, Smoker M, de Vroomen MJ et al. (2003) Multi-functional T-DNA/Ds tomato lines designed for gene cloning and molecular and physical dissection of the tomato genome. *Plant Mol Biol* 51: 83–98
105. Mathews H, Clendennen SK, Caldwell CG, Connors K, Matheis N, Schuster DK, Menasco DJ, Wagoner W, Lightner J, Wagner DR (2003) Activation tagging in tomato identifies a transcriptional regulator of anthocyanin biosynthesis, modification, and transport. *Plant Cell* 15: 1689–1703
106. Emmanuel E, Levy AA (2002) Tomato mutants as tools for functional genomics. *Curr Opin Plant Biol* 5: 112–117
107. Li X, Song Y, Century K, Straight S, Ronald P, Dong X, Lassner M, Zhang Y (2001) A fast neutron deletion mutagenesis-based reverse genetics system for plants. *Plant J* 27: 235–242
108. McCallum CM, Comai L, Greene EA, Henikoff S (2000) Targeted screening for induced mutations. *Nat Biotechnol* 18: 455–457
109. Colbert T, Till BJ, Tompa R, Reynolds S, Steine MN, Yeung AT, McCallum CM, Comai L, Henikoff S (2001) High-throughput screening for induced point mutations. *Plant Physiol* 126: 480–484
110. Koornneef M, Dellaert LWM, van den Veen JH (1982) EMS- and radiation-induced mutation frequencies at individual loci in *Arabdospis thaliana* (L.) Heynh. *Mutat Res* 93: 109–123
111. Menda N, Semel Y, Peled D, Eshed Y, Zamir D (2004) *In silico* screening of a saturated mutation library of tomato. *Plant J* 38: 861–872
112. Li X, Zhang Y (2002) Reverse genetics by fast neutron mutagenesis in higher plants. *Funct Integr Genomics* 2: 254–258

113. Lindroth AM, Cao X, Jackson JP, Zilberman D, McCallum CM, Henikoff S, Jacobsen SE (2001) Requirement of CROMOMETHYLASE3 for maintenance of CpXpG methylation. *Science* 292: 2077–2080
114. Yeung AT, Hattangadi D, Blakesley L, Nicolas E (2005) Enzymatic mutation detection technologies. *Biotechniques* 38: 749–758
115. Winkler S, Schwabedissen A, Backasch D, Bökel C, Seidel C, Bönisch S, Fürthauer M, Kuhrs A, Cobreros L, Bran M, Gonzalez-Gaitan M (2005) Target-selected mutant screen by TILLING in *Drosophila*. *Genome Res* 15: 718–723
116. Greene EA, Codomo CA, Taylor NE, Henikoff JG, Till BJ, Reynolds SH, Enns LC, Burtner C, Johnson JE, Odden AR et al. (2003) Spectrum of chemically induced mutations from a large-scale reverse-genetic screen in *Arabidopsis*. *Genetics* 164: 731–740
117. Perry JA, Wang TL, Welham TJ, Gardner S, Pike JM, Yoshida S, Parniske M (2003) A TILLING reverse genetics tool and a web-accessible collection of mutants of the legume *Lotus japonicus*. *Plant Physiol* 131: 866–871
118. Caldwell DG, McCallum N, Shaw P, Muehlbauer GJ, Marshall DF, Waugh R (2004) A structured mutant population for forward and reverse genetics in Barley (*Hordeum vulgare* L.). *Plant J* 40: 143–150
119. Till BJ, Reynolds SH, Weil C, Springer N, Burtner C, Young K, Bowers E, Codomo CA, Enns LC, Odden AR et al. (2004) Discovery of induced point mutations in maize genes by TILLING. *BMC Plant Biology* 4: 12
120. Slade AJ, Fuerstenberg SI, Loeffler D, Steine MN, Facciotti D (2005) A reverse genetic, non transgenic approach to wheat crop improvement by TILLING. *Nat Biotechnol* 23: 75–81
121. Henikoff S, Comai L (2003) Single-nucleotide mutations for plant functional genomics. *Annu Rev Plant Biol* 54: 375–401
122. Till BJ, Reynolds SH, Greene EA, Codomo CA, Enns LC, Johnson JE, Burtner C, Odden AR, Young K, Taylor NE et al. (2003) Large-scale discovery of induced point mutations with high-throughput TILLING. *Genome Res* 13: 524–530
123. Gilchrist EJ, Haughn GW (2005) TILLING without a plough: a new method with applications for reverse genetics. *Curr Opin Plant Biol* 8: 211–215
124. Comai L, Henikoff S (2006) TILLING: practical single-nucleotide mutation discovery. *Plant J* 45: 684–694
125. Wienholds E, van Eeden F, Kosters M, Mudde J, Plasterk RHA, Cuppen E (2003) Efficient target-selected mutagenesis in zebrafish. *Genome Res* 13: 2700–2707
126. Wu JL, Wu C, Lei C, Baraoidan M, Bordeos A, Madamba MRS, Ramos-Pamplona R, Mauleon R, Portugal A, Ulat J et al. (2005) Chemical- and irradiation-induced mutants of Indica Rice IR64 for forward and reverse genetics. *Plant Mol Biol* 59: 85–97
127. Tanksley SD, McCouch SR (1997) Seed banks and molecular maps: Unlocking genetic potential from the wild. *Science* 277: 1063–1066
128. Comai L, Young K, Till BJ, Reynolds SH, Greene EA, Codomo CA, Enns LC, Johnson JE, Burtner C, Odden AR et al. (2004) Efficient discovery of DNA polymorphisms in natural populations by Ecotilling. *Plant J* 37: 778–786
129. Qiu P, Shandilya H, D'Alessio JM, O'Connor K, Durocher J, Gerard GF (2004) Mutation detection using Surveyor nuclease. *Biotechniques* 36: 702–707
130. Urbanczyk-Wochniak E, Luedemann A, Kopka J, Selbig J, Roessner-Tunali U, Wilmitzer L, Fernie AR (2003) Parallel analysis of transcript and metabolic profiles: a new approach in systems biology. *EMBO rep* 4: 1–5
131. Thimm O, Blasing O, Gibon Y, Nagel A, Meyer S, Kruger P, Selbig J, Muller LA, Rhee SY, Stitt M (2004) MAPMAN: a user-driven tool to display genomics data sets onto diagrams of metabolic pathways and other biological processes. *Plant J* 37: 914–939

132. Sweetlove LJ, Fernie AR (2005) Regulation of metabolic networks: understanding metabolic complexity in the system biology era. *New Phytol* 168: 9–24
133. Fridman E, Pichersky E (2005) Metabolomics, genomics, proteomics and the identification of enzymes and their substrates and products. *Curr Opin Plant Biol* 8: 242–248
134. Blank LM, Kuepfer L, Sauer U (2005) Large-scale ^{13}C-flux analysis reveals mechanistic principles of metabolic network robustness to null mutation in yeast. *Genome Biol.* 6: R49
135. Tewari M, Hu PJ, Ahn JS, Ayivi-Guedelhoussou N, Vidalain P-O, Li S, Milstein S, Armstrong CM, Boxem M, Butler MD et al. (2004) Systematic interactome mapping and genetic perturbation analysis of a *C. elegans* TGF-β signalling network. *Mol Cell* 13: 469–482
136. Enns LC, Kanaoka MM, Torii KU, Comai L, Okada K, Cleland RE (2005) Two callose synthases, GSL1 and GSL5, play an essential and redundant role in plant and pollen development and in fertility. *Plant Mol Biol* 58: 333–349
137. Weckwerth W, Loureiro ME, Wenzel K, Fiehn O (2004) Differential metabolic pathways unravel the effect of silent plant phenotypes. *Proc Natl Acad Sci USA* 101: 7809–7811
138. Scholtz M, Gatzek S, Sterling A, Fiehn O, Selbig J (2004) Metabolite fingerprinting: detecting biological features by independent component analysis. *Bioinformatics* 20: 2447–2454
139. Fernie AR, Trethevey RN, Krotzky AJ, Wilmitzer L (2004) Metabolite profiling: from diagnostic to system biology. *Nat Rev* 5: 1–7
140. Sweetlove LJ, Last RL, Fernie AR (2003) Predictive metabolic engineering: a goal for systems biology. *Plant Physiol* 132: 420–425
141. Rontein D, Dieuaide-Noubhani M, Dufourc EJ, Raymond P, Rolin D (2002) The metabolic architecture of plant cells: Stability of central metabolism and flexibility of anabolic pathways during the growth cycle of tomato cells. *J Biol Chem* 277: 43948–43960

Plant Systems Biology
Edited by Sacha Baginsky and Alisdair R. Fernie
© 2007 Birkhäuser Verlag/Switzerland

Transcriptional profiling approaches to understanding how plants regulate growth and defence: A case study illustrated by analysis of the role of vitamin C

Christine H. Foyer[1], Guy Kiddle[1] and Paul Verrier[2]

[1] *Crop Performance and Improvement Division, and* [2] *Biomathematics and Bioinformatics Division, Rothamsted Research, Harpenden, Hertfordshire AL5 2JQ, UK*

Abstract

In this chapter, basic technical aspects concerning the design of DNA microarray experiments are discussed including sample preparation, hybridisation conditions and statistical significance of the acquired data are detailed. Given that microarrays are perhaps the most used tool in plant systems biology there is much experience in the pitfalls in using them. Herein important considerations are presented for both the experimental biologists and data analyst in order to maximise the utility of these resources. Finally a case study using the analysis of vitamin C deficient plants is presented to illustrate the power of this approach in enhancing comprehension of important and complex biological functions.

Introduction

Vitamin C (vtc, ascorbic acid, AA) is a highly abundant, multifunctional metabolite in plants [1–4]. Low AA levels trigger programmed cell death (PCD) and promote early senescence [5, 6]. While cellular oxidation increases during leaf senescence [7] there is no evidence to suggest that progressive increases in oxidative damage to macromolecules causes ageing in plant cells as is the case in animal ageing [8]. AA is a key antioxidant vitamin in primates that is implicated in healthy ageing [9, 10]. It is therefore important to gain a comprehensive understanding of the diverse roles of AA in plant biology as well as knowledge of factors that limit AA production and accumulation in different plant organs.

The major pathway of AA synthesis in leaves occurs via GDP-D-mannose, GDP-L-galactose and L-galactose [11] but other entry points into the AA synthetic networks have been suggested [12, 13]. The mannose pathway branches at GDP-mannose, where an epimerase can form either GDP-gulose or GDP-L-galactose. Low shoot AA can be induced by perturbations in L-galactose metabolism, for example in the *Arabidopsis thaliana* low AA (*vtc*) *vtc4* mutant, which has decreased l-galactose 1-P phosphatase activity or in *vtc1*, which lacks GDP-mannose pyrophosphorylase (GMPase) [14] or in transformed plants with much reduced activity of this enzyme [15]. Decreases in L-galactose dehydrogenase [16] and L-galactono-1, 4-lactone dehydrogenase (GalLDH) activity [17] however, have less effect on AA content.

AA synthesis and accumulation in leaves is regulated by light and responds to both developmental and environmental triggers [18, 19]. High light grown plants have more AA than those grown with less irradiance [19] and AA levels are low basal senescent leaves [20]. Light exerts effects through control of respiration [19] and through altered gene expression [21]. In some species leaf AA accumulation fluctuates on a diurnal basis being lowest at night and increasing throughout the day [22, 23] but in other species no diurnal changes in leaf AA can be observed [18]. The capacity of AA re-generation from its oxidised forms also impacts on AA abundance [19, 24].

Several types of *A. thaliana vtc* mutants having low AA have been isolated [14, 25]. They have been useful in analysing the pathway of AA synthesis as well as in elucidating the roles of AA. The *vtc1* mutant was selected via its high sensitivity to ozone and it also has enhanced sensitivity to other abiotic stresses such as freezing and UV-B irradiation [14, 26]. This mutant has a single point mutation in the gene encoding GMPase, causing the conversion of a highly conserved proline to a serine at position 22. Hence, while the *vtc1* plants contain similar amounts of GMPase mRNA to the wild type, the GMPase protein in the mutant has a substantially lower enzyme activity. As a result the mutant rosette leaves have only about 30% of the leaf AA than that found in the wild type [25]. When grown in optimal growth conditions, *vtc1* has similar rates of photosynthesis to the wild type [27]. However, *vtc* leaves generally have a decreased capacity to accumulate zeaxanthin and as a result photosynthesis is more susceptible to inhibition by abiotic stress [28]. The *vtc2* mutants, which have even less ascorbate (15–20% [6]) are deficient in GDP-L-galactose phosphorylase, an enzyme that is at a branch point between AA synthesis and incorporation of L-galactose into polysaccharides. Ectopic expression of the animal AA biosynthetic enzyme L-gulono-1, 4-lactone oxidase, restores wild type AA levels and the wild type phenotype to the *vtc1* and *vtc2* mutants suggesting that the *vtc1* and *vtc2* phenotypes are caused largely by low AA alone [29]. The *vtc3* and *vtc4* mutants have about 50% of the wild type leaf AA levels [25].

Genetic screens based on either sugar-regulated gene expression or the arrest of development by high concentrations of sugar has led to the isolation of a number of sugar sensing mutants [30–34]. Much evidence has shown that abscisic acid (ABA) and some components of the ABA signal transduction cascades are involved in sugar signalling in higher plants [35, 36]. The *abi4* mutant for example plays a crucial role in ABA signalling and is also important in detecting both sucrose and glucose and in mediating the inhibitory effect of nitrate on LR development [37, 38].

Strategy and approach

In the following studies we have used the *vtc1* mutant and the *abi4* mutant. By comparing the transcriptome of the *vtc1* mutant with that of the wild-type we were able to explore the effects of low AA on the *A. thaliana* leaf transcriptome [6, 39]. Similarly, by feeding AA, we were able to greatly enhance tissue AA levels and thus compare the high AA transcriptome to that of controls. Expression analysis techniques were compared using the results from three to five pairs of array plates. Three to five independent samples of mutant and wild-type leaves were harvested from 5–6 week-old plants. Furthermore, comparisons of the *vtc1* mutant transcriptome with that of the *abi4* mutant that is unable to sense ABA has enabled elucidation of relationships between ABA and AA signalling.

Micro-array analysis

The gene expression microarray is a very powerful tool for exploring the expression level of large numbers of transcripts in a single experiment. Currently available commercial microarrays can be used to track the expression levels of 60,000 or more transcripts. RNA extracted from whole plants, specific tissues or specific cells is normally used in a hybridisation process to compare the expression levels in one system to that of another. Where an organism has been well studied and the gene responses are well understood, microarray technology can be used as a diagnostic tool to determine when samples are behaving abnormally. The majority of microarray experiments follow a similar set of procedures. Assuming that a suitable microarray is available, the selected material is prepared for hybridisation with the microarray slide and is inoculated with a number of control RNAs. Dependant on the type of the experiment, the sample RNA may be labelled with Cy3 or Cy5 dyes. Following hybridisation, the microarray is then scanned with a high-precision laser scanning device to provide a measure of the quantity of material hybridising to each probe cell of the microarray. The data is then processed with appropriate statistically sound analysis software to derive the comparative levels of the microarray probes, either within a single microarray slide or across multiple slides that comprise the experiment. Having obtained the levels of the represented genes, the task is then one of identifying relationships between the genes under the conditions of the experiments which normally involves the application of considerable biological knowledge.

The process briefly outlined above describes purely the basic processes that have to be undertaken. In reality, each step requires the adherence to appropriate protocols, enormous care when undertaking the RNA extraction, inoculation, labelling, hybridisation, scanning setup and data analysis stage. The analysis stage brings with it a further complication in that the quantity of data can be overwhelming. An experiment utilising a 40,000 spot microarray with five experimental conditions and three replicates will generate 600,000 expression levels (and many other values that need to be considered when looking at the statistical significance), which stretches the capabilities of the most able spreadsheet manipulator. Therefore, it is necessary to deploy appropriate software packages to analyse the basic information coming from the experiments. The package has to pull out the pertinent points of interest for the biologist to then examine the system and obtain some insight as to the processes involved in the organism.

The microarray slide

The majority of microarray users will obtain their slides from one of the commercial providers or from a collaborating research group that produces a volume of slides. While it is not essential to know the detail as to how a microarray slide is produced, there are certain issues that should be considered when preparing an experiment and when analysing the results. There are two main types of microarray. A portion of a 'spotted' array is shown in Figure 1. This is produced by a robot that deposits small quantities of each of the cDNA or oligonucleotide target probe onto the array slide. This process is termed printing and the probe spots are produced in blocks. The number of spots in each block normally depends on the design of the array. Each block is printed by one print needle. For example, a slide may have 48 blocks, each comprising 20 columns, by 25 rows. While the robot tries to align the spots so that they have the same size with similar spacing, in practice, the spots tend to vary both in size and alignment. In addition, while alignment within a block may be quite good, alignment between blocks is not as precise. The probe spot alignment, or misalignment is an important issue when preparing to scan the image. No two print needles are exactly the same and each wears in use so that spots vary in size. The process of production can be imperfect and sometimes the probe spot dries too quickly leading to doughnut or cusp shaped hybridisation to the probe. Thus, the problem of variations in density over the spots and between slides requires resolution by analysis software [40].

Figure 1. Two blocks of a cDNA microarray after hybridisation showing the Cy5 response. Note the contamination at top right and the large spot sizes in the left hand block. Note also that some spots exhibit doughnut and cusp-like hybridisation.

Figure 2. A section of an Affymetrix ath0 oligonucleotide microarray chip after hybridisation. This shows the typically rectangular probe spots found in this production type. Note that this slide has some obvious contamination that affects a large number of probe cells, and although not easily discernable in this image, there is a general degradation of image quality to the left of the obvious white-out.

The second type of array is the manufactured oligonucleotide array. The most commonly found arrays of this form are geneChip™ proprietary products of Affymetrix. In this process, each probe is generated through the application of a sequence of printed masks, nucleotide washes and etching washes to generate the appropriate short cDNA fragments in each array cell. These arrays have very regular spacing and can be produced at very high densities. Figure 2 shows a portion of an Affymetrix geneChip array with some damage.

The approach to the creation of cDNA fragments for each cell in the Affymetrix arrays is interesting in itself. Cells are created as perfect cDNA matches and as mismatch cells, where the mis-match cells will have one nucleotide being altered from the perfect match. To determine if a hybridisation match occurs, account can be taken of the perfect matches and the mis-matches. This approach has sound groundings but it does mean that an analysis package has to be used to determine the hybridisation levels.

For the spotted array, the control is labelled with one dye and the experiment with the other dye. One slide is then hybridised with both the control and the experiment. Often the process is repeated with the dyes reversed and hybridised onto a second plate. This is known as a dye-swap experiment. This helps in providing a better framework for experiment analysis when determining the relative expression levels. Both types of microarray suffer from the problem that the hybridisation process is not perfect and there is invariably a gradient of hybridisation quality across the slide. In addition, there are frequently found contaminations which may affect a few or many probes, due to air bubbles, dust or imperfect drying of the slide or some other mechanism that results in streaks, blotches and non-uniform slide density. Figures 1 and 2 show portions of hybridised arrays where not everything went perfectly. The process of creating a hybridised array can be fraught. Thus, it is wise to invest time in practising the technique before investing in the use of hard-won experimental material which may not easily be re-created.

Types of experiment

The type of experiment that could be conducted using microarray technologies is limited perhaps only by imagination. However, *dye-swap experiments* are most commonly used on spotted arrays. Here, a control sample is labelled with, e.g., Cy3 and the sample which it is to be compared with is labelled with Cy5. Both labelled samples are hybridised to the same array. When scanned, a value for the Cy3 and the Cy5 labelled signal levels is obtained. A comparison is then made between the signal levels of the Cy3 and Cy5 labelled expressions. The resultant comparison gives a value for the difference between the expression levels of the two samples for each DNA or RNA fragment in the probe set. The experiment is then repeated but the control is labelled with Cy5 and the experiment sample labelled with Cy3. A comparison can then be made between the two sets of data to obtain a better estimate of the expression levels. Comparisons are often made as ratios of one level against the other. Sometimes $\log_2$ values of the ratio are used and the straight ratio may be called the fold level, so it is best to check the definitions that are being used.

Another common experiment, particularly with Affymetrix labelling style arrays, involves no labelling with one sample only hybridised per array slide. This requires that good analysis technologies are available when comparing data across slides. Single array techniques are often used in diagnostic experiments. Time series experiments utilise either single sample slides or dye swap pairs with material being taken from a sequence of samplings, the timing varying according to the purpose of the experiment. Sample times may vary from minutes to days according to the underlying process being investigated. In all experiments, consideration must be given to experiment replication. Normally, costs prohibit large quantities of replication, but normal practice is to make three biological replicates to ensure that unusual biological variation is masked and that the unfortunate appearance of slide damage/contamination does not totally ruin the complete experiment. Where it is known that the biological sample is likely to exhibit a very noisy response, then additional replicates may well be required. There is normally no need to make technical replicates of the same biological sample unless they are to experiment with the procedures or to gain experience.

Scanning a slide

After hybridisation, each slide must be scanned with an appropriate laser scanner and its associated software package. The quality of the scan setup will have an impact on the subsequent analysis and quality of the results. Normally, the process is to identify each spot or cell with a mask. This mask is intended to enclose the printed spot or oligonucleotide cell (easier with the manufactured cell chip which has regular spacing). In the case of printed spot arrays, the alignment of the spots is not always perfect and the microarray for any printed batch will need the mask to be checked and manually aligned to ensure that spots are not missed or cut by the spot mask area. Once the mask has been aligned and the mask saved, the scanned levels of the control probes can be assessed. An analysis of the levels of the controls will give an indication of both the hybridisation quality and will indicate any serious change of levels across the slide. If the non-control probes show overall low signal strength, the illumination may be increased. However, this might make some very strong hybridisation levels to become saturated (reach the peak recordable light intensity). While this may not be a problem, it does mean that relative hybridisation strengths cannot be compared with the saturated probes. Most scanners allow saturated spots and contaminated spots to be flagged with an appropriate value to record the problem. If a relative level is required for spots that have to be saturated to lift lower intensity spots out of the background, a second scan can be made where the signal level is lower. This second scan can be used to gain a higher level of knowledge of the relative expressions and will require additional steps in the expression level analysis to merge the multiple scans of one slide.

The scanning process divides the slide into pixels, where a pixel represents the resolution of the scanner. A single spot will typically be divided into 25 or more pixels. The scanning software will normally calculate the mean and median intensities of the pixels in each spot mask and various statistical measures of the intensity distribution. It will also determine similar values for the areas outside the spot mask to derive background intensity. Most scanning software will calculate a number of additional measures and output these to a file in a standard format suitable for reading by an analysis package. Values to be output are often user selectable and care must be taken to ensure that all the required values are output and it is good practice to keep the output format and order the same for all scans in the same experiment to avoid subsequent errors or misunderstanding in the data organisation.

Experimental design issues

When planning a microarray experiment, it is just as important to consider the design of the experiment in terms of sample collection as is the design of the way the microarrays will be hybridised. For example, where RNA extraction leads to low volumes of RNA, it is often necessary to produce many numbers of plants to obtain the appropriate material quantities. If these plants are grown in a glasshouse, then normal methods of randomisation should be employed to ensure that the growing environment is not placing undue emphasis on the outcome in that one location may be receiving undue water, fertilisation, drought, heat etc.

As an example, suppose we have an experiment where we wish to compare the tissue sampled from 20 isogenic and 20 transgenic plants and we wish to have three replicates. The glasshouse space available is of necessity limited and there may be slight differences in the basic treatments of the plants in different parts of the glass-

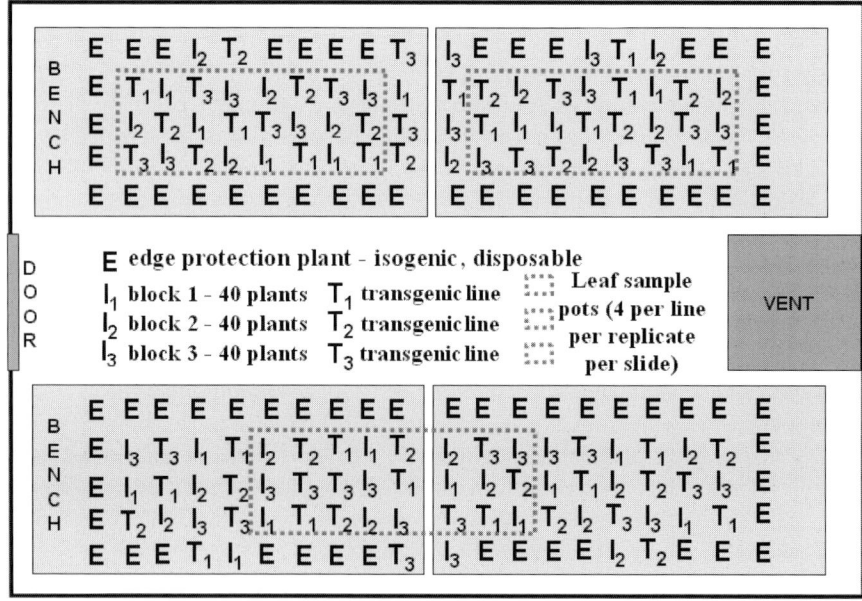

Figure 3. Random block layout for wheat plant pots to accommodate an experiment with three replicates, the control isogenic line and the experiment, the transgenic line; each experiment requiring 10 plants to gather sufficient material for RNA extraction of seed embryo. An additional experiment using the same plants has been accommodated in which leaf tissue is sampled with three replicates.

house benching as well as possible effects of the proximity to the glass, proximity to the edge of the bench and proximity to a neighbour. To provide a suitable random placement of the plants in the growing environment a random block design was produced using a standard statistical package (Genstat (http://www.vsn-intl.com/genstat/)) as shown in Figure 3. While the experiment could have proceeded without the random block design, the use of it will ensure that any systematic effects of the environment will not unduly weight the results.

The design of the microarray itself requires consideration. In most cases, this will be outside the control of the experimenter, but a microarray should have controls placed at random locations across the slide and be of sufficient number to provide an indication of poor hybridisation technique. In addition, each probe fragment should be repeated at least once on the slide, preferably the copies should not be close to each other. Probe copies provide a better estimate and also give further indication of non-uniform hybridisation. The comparison of one sample against another is straightforward in experiments where there is just one control and one sample, but where there are several experimental samples (with perhaps different treatments, or different phenotypes), consideration should be made of the design of the experiment in terms of which sample is hybridised with which for dye-swap experiments. For example comparing Control C against experiment A and Control

C against experiment B does enable an analysis to be made to compare A against B and this will potentially increase the error over a straight microarray experiment in which A is compared to B. The problem becomes more severe when there are several comparisons being made. The further the comparison gets from being an experimental comparison, the more error will creep into the results. The basic principle for a sound analysis is to minimise the distance between comparisons. A distance, of two plates between comparisons is acceptable, while distances greater than three may lead to misleading results. However, this has to be moderated with the cost of the experiment. For multi-dye hybridisations, it is sensible to use a single control over all plates within one experiment. Control *versus* A plus Control *versus* B enables a comparison of A *versus* B to be made through an analysis. The careful designing of an array experiment can save a lot of money by reducing the number of slides required and it is always wise to consult a statistician for the design.

Analysing the scans

The problem of extracting expression levels from the scanned microarrays has been exercising mathematicians and statisticians for several years. The methodologies are numerous and the techniques are improving [41]. Novel and revised methodologies appear in the literature at an alarming rate. The practicing biologist is faced with the major dilemma of how to proceed and with which methodology. The quickest approach in the laboratory would be to use a package such as the proprietary GeneSpring product which can be used to analyse large numbers of microarrays of both the spotted type and the manufactured Affymetrix style. The more adventurous could well make use of the R package which is a public domain statistical package coupled with the growing library of microarray analysis tools prepared for the R environment and GeneSpring users can import some R procedures into the package. This may be driven directly or through the Bioconductor suite. There are also many public domain packages that are suitable for handling either spotted arrays or the Affymetrix style arrays. In any event, the successful use of any of these software tools requires that the user understands how to use the tool and the process that he is trying to achieve. It is important to understand that the whole basis of the analysis of the expression levels is that almost all the expression levels will be similar when two samples are compared. The analysis packages make use of this fact. The significantly differentially expressed spots may number a few hundred to a few thousand. An initial check of the overall spot levels can be made by producing a scattergram and a frequency histogram without any adjustments to values other than grouping. This will give an immediate impression of potential bias in the slide and an indication of the amount of high expressers as well as a possible indication of hybridisation problems. A simple regression of two slides against each other will also show a broad comparison between the expression levels.

The analysis process is straightforward in terms of the overall strategy, but complex in terms of some of the techniques. A simple dye swap experiment using a pair of microarrays without replication (for the sake of clarity) will be used to illustrate the type of analysis required. The following is not intended to be a definitive approach to the analysis of a dye-swap spotted array experiment, but rather a description of a frequently used technique. The example is given to illustrate the type of steps that have to be taken by the analysis software. Other experiment types are not covered in such depth. In this experiment we have as a control, leaf tissue of *Arabidopsis thaliana* wild strain Col0. The experiment plant is leaf tissue of the same species but the *vtc1* mutant, an AA-deficient strain. The microarray being used is the Stanford University cDNA chip which comprises about 4,500 spots representing 7,800 genes. The chip was produced by a robot with 48 printing needles giving rise to 48 blocks of 18 cols by 18 rows of spots. Not all spot locations contain a probe. Following hybridisation and labelling with Cy3 and Cy5, the two microarrays were scanned and the resultant file provided the mean and median spot pixel intensities and variance for both foreground and background levels. The single dye-swap experiment resulted in four images and scan files, a Cy3 and Cy5 for each plate. The dye was reversed on the second plate. The analysis mechanism described by Yang et al. [41] (Normalization for cDNA Microarray Data, Berkeley Technical Report; http://citeseer.nj.nec.com/406329.html) has been followed to undertake a print-tip normalisation with robust smoothing. For any spot j, j = 1,...,p where p is the number of spots, the measured fluorescent intensities R_j and G_j are the Red and Green dye values respectively. Background intensities could be subtracted but have not been in this analysis on the assumption that the equipment setup stabilised the background, but a recalculation with background removed should perhaps be undertaken for comparison, but note that the background determination used by the scanner will possibly include many small blemishes and if the mask is not properly set on scanning, will also include some hybridised spot pixels making the background levels rather misleading. This method is also sensitive to very low intensity levels (often found after background levels are subtracted from the individual spot intensities) where misleading results can be obtained

The log intensity ratio $M = \log_2(R/G)$ gives a useful measure of the changes in expression level. Plotting M against $A = \log_2(\sqrt{RG})$ assists in the identification of spot artefacts and intensity dependent patterns, since A is proportional to intensity and it is known that the dyes fluoresce differently at different intensities. The M *versus* A plots can give an immediate indication of the overall expression levels. Figure 4(A) shows a typical M-A plot of raw data. In this case there is a marked tendency to favour the green dye at the lower intensity levels. Note that in producing this M-A plot, the controls and empty spot cells have been removed from the analysis. Note that the horizontal line is M=0, thus any spot above the line is expressing more in Red and below the line, more in Green. One would expect that the majority of probes will show very little change between the two samples. It is useful to view the distribution of M values, Figure 4(D) shows the distribution density of M values, the tick mark representing M=0. From this it can be seen that the distribution is skewed and while there are plenty of outliers, the majority of spots appear in a broad central distribution. The statistical task is to look at the causes of the distribution being non-normal and to implement methods to sharpen this distribution so that the majority of spots fall into a tight cluster around M=0. Yang et al. recom-

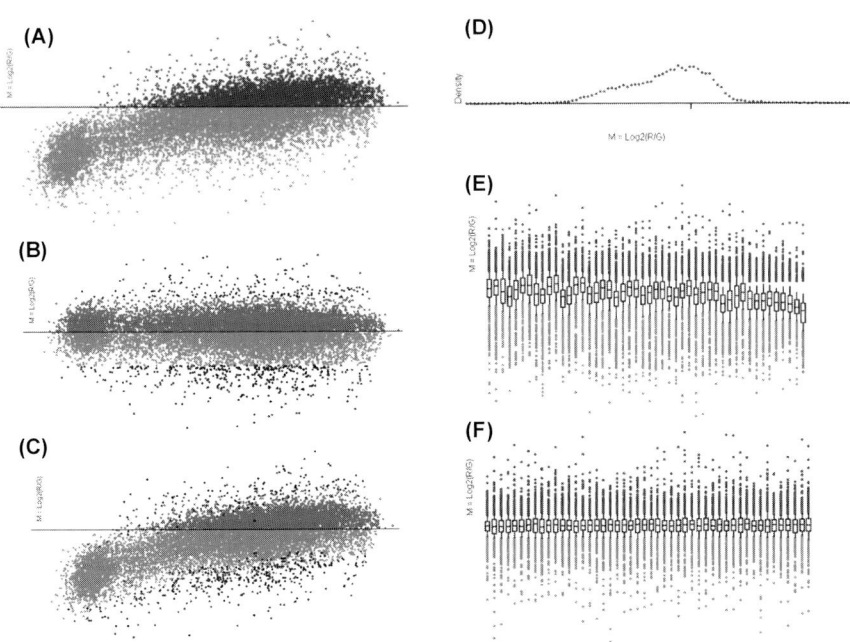

Figure 4. M-A plots (A) showing the typical spread of spot intensities for the raw data. Background has not been removed. The horizontal line represents M=0. Any spot above this line indicates a relative over-expression of the red (Cy5) labelled sample and anything below the line represents a relative over-expression of the green (Cy3) labelled sample. (B) After Lowess normalisation and block scaling, the spots above the cut=off of |0.95| show the selected differentially expressed genes. (C) the selected genes thrown back onto the raw data distribution. (D) the raw data distribution of M, (E) the distribution of M values in each group in the raw data. (F) the distribution of M in each group after applying a Loweess normalisation.

mend that the data should be normalised within each print-tip group because each print tip has different properties and this leads to variation between the printed spots between the tips, but similar spot profiles should be seen for a single print tip. The Stanford arrays can be taken to have 48 print tips, with at least 48 blocks, which may each differ in their characteristics. Print tips are identified by the on-plate Block numbers in the scanner output. Even if the utilisation of print tips in the blocks is not known, any one block can be treated as a separate group and this method would then treat systematic variations across the plate.

Within a print tip, Yang et al. [40] perform a transformation of the data using a robust Lowess smoothing, although a smoothed spline approach could also be used. The Lowess method performs a robust local linear fit to the data. Since the majority of M values should be expected to be similar (little or no change in expression levels with a value close to 1, the Lowess is made robust by disregarding points that lie outside five standard deviations adjustable from the mean value. Lowess takes a percentage of the points near the x-value (A in our case) to create a localised linear regression fit to the data, having due regard to robustness. The fraction used is typically 20%. The Lowess fit thus gives a modified distribution of data for each print tip. The mean of this distribution can then be used to normalise the data within each print tip group. We thus have the Lowess transformation of

$$M \rightarrow M - c_i(A)$$

where $c_i(A)$ is the Lowess fit to the M *versus* A plot for the i_{tu} grid, where i = 1,…,I, and I represents the number of print-tips. Figure 4(E) shows the separated print tip groups before normalisation against a Lowess fit. This should be compared with Figure 4(F) that shows the distribution after Lowess fitting has been performed: This transformation improves the distribution of the data, making for better comparisons. It can be improved further by scaling each print tip group with the others to remove cross-plate variation in the hybridisation process. This method is not necessarily the best approach to across plate normalisation, but is reasonably sound. This, then, provides a full plate normalisation enabling comparisons of individual spot intensities to be made across the whole plate. Yang et al. found that appropriate robust scale factor to apply is a_i^2, where

$$q_i = MAD_i / {}^I\sqrt{[\Pi_{i=1}^I MAD_i]}$$

where MAD is the median absolute deviation, defined by

$$MAD_i = median_j \{ | M_{ij} - median_j(M_{ij}) | \}$$

Where, I denotes the total number of print-tip groups and M_{ij} denotes the i_{tu} log ratio in the i_{tu} print tip group, j = 1,…,n_j. This robust MAD statistic will not be affected by the small percentage of differentially expressed genes which will appear as outliers in the M *versus* A plots. The resultant scaled distribution, is now sharper and centred on M=0 as can be seen in Figure 4(B). In this case, the outliers, where | M | > 0.95 have been highlighted. These spots are considered to be worthy of further investigation. The cut-off point for | M | is somewhat arbitrary and can be determined by selective PCR. In reality, the number of expressed genes to be investigated will limit the positioning of the base-line cut-off. To continue the analysis, the marked spots are saved along with their original ID's for later comparison with other data. In the dye-swap experiments, Yang et al. have suggested that a between plate normalisation of 0.5 (M + M') *versus* 0.5 (A + A') will provide an immediate comparison between the plates. In this case A and M are for one plate and the A' and M' are for the dye swapped plate.

It has been found that the method of normalising and scaling each plate without background removal leads to less error. Each set of expressed spots fitting the criteria | M | > 0.95 are then compared. Spots appearing in both the un-swapped and dye-swapped plates with these high expression level changes are then considered as likely candidates for function investigation. Spots that do not appear in both lists are considered as dubious spots which may be worth following up, but there is insufficient evidence to include them in the likely spots list. The un-swapped array is shown in Figure 4(C) with the raw data highlighted spots marking the spots meeting the | M | > 0.95 criteria following normalisation and scaling. The dye-swapped plate must undergo a similar analysis. Following some manipulation, a set of spots was found to match the selection criteria. Consideration must be given to any spots that are flagged as damaged or are saturated. Only by examining the original image can the damaged spots be declared as possible for inclusion or must be excluded from the analysis. Saturated spots should be noted in order that later comparisons are informed of the artificially low intensity value being recorded. Should the experiment include replicates, the mean plates should be further normalised between them to obtain comparable values. This is normally performed by taking the plate with median spot expression level of each plate and using this plate as a normalising factor for all plates in the experiment. However, with a dye swap experiment using the above print-tip analysis, the result is a set of ratios. The ratios should not change significantly if all the values are raised or lowered in a broad spectrum spot normalisation process. If there are a number of spots exhibiting low intensities, and there will normally be many of these, the ratios of these intensities may be overemphasised by the analysis process. Therefore it is recommended that a small intensity value be added to all spot intensities prior to analysis, typically this will be a value of around 50. This 'trick' to avoid artefacts of the analysis process is particularly important if the background intensity level is subtracted from the foreground spot intensity. After reversing the results of the second dye-swap analysis, the two sets of results can be combined, usually as a mean value of the spot ratios and with replicate plates, a similar combination taken. Statistical considerations should be made and the variance used to give some confidence to the values obtained. For the cut-off of $|M| >= 0.95$, we obtained 255 spots with differential expression levels. But what does it tell us and how do we proceed? The first step in the further analysis is to identify the gene related to the probe fragment. This may be provided by the microarray supplier as an EST or gene accession number or loci. Alternatively, only the sequence may be known. Whatever is the given information; this must be used to seek appropriate annotation for the selected probes. For the print-tip analysis above, the plain results are given in Table 1 where the spots giving an absolute log fold change of ≥ 1.5 is shown. The interpretation of these results is given in later sections of this chapter. Note that the expression levels are often referred to as 'fold-change' and some authors use Log base 2 to express the change, where others show the actual change. In the former, a negative value indicates the divisor spot is expressing more and a value of 0 means they are equal. The data is now ready for exploration and this normally requires several steps:

a) Check the identity of the probes of interest and if possible check the sequence used is functionally equivalent to the target
b) Check for recent annotations of the probes of interest
c) Compare with similar or related experiments for additional hints of activity levels
d) Explore related biology/processes etc

Table 1. Results of print tip analysis showing $|\log_2(R/G)| \geq 1.5$. The annotation given is that located at the time for the experiment (2004) and includes several unknown functional equivalents

			results for 1–50	
SpotName	ID	Loci	Log2-RbyG	Annotation
N96309	G8C11T7	At3g45780	–2.09	nonphototropic hypocotyl 1
T45480	132I17T7		2.05	UDP-glucoronosyl/UDP-glucosyl transferase family protein contains Pfam profile: PF00201 UDP-glucoronosyl and UDP-glucosyl transferase
BE521605	M20E9STM		–2.04	
T13744	38C12T7		–2.02	expressed protein contains similarity to cotton fiber expressed protein 1 [*Gossypium hirsutum*] gi|3264828|gb|AAC33276
N65691	229K3T7		–2.01	expressed protein contains similarity to cotton fiber expressed protein 1 [*Gossypium hirsutum*] gi|3264828|gb|AAC33276
T20589	88I21T7	At1g09310	–2.01	expressed protein contains Pfam profile PF04398: Protein of unknown function, DUF538
M90508	PR-1		–1.98	Not found in TAIR. EMBL: *Arabidopsis thaliana* PR-1-like mRNA, complete cds.
M90508	PR-1		–1.98	Not found in TAIR. EMBL: *Arabidopsis thaliana* PR-1-like mRNA, complete cds.
H76907	205J15T7		–1.95	nonspecific lipid transfer protein 1 (LTP1) identical to SP|Q42589
T41722	65F10T7		1.92	zinc finger (C2H2 type) family protein (ZAT12) identical to zinc finger protein ZAT12 [*Arabidopsis thaliana*] gi|1418325|emb|CAA67232
R86807	124I15T7		–1.89	expressed protein
T22117	96O24T7		–1.88	expressed protein
N37319	209K19T7		–1.87	long hypocotyl in far-red 1 (HFR1) / reduced phytochrome signalling (REP1) / basic helix-loop-helix FBI1 protein (FBI1) / reduced sensitivity to far-red light (RSF1) / bHLH protein 26 (BHLH026) (BHLH26) identical to SP|Q9FE22 Long hypocotyl in far-red 1 (bHLH-like protein HFR1) (Reduced phytochrome signalling) (Basic helix-loop-helix FBI1 protein) (Reduced sensitivity to far-red light) [*Arabidopsis thaliana*]
T43374	118F16T7	At2g38540	–1.86	nonspecific lipid transfer protein 1 (LTP1) identical to SP|Q42589
AA395470	94E10XP	At3g21760	–1.84	glycosyltransferase family; contains Pfam profile: PF00201 UDP-glucoronosyl and UDP-glucosyl transferase

Table 1 (continue)

				results for 1–50
SpotName	ID	Loci	Log2-RbyG	Annotation
H37424	181F10T7	At2g44790	1.83	uclacyanin II; almost identical to uclacyanin II GI:3399769 from [Arabidopsis thaliana]
BE521509	M20A8XTM		–1.8	
AA721829	126C9T7		–1.78	
H75999	193C17T7	At1g11210	–1.75	expressed protein; similar to hypothetical protein GB:AAD50003 GI:5734738 from [Arabidopsis thaliana]
R90351	192M4T7	At2g22125	–1.73	C2 domain-containing protein; contains Pfam profile PF00168: C2 domain
T75691	142K12T7		–1.72	expressed protein contains Pfam profile PF04862: Protein of unknown function, DUF642
AA650788	283D6T7		1.69	glutathione S-transferase, putative similar to glutathione transferase GB:CAA09188 [Alopecurus myosuroides]
N37141	208H21T7		–1.68	alpha-xylosidase (XYL1) identical to alpha-xylosidase precursor GB:AAD05539 GI:4163997 from [Arabidopsis thaliana]; contains Pfam profile PF01055: Glycosyl hydrolases family 31; identical to cDNA alpha-xylosidase precursor (XYL1) partial cds GI:4163996
AA395252	119G10XP		1.67	glycerophosphoryl diester phosphodiesterase family protein weak similarity to SP|P37965 Glycerophosphoryl diester phosphodiesterase (EC 3.1.4.46) [Bacillus subtilis]; contains Pfam profile PF03009: Glycerophosphoryl diester phosphodiesterase family
AI100032	149E11XP	At2g08383	–1.65	predicted protein
H36203	175O18T7	At3g16370	–1.64	GDSL-motif lipase/hydrolase protein; similar to family II lipases EXL3 GI:15054386, EXL1 GI:15054382, EXL2 GI:15054384 from [Arabidopsis thaliana]; contains Pfam profile: PF00657 Lipase Acylhydrolase with GDSL-like motif
AA605360	185F1XP	At1g49750	–1.63	leucine rich repeat protein family; contains leucine-rich repeats, Pfam:PF00560
N38199	220N21T7		–1.61	defective chloroplasts and leaves protein-related / DCL protein-related similar to defective chloroplasts and leaves (DCL) protein SP: Q42463 from [Lycopersicon esculentum]

Table 1 (continue)

SpotName	ID	Loci	Log2-RbyG	Annotation
			results for 1–50	
T22370	104E20T7		−1.6	germin-like protein (GER1) identical to germin-like protein subfamily 3 member 1 SP|P94040; contains Pfam profile: PF01072 Germin family
AA394884	314A10T7	At1g75540	−1.6	diadenosine 5',5'''-P1,P4-tetraphosphate hydrolase, putative; similar to diadenosine 5',5'''-P1,P4-tetraphosphate hydrolase GI:1888556 from [*Lupinus angustifolius*], [*Hordeum vulgare* subsp. *vulgare*] GI:2564253; contains Pfam profile PF00293: NUDIX domain
T21853	103M21T7	At4g21960	−1.59	peroxidase, putative; identical to peroxidase [*Arabidopsis thaliana*] gi|1402904|emb|CAA66957
N38263	222A6T7	At3g10490	−1.59	expressed protein; N-terminus similar to unknown protein GB:AAD25613 [*Arabidopsis thaliana*]
N65640	240K8T7	At2g39530	1.57	expressed protein
AA712435	190N22T7	At5g38980	−1.55	expressed protein
BE520960	M15H9STM		1.55	
R90675	191G3T7	At1g22500	−1.52	RING-H2 zinc finger protein ATL5 -related; similar to RING-H2 zinc finger protein ATL5 GI:4928401 from [*Arabidopsis thaliana*]
H37681	185B17T7	At4g29510	−1.5	protein arginine N-methyltransferase, putative; similar to protein arginine N-methyltransferase 1-variant 2 (Homo sapiens) GI:7453575

Affymetrix style microarray analysis

For our next example, we consider an experiment using a number of microarrays produced by Affymetrix, the Ath0 (also known as the AG-8K) chip which contains spots representing some 30,000 *Arabidopsis thaliana* genes. The Affymetrix chip contains multiple repeats of 'perfect match' (PM) oligonucleotide fragments for each target sequence together with a similar number of 'mis-match' (MM) fragments, where each MM spot differs in one base. The various PM's and MM's are dispersed across the physical plate. These arrays require a different type of analysis. Some approaches to the analysis make a comparison of the PM and MM values to determine if true hybridisation has been detected at a given target. Other packages ignore the MM values and simply determine the hybridisation levels through the PM probes alone. Among the former are the Affymetrix (GCOS) and dChip. Techniques such as Robust Multichip Average (RMA) [42] and gcRMA (available in Bioconductor (http://www.bioconductor.org/)) ignore the MM probes and consider only the normalisation of the PM probes. The RMA approach is gaining in popularity and while gcRMA is considered the best of these approaches as it includes a Bayes empirical GC content correction on the basic RMA methodology. A quick analysis may readily be conducted using RMAExpress which is a standalone, public domain package purely for the fast application of RMA to extract the expression levels over many chips. GeneSpring can analyse Affymetrix scans using their in-built algorithms or can be used to analyse with RMA or gcRMA by importing the appropriate R-library. Bioconductor can of course be used for the application of these techniques as well. The proprietary packages provide many features for exploring the data before and post processing and the reader is referred to the documentation of such packages to see how this may be performed.

The use of RMA Express (http://stat-www.berkeley.edu/users/bolstad/RMA Express/RMAExpress.html), R (http://www.r-project.org/) and many other algorithm collections requires that the user work hard and have some experience in the collection and analysis of post-processed data. Experienced users will make use of available database systems (MySQL, MS Access, ORACLE or Postgres for example) and statistical engines (R, Genstat etc.) to import (and in some instances determine expression levels) the normalised data collection and perform the appropriate calculation of confidence levels, spot-level comparisons, linking to annotation and selection/export of results of interest. The visualisation of features is often a most valuable exploration tool and the methods of distance clustering for the production of 'heat maps' which shows the 'nearness' of plates to each other along with the levels of expression and the use of Principal Component Analysis which separates the main causes of differences between the experiments and helps to identify significantly distinct gene sets across experiments are two primary methods available for the exploration of the data. Such methods are available in the larger packages and in the many public domain tools.

Time series microarray analysis

One frequent class of microarray experiment is that of the time series. In this case, there is usually a biologically replicated series of microarrays taken at intervals of hours, days or weeks according to the organism being studied. Often these form a series of about five time steps. The objective being to determine how an organism responds to various stimuli over time compared to an appropriate control. The analysis of such short time series requires the use of appropriate techniques [43] due to the large number of genes and the small number of time steps where many patterns are expected to arise at random. One implementation for the analysis of short time series is available in the public domain package STEM (http://www.cs.cmu.edu/~jernst/stem/). Figure 5 shows the results of the STEM-based analysis of a time series of an experiment on *Arabidopsis*, using the Affymetrix ath1 chip for which the scanned data was analysed using RMA Express to obtain the expression levels

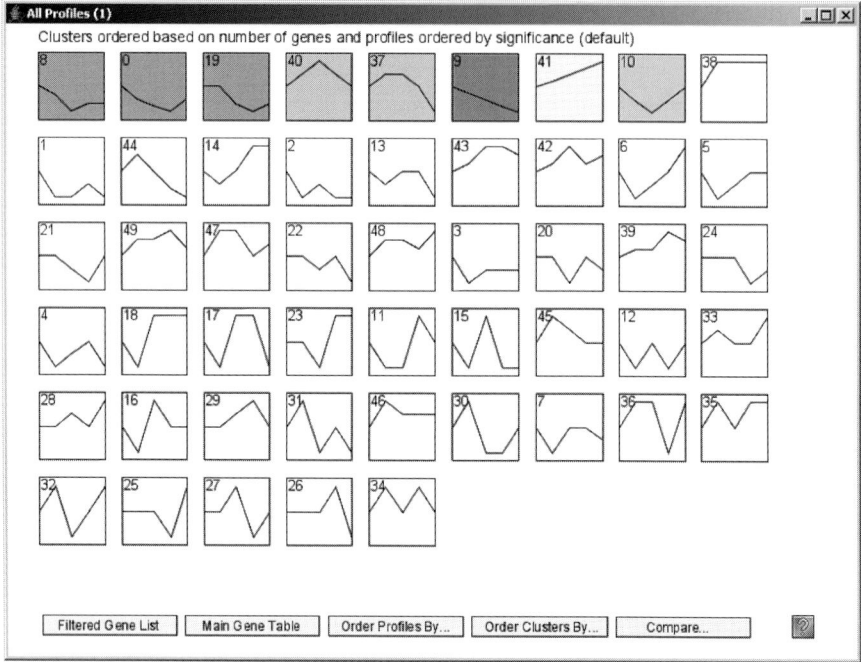

Figure 5. STEM package cluster groups. The greyed cells identify the significant clusters.

normalised across all plates in the experiment. The expression levels were then exported in a suitable format for STEM, along with the available GO ontology for *Arabidopsis*. In this experiment there are five time steps available. Figure 5 shows the resultant set of distinct time-series clusters found by the package. The greyed boxes indicate the clusters of statistical significance. Examination of the first group shows (Fig. 6) that 65 genes on the arrays follow this specific expression level change over the time series. With the associated Go annotation, STEM also provides the gene annotation sorted by function enabling a rapid assimilation to be made of the activities taking place and also often shows the appearance of genes of unknown function following this same pattern. The problem for the biologist is to interpret the different clusters and to perhaps locate causal genes for which one cluster might follow the activity of another.

Significance levels

The analysis of an array would not be complete without some form of measure of the confidence level of any given spot value or cluster. Essentially, there are two levels of significance that require to be considered. Firstly, the actual spot levels and the values of the pixels that makes up these spots. There may be considerable variation in the pixel intensity for a single spot (e.g., in the case of a 'doughnut' or 'cusp'-like spot) and this will have an impact on the quality of the signifi-

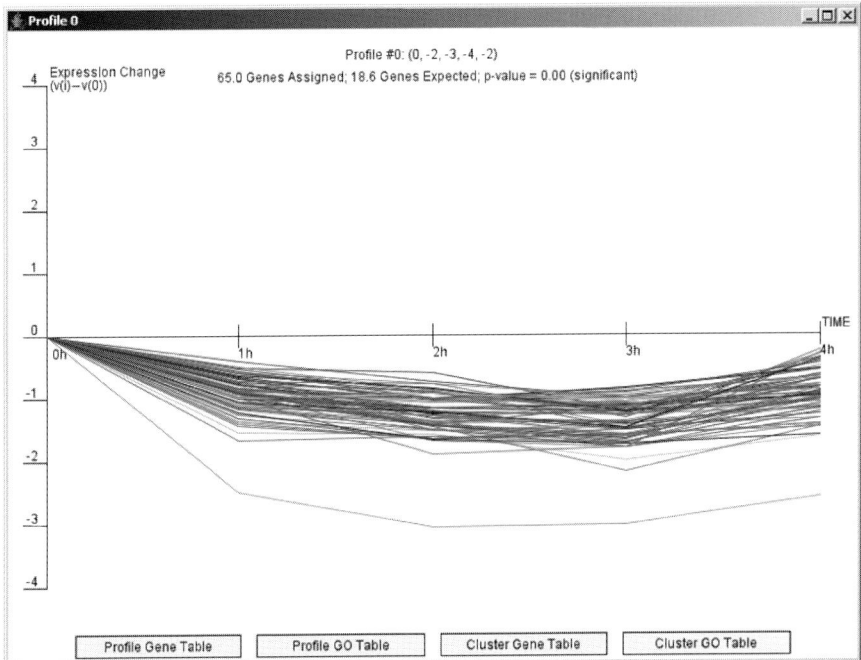

Figure 6. The first cluster profile from STEM showing 65 genes that fit this expression pattern.

cance of the spot value. There may also be 'missing' spots across an experiment, where one spot is damaged in a set of replicates and there may be a large variance in the intensities of one probe across the replicates. These sources of uncertainty need to be considered in the analysis. In addition, it is possible to assess the probability of a selected spot being present at high intensities through chance in these experiments. Both these measures are frequently produced by the various analysis packages, but not all. This chapter cannot deal with the methods used to describe such statistics, and reference should be made to an appropriate text such as that of Wit and McClure [44] which also gives a very thorough review of analysis approaches.

Resources

A large experiment with 150 arrays each representing say 30,000 genes will eat away the average resources of the normal computer user. Many analysis packages are memory hungry and the volume of calculations is sufficiently large to strain the smaller desktop computers. As an illustration, a typical PC running under the Microsoft Windows XP operating system configured for analysing this large number of arrays is likely to have 300 Gb of local disc, 4 Gb local memory, 3 GHz CPU chip and a large size monitor. Be prepared to handle the disk store back-up requirement.

Discernable signatures within the *vtc* transcriptome

Using the Affymetrix Ath1 (AG-8K) array, we found that AA deficiency in the *vtc1* mutant led to the differential expression of 171 genes, of which 97 genes were induced and 74 genes were repressed. A comparable experiment conducted using the Affymetrix ATH1-22K full genome array yielded 821 differentially expressed genes of which 249 were induced and 572 were repressed. In comparison, the *abi4* mutant leaves yielded 535 differentially expressed genes compared to the wild type control leaves using the Affymetrix ATH1-22K array. Of these 149 genes were induced and 386 were repressed. From analysis of the gene expression patterns we were able to determine that AA content influences the following processes.

Innate immune resistance to pathogens

One of the most interesting features of the *vtc1* transcriptome is the synchronised accumulation of transcripts encoding pathogenesis resistance (PR) proteins [39, 45]. These results suggested that low AA might confer enhanced basal resistance to pathogen attack. This hypothesis was confirmed in experiments using a number of pathogens such as *Pseudomonas syringae* [5, 6]. In contrast to low symplastic AA, which enhances pathogen resistance [6], low abundance of AA specifically in the apoplast as a result of high ascorbate oxidase (AO) activity, decreases pathogen resistance [46].

Effects on growth and development

AA and AO have long been considered to influence cell expansion [23, 46–48] and mitosis [4, 49]. The low AA transcriptome revealed effects of AA on plant hormone metabolism that indicate how AA can influence growth. AA-modulated transcripts that have the potential to influence plant growth and development are listed in Tables 2–5. Some of the implications of these results are as follows.

Effects on ABA and giberrellic acid

The *vtc1* signature contained transcripts indicating an increased abundance of ABA in the *vtc* mutants, a feature confirmed by measurements of leaf ABA contents [45]. The upregulation of this plant hormone in *vtc* leaves coincides with enhanced pathogen resistance and slowed growth [6, 50]. We therefore considered whether at least a part of AA signalling in leaves proceeds via ABA-dependent pathways. We thus examined whether ABA signalling events were also involved in AA-signalling. A comparison of the transcriptome of *abi4* and *vtc1* leaves relative to that of the wild type leaves revealed that a large number of transcripts were modified in a similar manner in *abi4* and *vtc1* leaves. A comparison of the data given in Table 4 for *vtc1* and Table 5 for *abi4*, illustrates this point well for transcripts concerned with cell cycle regulation, development and hormone and cell signalling. The extent of cross talk between ABA and AA signalling pathways is now under further investigation.

Table 2. Comparisons of key transcripts related to plant growth and development modified in *vtc1* leaves relative to wild type using the Affymetrix GeneChip Arabidopsis Genome Array (AG-8K array; [45])

Fold	Gene ID	Description	Function
–1.45	At5g44290	CDC2a type cyclin (AK23; G1→S)	cell cycle
–1.31	At1g30690	patellin-4(cytokinesis)	cell cycle
+1.22	At4g39180	Putative SEC14 protein (cytokinesis)	cell cycle
+1.48	At2g23430	cyclin-dependent kinase inhibitor (KRP1; G1→S)	cell cycle
+2.16	At2g18050	histone H1-3 (HIS1-3)	cell cycle
–1.3	At1g01720	ATAF1 Mrna (NAM)	development
–1.26	At4g20370	twin sister of FT (TSF)	development
–1.23	At4g33680	Abarrent growth and death 2	development
–1.21	At2g02450	Putative no apical meristem (NAM) protein	development
+1.33	At5g41410	homeobox protein (BEL1; NAM)	development
+1.53	At2g17040	putative no apical meristem (NAM) protein	development
+1.65	At4g26850	vitamin C defective 2 (VTC 2)	development
+1.2	At2g36690	putative giberellin beta-hydroxylase	hormone
+1.57	At4g00700	putative phosphoribosylanthranilate	hormone
+1.7	At4g19170	9-cis neoxanthin cleavage enzyme	hormone

Fold: – ve fold change (repressed); + ve fold change (induced);
Gene ID *A. thaliana* gene identifier;
Description: name of protein encoded by transcript modified;
Function: functional classification of each encoded protein was obtained from the Protein Families Data Base (Pfam; http://www.sanger.ac.uk/Software/Pfam/).

ABA and gibberellic acid (GA) often act antagonistically to modulate plant growth and defence. An interesting example of this antagonistic behaviour in relation to antioxidant defence concerns the regulation of PCD in the aleurone layer of seeds. ABA increases antioxidant gene expression and decreased sensitivity to H_2O_2 and susceptibly to PCD [51, 52] while application of GA decreased antioxidant gene expression and increased sensitivity to H_2O_2 and susceptibly to cell death [51, 52]. AA is a co-factor for the 2-oxoacid-dependant dioxygenase (2ODD) family of enzymes [47]. These enzymes are responsible for the synthesis of a wide range of crucial secondary metabolites including hormones [47]. One example is the aminocyclopropane-1-carboxylate (ACC) oxidase that is involved in ethylene synthesis. The ACC oxidase requires AA and Fe^{2+} for optimal rates of catalysis [53]. Furthermore cytosolic 2ODD's catalyse the final stages of GA synthesis, where GA12-aldehyde is converted to bioactive GA [54, 55]. In *in vitro* assays, 2ODD activities can often be enhanced by AA [54]. The KNOX family of transcription factors exert control over GA synthesis. Interestingly, transcripts encoding the homeodomain transcription factor BEL1 which activate the KNOX transcription factors [56, 57] are modulated by AA. Cellular AA availability may therefore contribute to the control of the *BEL1* and KNOX proteins.

Table 3. Comparisons of key transcripts related to plant growth and development modified in wild type *A. thaliana* leaves as a result of ascorbate feeding, using data obtained from the Stanford Universities cDNA microarrays [41]

Fold	Gene ID	Description	Function
−2.62	At1g12430	kinesin-like protein (cytokinesis)	cell cycle
−2.58	At4g39050	kinesin like protein (MKRP2; cytokinesis)	cell cycle
−2.35	At1g52740	putative histone H2A	cell cycle
−2.22	At1g47210	putative cyclin-A (CYCA3.2; G1→S)	cell cycle
−2.1	At4g08950	putative phi-1-like phosphate-induced protein	cell cycle
+1.99	At5g03340	cell division control protein (CDC48E; cytokinesis)	cell cycle
+2.03	At3g28780	histone-H4-like protein	cell cycle
−2.79	At2g29890	putative villin (actin binding)	development
−2.62	At1g57720	similarity to elongation factor 1-gamma 2	development
−2.57	Atg73680	similarity to feebly-like protein	development
−2.51	At1g09640	eukaryotic translation elongation factor 1 complex	development
−2.31	At3g23550	aberrant lateral root formation 5	development
−2.03	At1g69490	NAC-like, activated by AP3/PI protein	development
−1.97	At4g12420	putative pollen-specific protein	development
−1.95	At5g41410	homeotic protein (BEL1;NAM)	development
+2.14	At3g57520	imbibition protein homolog	development
+2.17	At5g44120	similarity to legumin-like protein	development
−2.89	At1g05180	auxin-resistance protein (AXR1; IAA)	hormone
−1.96	At4g19170	9-*cis* neoxanthin cleavage enzyme (ABA)	hormone
+2.38	At4g37390	Indole-3-acetic acid-amido synthetase (GH3.2; IAA)	hormone
−2.89	At4g29810	MAP kinase kinase 2 (MAPKK2; MK1)	signalling
−2.32	At3g59220	pirin-like protein	signalling
−2.08	At3g18820	putative GTP binding protein	signalling
−2.01	At4g09720	rab7-like protein (GTP-binding protein)	signalling

Fold: − ve fold change (repressed); + ve fold change (induced);
Gene ID *A. thaliana* gene identifier;
Description: name of protein encoded by transcript modified;
Function: functional classification of each encoded protein was obtained from the Protein Families Data Base (Pfam; http://www.sanger.ac.uk/Software/Pfam/).

Table 4. Comparisons of key transcripts related to plant growth and development modified in *vtc1* leaves relative to wild type using the Affymetrix ATH1-22K arrays

Fold	Gene ID	Description	Function
−0.90	At1g75780	tubulin beta-1 chain	Cell cycle
+0.90	At3g53230	cell division control protein (CDC48E; cytokinesis)	Cell cycle
+0.90	At5g10400	*histone H3-like protein	Cell cycle
+0.99	At3g46030	*histone H2B-ike protein	Cell cycle
−1.69	At5g24780	vegetative storage protein (Vsp1)	development
−1.22	At1g28330	dormancy-associated protein	development
−0.97	At5g62210	embryo-specific protein	development
−0.94	At4g13560	*putative protein LEA protein	development
+0.87	At5g33290	putative protein EXOSTOSIN-1	development
+0.89	At4g02380	*late embryogenesis abundant 3 family protein / LEA3	development
+1.00	At3g49530	NAC2-like protein	development
+1.06	At3g44350	*NAC domain-like protein	development
+1.08	At1g61340	late embryogenesis abundant protein (LEA)	development
+1.23	At3g54150	embryonic abundant protein	development
+1.33	At3g25290	auxin-responsive family protein	development
+1.58	At5g22380	*NAC-domain protein-like	development
+1.75	At2g43000	NAM (no apical meristem)-like protein	development
+1.84	At2g17040	*NAM (no apical meristem)-like protein	development
−0.96	At1g78440	*gibberellin 2- oxidase	hormone
−0.89	At1g05560	indole-3-acetate beta-D-glucosyltransferase	hormone
+1.02	At4g29740	cytokinin dehydrogenase 4	hormones
+1.18	At5g20400	ethylene-forming-enzyme-like dioxygenase	hormone
+2.02	At5g13320	auxin-responsive GH3 family protein	hormone
+0.89	At4g08470	putative mitogen-activated protein kinase	signalling
+0.91	At3g45640	mitogen-activated protein kinase 3 (MAP kinase 3; AtMPK3)	signalling
+1.01	At1g73500	*mitogen-activated protein kinase kinase (MAPKK; MKK9)	signalling

Transcriptome comparison acquired using the Affymetrix GeneChip Arabidopsis Genome Array (ATH1-22K).

Fold: – ve fold change (repressed); + ve fold change (induced);
Gene ID *A. thaliana* gene identifier;
Description: name of protein encoded by transcript modified;
Function: functional classification of each encoded protein was obtained from the Protein Families Data Base (Pfam; http://www.sanger.ac.uk/Software/Pfam/).
* Transcript abundance also changed in abi4-102 leaves (Tab. 4), identified using the same technology.

Table 5. Comparisons of key transcripts related to plant growth and development modified *abi4-102* leaves relative to wild type using the Affymetrix GeneChip Arabidopsis Genome Array (ATH1-22K)

Fold	Gene ID	Description	Function
–1.09	At3g50240	Kinesin-like protein (KIF4; cytokinesis)	cell cycle
–0.86	At4g27180	Kinesin-related protein katB (ATK2; cytokinesis)	cell cycle
–0.85	At3g16000	myosin heavy chain-like protein (cytokinesis)	cell cycle
+0.86	At2g38810	histone H2A	cell cycle
+0.92	At5g22880	histone H2B-like protein	cell cycle
+1.00	At3g45930	histone H4-like protein	cell cycle
+1.24	At3g46030	*histone H2B-like protein	cell cycle
+1.44	At5g10400	*histone H3-like protein	cell cycle
–1.61	At4g13560	*putative protein LEA protein	development
–1.28	At1g34180	similar to NAM-like protein	development
–1.17	At1g52690	late embryogenesis-abundant protein (LEA76)	development
–1.04	At5g55400	fimbrin (actin binding)	development
–0.99	At3g13470	putative chaperonin 60 beta	development
–0.93	At1g72030	GCN5-related N-acetyltransferase (GNAT)	development
+0.85	At3g44350	*putative NAC-domain containing protein 61	development
+0.88	At4g02380	*'embryogenesis abundant 3 family protein / LEA3 family protein	development
+0.91	At1g01720	similar to NAC domain protein	development
+1.33	At2g39030	GCN5-related N-acetyltransferase (GNAT)	development
+1.40	At1g52890	similar to NAM (no apical meristem) protein	development
+1.44	At2g17040	*NAM (no apical meristem)-like protein	development
+1.58	At5g22380	*NAC-domain protein-like	development
–0.94	At1g15550	putative similar to gibberellin 3 beta-hydroxylase	hormone
–0.89	At1g78440	*gibberellin 2-oxidase	hormone
+0.96	At4g11280	1-aminocyclopropane-1-carboxylate synthase 6	hormone
+1.40	At1g73500	*putative mitogen-activated protein kinase kinase (MKK9)	signalling

Fold: – ve fold change (repressed); + ve fold change (induced);
Gene ID *A. thaliana* gene identifier;
Description: name of protein encoded by transcript modified;
Function: functional classification of each encoded protein was obtained from the Protein Families Data Base (Pfam; http://www.sanger.ac.uk/Software/Pfam/).
* Transcript abundance also changed in *vtc1-1* leaves (Tab. 3), identified using the same technology.

The synthesis of biologically active GAs (GA_1 and GA_4) is dependent upon the activities of the GA 20 oxidase (GA20OX/GA5) enzymes. The expression of the GA20OX genes is regulated by feedback inhibition by GA [58, 59]. For example, *GA5* transcripts accumulate in gibberellin-deficient plants [60]. Furthermore, sense and antisense expression of *GA5* has direct effects on the bioactive gibberellin content of transformed *A. thaliana* plants and also effects growth [61]. The expression of GA5 can therefore be used as a physiological marker for bioactive GA. GA5 transcripts were much more abundant in *vtc*1 leaves than those of the wild type, suggesting that bioactive GAs were much lower in *vtc*1 leaves.

Affects on two mitogen activated protein kinase cascades
Mitogen activated protein kinase (MAPK) cascades are also involved in redox signal transduction [62]. It is therefore not surprising that leaf AA abundance influenced the mRNAs encoding a MAPK (AtMPK3; At3g45640) and a MAPK kinase (MAPKK9; At1g73500; Tab. 5), which were increased in *vtc*1 shoots. The expression of AtMPK3 is regulated by ABA and it is thought to act by phosphorylation of the ABI5 transcription factor [63]. We have also shown that the amount of AA in the apoplast specifically also responses to auxin and GA through effects on MAP kinase activity [46].

Effects on the cell cycle
Cell cycle regulation involves components that respond to signals from the external environment as well as intrinsic developmental programmes and it ensures that DNA is replicated with high fidelity within the constraints of prevailing environmental conditions [64, 65]. *Arabidopsis* has two A1-type (CYCA1;1 and CYCA1;2), four A2-type (CYCA2;1, CYCA2;2, CYCA2;3, and CYCA2;4) and four A3-type (CYCA3;1, CYCA3;2, CYCA3;3, and CYCA3;4) cyclins. In synchronised tobacco BY2 cells, different A-type cyclins are expressed sequentially at different time points from late G1/early S-phase through to mid M-phase [66]. The alfalfa A2-type cyclin Medsa; CYCA2;2 is expressed during all phases of the cell cycle, but its associated kinase activity peaks both in S-phase and during the G2/M transition [67]. Cyclin-dependent kinases (CDKs) play a central role in cell cycle regulation, with negative kip-related proteins (KRP) and positive (D-type cyclins) regulators acting downstream of environmental inputs at the G1 checkpoint [65, 68].

The components that are modulated by AA in the control on the cell cycle remain to be characterised but effects of AA are independent of glutathione another abundant cellular antioxidant [49]. The expression of a number of genes encoding kinases were altered in *vtc*1 leaves compared to the wild type [45] (Tabs 2–4). A number of transcripts that transcripts are either known to be cell cycle regulated or could be associated with progression through the cell cycle are shown in Figure 7. At this stage we can only draw tentative conclusions from the transcriptome results as changes in gene expression can be an indirect effect of arrest in cell cycle phases, rather than being direct targets of AA signalling. Here we consider the changes in expression as a molecular footprint revealing the points of cell cycle arrest (providing that the transcripts are indeed cell cycle regulated). They are thus putative targets which will induce arrests at specific phases of the cell cycle. While transcripts encoding D-type cyclins were similar in *vtc1* and wild type leaves and they were not changed by feeding AA, the expression of KRP1, a cyclin dependant kinase inhibitor (ICK1; At2g23430) was upregulated in the *vtc1* transcriptome suggesting that low AA favours decreased D-type cyclin expression.

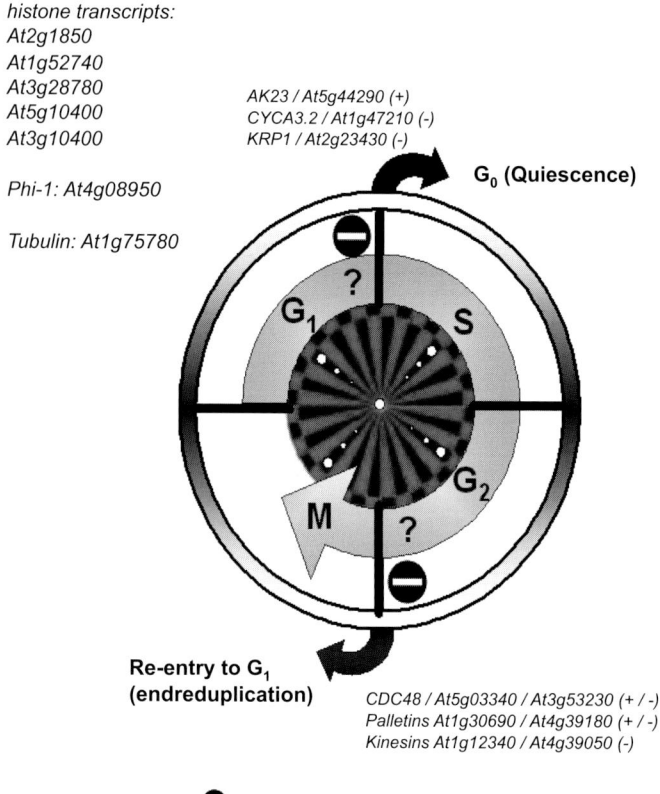

Figure 7. Ascorbate-modulated cell cycle genes. Classification: plus indicates induced by high ascorbate/redox state while minus indicates repressed by high ascorbate/redox state. Thus genes that are induced by low ascorbate should be repressed by high ascorbate. Genes decreased by low ascorbate have a (plus) and genes increased by low ascorbate have a (minus) and genes that are decreased by high ascorbate thus have a (plus) while those increased by high ascorbate have a (minus).

While A cyclins and KRP function in the G1/S transition, changes in histone transcripts are related to S-phase progression. Leaf AA content has a large effect on the abundance of tubulin transcripts. Changes in tubulin configuration occur during G2/M. However, tubulin contents are also influenced by other events such as the exit from the cell cycle and elongation, as well as the transport of protein complexes throughout the cell cycle. Kinesins are required at the G2/M phase. While a number of issues have to be considered during the interpretation of these data, it would appear that that AA exerts effects at several points in the cell cycle and not just the G1/S transition. Some of the observed changes in transcripts could be due to knock on effects caused by a primary block or delay during cell cycle progression by inducing a even a partial restriction at any one cell cycle checkpoint. This will affect the expression of genes involved at the next checkpoint. Hence, having more proliferating cells lingering longer in G1 will reduce the population of cells in G2, and therefore the levels of G2/M associated transcripts. It is therefore important to verify these findings using flow cytometrical analysis data. If the AA-modulated arrest occurs at both checkpoints, one would expect to find no changes in the balance of cells in G1 and G2. However, such analyses might be complicated by the superimposed effects of endoreduplication. The nuclear location of the non expressor of PR proteins (NPR1) in *vtc1* leaves [6] may also suggest effects of low AA on endoreduplication levels in *Arabidopsis* [69]. The transcriptome data may suggest that expression of transcripts associated with cytokinesis are modified in vtc1 leaves and this may affect endoreduplication levels. For example, the expression of two *Arabidopsis* CDC48 proteins known to regulate cell plate turnover and endoplasmic reticulum assembly during cytokinesis are modified in *vtc1* rosettes compared to the wild type and these are also modified in *vtc1* leaf discs following AA feeding (At5g03340). The expression of patellin genes (At4g39180 and At1g30690) was also modified in *vtc1* shoots. Patellins have been associated with membrane trafficking events during cell plate formation [70]. The decreased abundance of kinesin transcripts (At1g12430 and At4g39050) in *vtc1* leaves compared to those of the wild type suggests that AA could influence the cell cycle through the various roles of these proteins in centromere separation; chromosome attachment to microtubules; and aggregation to the cell plate during metaphase. It is of interest to note that one of the kinesins (MKRP2; At4g39050) whose mRNA abundance is deceased in *vtc1* is targeted to mitochondria [71].

Conclusions and perspectives

Plants created the aerobic world in which we live and hence they have already tackled the key problems of living with oxygen and found solutions in antioxidants and in redox signalling. The above discussion illustrates how combined physiological and genetic approaches can be used to identify relevant transcripts and genes for further analysis and how such data can be used to form testable hypotheses regarding metabolite signalling functions. The results show that AA is not only integral to the redox regulation of plant cells [1, 2] but that it is also a crucial metabolic regulator influencing plant growth and development. Much of the information that has allowed the development of current concepts concerning the central role of AA has come from transcript data. The evidence discussed here illustrates how microarray analysis can be used to give a comprehensive perspective of the influence of a metabolite such as AA on the leaf transcriptome and hence plant metabolism, physiology and development. The underpinning technologies have become routine and reliable while the methods of transcriptome analysis and data mining have become increasingly more sophisticated, useful and informative. Thus, we consider that microarray approaches and transcriptomics are the most easily accessible and user-friendly of all the information-rich –omics technologies available to help the plant scientist advance current knowledge.

For simplicity, in this discussion we have considered only certain of the AA transcriptome and how these features have enabled us to develop hypotheses for further testing by more classic physiology and molecular genetic approaches. In this way, the microarray analysis has provided a much deeper understanding of the interactions between AA and plant hormones that underpin key aspects plant biology than could have been gleaned by other approaches. With regard to the regulation of the cell cycle, we can only draw tentative conclusions at present but the transcriptome results suggests at least two redox regulated sites influenced by AA availability. We can use this information to test whether AA-dependent changes in component gene expression are direct targets of AA signalling or indirect effects of for example, arrest in cell cycle phases.

Acknowledgements

Rothamsted Research and the Institute of Biotechnology receive grant-aided support from the Biotechnology and Biological Sciences Research Council of the UK (BB/C51508X/1, [C.F]). We thank Dr Spencer Maughan and Walter Dewitte for helpful discussions concerning putative cell cycle components and for critical reading of the manuscript.

References

1. Foyer CH, Noctor G (2005) Redox homeostasis and antioxidant signalling: a metabolic interface between stress perception and physiological responses. *Plant Cell* 17: 1866–1875
2. Foyer CH, Noctor G (2005) Oxidant and antioxidant signalling in plants: a re-evaluation of the concept of oxidative stress in a physiological context. *Plant Cell Environ* 28: 1056–1071
3. Fry SC (1998) Oxidative scission of plant cell wall polysaccharides by ascorbate-induced hydroxyl radicals. *Biochem J* 332: 507–515
4. Potters G, Horemans N, Caubergs R J, Asard H (2000) Ascorbate and dehydroascorbate influence cell cycle progression in tobacco cell suspension. *Plant Physiol* 124: 17–20
5. Barth C, Moeder W, Klessig DF, Conklin PL (2004) The timing of senescence and response to pathogens is altered in the ascorbate-deficient mutant vitamin C-1. *Plant Physiol* 134: 178–192
6. Pavet V, Olmos E, Kiddle G, Mowla, S, Kumar S, Antoniw J, Alvarez ME, Foyer CH (2005) Ascorbic acid deficiency activates cell death and disease resistance responses in *Arabidopsis thaliana*. *Plant Physiol* 139: 1291–1303
7. Thomas H, Ougham HJ, Wagstaff C, Stead AD (2003) Defining senescence and death. *J Expt Bot* 54: 1127–1132
8. Finkel T, Holbrook NJ (2003) Oxidants, oxidative stress and the biology of ageing. *Nature* 408: 239–247
9. Partridge L, Gems D (2002) Mechanism of ageing: public or private. *Nature Rev Genetics* 3: 165–175
10. Kurzweil R, Grossman T (2005) Fantastic voyage: Live long enough to live forever. Emmaus, Pennsylvania, Rodale Press, 1–452
11. Wheeler GL, Jones MA, Smirnoff N (1998) The biosynthetic pathway of vitamin C in higher plants. *Nature* 393: 365–369

12. Smirnoff N, Running JA, Gatzek S (2004) Ascorbate biosynthesis: a diversity of pathways. In: H Asard, JM May, N Smirnoff (eds.) *Vitamin C. Functions and Biochemistry in Animals and Plants.* Bios Scientific Publishers, Oxon, UK, Chapter 1: 7– 29
13. Agius F, González-Lamothe R, Caballero JL, Muñoz-Blanco J, Botella MA, Valpuesta V (2003) Engineering increased vitamin C levels in plants by overexpression of a D-galacturonic acid reductase. *Nature Biotechnol* 21: 177–181
14. Conklin PL, Norris SR, Wheeler GL, Williams EH, Smirnoff N, Last RL (1999) Genetic evidence for the role of GDP-mannose in plant ascorbic acid (vitamin C) biosynthesis. *Proc Natl Acad Sci USA* 30: 4198–4203
15. Keller R, Springer F, Renz A, Kossmann J (1999) Antisense inhibition of the GDP mannose pyrophosphorylase reduces the ascorbate content in transgenic plants leading to developmental changes during senescence. *Plant J* 19: 131–141
16. Gatzek S, Wheeler GL Smirnoff N (2002) Antisense suppression of L-galactose dehydrogenase in *Arabidopsis thaliana* provides evidence for its role in ascorbate synthesis and reveals light-modulated L-galactose synthesis. *Plant J* 30: 541–553
17. Tabata K, Ôba K, Suzuki K, Esaka M (2001) Generation and properties of ascorbic acid-deficient transgenic tobacco cells expressing antisense RNA for L-galactono-1,4-lactone dehydrogenase. *Plant J* 27: 139–148
18. Bartoli CG, Guiamet JJ, Kiddle G, Pastori G, Di Cagno R, Theodoulou FL, Foyer CH (2005) The relationship between L-galactono-1, 4-lactone dehydrogenase (GalLDH) and ascorbate content in leaves under optimal and stress conditions. *Plant Cell and Environment* 28: 1073–1081
19. Bartoli CG, Yu J, Gómez F, Fernández L, Yu J, McIntosh L, Foyer CH (2006) Inter-relationships between light and respiration in the control of ascorbic acid synthesis and accumulation in *Arabidopsis thaliana* leaves. *J Exp Bot* 57: 1621–1631
20. Bartoli CG, Pastori GM, Foyer CH (2000) Ascorbate biosynthesis in mitochondria is linked to the electron transport chain between complexes III and IV. *Plant Physiol* 123: 335–343
21. Tabata K, Takaoka T, Esaka M (2002) Gene expression of ascorbic acid-related enzymes in tobacco. *Phytochemistry* 61: 631–635
22. Tamaoki M, Mukai F, Asai N, Nakajima N, Kubo A, Aono M, Saji H (2003) Light-controlled expression of a gene encoding L-galactono-γ-lactone dehydrogenase which affects ascorbate pool size in *Arabidopsis thaliana*. *Plant Sci* 164: 1111–1117
23. Pignocchi C, Fletcher JM, Wilkinson JE, Barnes JD, Foyer CH (2003) The function of ascorbate oxidase in tobacco. *Plant Physiol* 132: 1631–1641
24. Chen Z, Young TE, Ling J, Chang SCh, Gallie DR (2003) Increasing vitamin C content of plants through enhanced ascorbate recycling. *Proc Natl Acad Sci USA* 100: 3525–3530
25. Conklin PL, Saracco SA, Norris SR, Last RL (2000) Identification of ascorbic acid deficient *Arabidopsis thaliana* mutants. *Genetics* 154: 847–856
26. Conklin PL, Williams EH, Last RL (1996) Environmental stress sensitivity of an ascorbic acid-deficient *Arabidopsis* mutant. *Proc Natl Acad Sci USA* 3: 9970–9974
27. Veljovic-Jovanovic SD, Pignocchi, C, Noctor G, and Foyer CH (2001) Low ascorbic acid in the *vtc-1* mutant of *Arabidopsis* is associated with decreased growth and intracellular redistribution of the antioxidant system. *Plant Physiol* 127: 426–435
28. Mulle-Moule P, Conklin PL, Niyogi KK (2002) Ascorbate deficiency can limit violaxanthin de-epoxidase activity *in vivo*. *Plant Physiol* 128: 970–977
29. Radzio A, Lorence A, Chevone BI, Nessler C L (2003) L-Gulono-1, 4-lactone oxidase expression rescues vitamin-C deficient *Arabidopsis* (vtc) mutants. *Plant Mol Biol* 53: 837–844
30. Dijkwel PP, Huijser C, Weisbeek P, Chua N-M, Smeekens SCM (1997) Sucrose control of phytochrome A signaling in *Arabidopsis*. *Plant Cell* 9: 583–595

31. Martin T, Hellmann H, Schmidt R, Willmitzer L, Frommer WB (1997) Identification of mutants in metabolically regulated gene expression. *Plant J* 11: 53–62
32. Arenas-Huertero F, Arroyo, A, Zhou L, Sheen J, Leon P (2000) Analysis of *Arabidopsis* glucose insensitive mutants, *gin5* and *gin6*, reveals a central role of the plant hormone ABA in the regulation of plant vegetative development by sugar. *Genes Dev* 14: 2085–2096
33. Sheen J, Zhou L, Jang JC (1999) Sugars as signaling molecules. *Curr Opin Plant Biol* 2: 410–418
34. Smeekens S, Rook F (1997) Sugar sensing and sugar-mediated signal transduction in plants. *Plant Physiol* 115: 7–13
35. Zhou L, Jang JC, Jones TL, Sheen J (1998) Glucose and ethylene signal transduction crosstalk revealed by an *Arabidopsis* glucose-insensitive mutant. *Proc Natl Acad Sci USA* 95: 10294–10299
36. Huijser C, Kortstee A, Pego J, Weisbeek P, Wisman E, Smeekens S (2000) The *Arabidopsis* SUCROSE UNCOUPLED-6 gene is identical to ABSCISIC ACID INSENSITIVE-4: involvement of abscisic acid in sugar responses. *Plant J* 23: 577–585
37. Signora L, De Smet I, Foyer CH, Zhang H (2001) ABA plays a central role in mediating the regulatory effects of nitrate on root branching in *Arabidopsis*. *Plant J* 28: 655–662
38. De Smet I, Signora L, Beeckman T, Inze D, Foyer CH, Zhang H (2003) An ABA-sensitive lateral root developmental checkpoint in *Arabidopsis*. *Plant J* 33: 543–555
39. Kiddle G, Pastori GM, Bernard B, Pignocchi C, Antoniw J, Verrier PJ, Foyer CH (2003) Effects of leaf ascorbate content on defense and photosynthesis gene expression in *Arabidopsis thaliana*. *Antioxidants and Redox Signalling* 5: 23–32
40. Yang YH, Dudoit S, Luu P, Speed T (2001) Normalization for cDNA microarry data. Berkley Technical report. http://www.stat.berkeley.edu/users/terry/zarray/Html/normspie.html
41. Allissul DA, Cui X, Page GP, Sabripour M (2005) Micoarray data analysis from disarry to consolidation and consensus. *Nature Rev Genetics* 7: 55–65
42. Bolstad BM, Irizarry RA, Astrand M, Speed TP (2003) A comparison of normalization methods for high density oligonucleotide array data based on bias and variance. *Bioinformatics* 19: 185–193
43. Ernst J, Nau GJ, Bar-Jospeh Z (2005) Clustering short time series gene expression data. *Bioinformatics* 21(suppl): i159–i168
44. Wit E, McClure J (2004) Statistics for microarrays, design, analysis and inference. John Wiley & Sons, Chichester, UK
45. Pastori GM, Kiddle G, Antoniw J, Bernard S, Veljovic-Jovanovic S, Verrier PJ, Noctor G, Foyer CH (2003) Leaf vitamin C contents modulate plant defense transcripts and regulate genes controlling development through hormone signaling. *Plant Cell* 15: 939–951
46. Pignocchi C, Kiddle G, Hernández I, Foster SJ, Asensi A, Taybi T, Barnes J, Foyer CH (2006) Ascorbate-oxidase-dependent changes in the redox state of the apoplast modulate gene transcription leading to modified hormone signaling and defense in tobacco. *Plant Physiol* 141: 423–435
47. Arrigoni O, de Tullio MC (2000) The role of ascorbic acid in cell metabolism: between gene-directed functions and unpredictable chemical reactions. *J Plant Physiol* 157: 481–488
48. Pignocchi C, Foyer CH (2003) Apoplastic ascorbate metabolism and its role in the regulation of cell signalling. *Curr Opin Plant Biol* 6: 379–389
49. Potters G, Horemans N, Bellone S, Caubergs R J, Trost P, Guisez Y, Asard H (2004) Dehydroascrobate influences the plant cell cycle through a glutathione-independent reduction mechanism. *Plant Physiol* 134: 1479–1487

50. Olmos E, Kiddle G, Pellny T, Kumar S, Foyer CH (2006) Modulation of plant morphology, root architecture and cell structure by low vitamin C in *Arabidopsis thaliana*. *J Exp Bot* 57: 1645–1655
51. Fath A, Bethke PC, Jones RL (2001) Enzymes that scavenge reactive oxygen species are down-regulated prior to gibberellic acid-induced programmed cell death in barley aleurone. *Plant Physiol* 126: 156–166
52. Fath A, Bethke P, Beligni V, Jones R (2002) Active oxygen and cell death in cereal aleurone cells. *J Exp Botany* 53: 1273–1282
53. Dong JG, Fernandez-Maculet JC, Yang SF (1992) Purification and characterization of 1-aminocyclopropane-1-carboxylate oxidase from apple fruit. *Proc Natl Acad Sci USA* 89(20): 9789–9793
54. Hedden P (1992) 2-Oxoglutarate-dependent dioxygenases in plants: mechanism and function. *Biochem Soc Trans* 20(2): 373–377
55. Hedden P, Kamiya Y (1997) Gibberellin biosynthesis: Enzymes, genes and their regulation. *Ann Rev Plant Physiol Plant Mol Biol* 48: 431–460
56. Quaedvlieg N, Dockx J, Rook F, Weisbeek P, Smeekens S (1995) The homeobox gene ATH1 of *Arabidopsis* is de-repressed in the hotomorphogenicmutants cop1 and det1. *Plant Cell* 7(1): 117–129
57. Bellaoui M, Pidkowich MS, Samach A, Kushalappa K, Kohalmi SE, Modrusan Z, Crosby WL, Haughn GW (2001) The *Arabidopsis* BELL1 and KNOX TALE homeodomain proteins interact through a domain conserved between plants and animals. *Plant Cell* 13(11): 2455–2470
58. Phillips AL, Ward DA, Uknes S, Appleford NE, Lange T, Huttly AK, Gaskin P, Graebe JE, Hedden P (1995) Isolation and expression of three gibberellin 20-oxidase cDNA clones from *Arabidopsis*. *Plant Physiol* 108(3): 1049–1057
59. Xu YL, Gage DA, Zeevaart JA (1997) Gibberellins and stem growth in *Arabidopsis thaliana*. Effects of photoperiod on expression of the GA4 and GA5 loci. *Plant Physiol* 114(4): 1471–1476
60. Chiang HH, Hwang I, Goodman HM (1995) Isolation of the *Arabidopsis* GA4 locus. *Plant Cell* 7(2): 195–201
61. Coles JP, Phillips AL, Croker SJ, Garcia-Lepe R, Lewis MJ, Hedden P (1999) Modification of gibberellin production and plant development in *Arabidopsis* by sense and antisense expression of gibberellin 20-oxidase genes. *Plant J* 17(5): 547–556
62. Kyriakis JM, Avruch J (1996) Sounding the alarm: protein kinase cascades activated by stress and inflammation. *J Biol Chem* 271: 24313–24316
63. Lu C, Man MH, Guevara-Garcia A, Fedoroff NV (2002) Mitogen-activated protein kinase signaling in postgermination arrest of development by abscisic acid. *Proc Natl Acad Sci USA* 99: 15812–15817
64. Dewitte W, Riou-Khamlichi C, Scofield S, Healy JM, Jacqmard A, Kilby NJ, Murray JA (2003) Altered cell cycle distribution, hyperplasia, and inhibited differentiation in *Arabidopsis* caused by the D-type cyclin CYCD3. *Plant Cell* 15: 79–92
65. de Jager SM, Maughan S, Dewitte W, Scofield S, Murray JA (2005) The developmental context of cell-cycle control in plants. *Semin Cell Dev Biol* 16: 385–396
66. Reicheld J-P, Venoux T, Lardon F, Van Montagu M, Inze D (1999) Specific checkpoints regulate plant cell cycle progression in response to oxidative stress *Plant J* 17: 647–656
67. Roudier F, Fedorova E, Györgyey J, Feher A, Brown S, Kondorosi A, Kondorosi E (2000) Cell cycle function of a *Medicago sativa* A2-type cyclin interacting with a PSTAIRE-type cyclin-dependent kinase and a retinoblastoma protein. *Plant J* 23 (1): 73–83
68. Dewitte W, Murray JAH (2003) The plant cell cycle. *Annu Rev Plant Biol* 54: 235–264

69. Vanacker H, Lu H, Rate DN, Greenberg JT (2001) A role for salicylic acid and NPR1 in regulating cell growth in *Arabidopsis*. *Plant J* 28: 209–216
70. Peterman TK, Ohol YM, McReynolds LJ, Luna EJ (2004) Patellin1, a novel sec14-like protein, localizes to the cell plate and binds phosphoinositides. *Plant Physiol* 136: 3080–3094
71. Lee YR, Liu B (2004) Cytoskeletal motors in *Arabidopsis*. Sixty-one kinesins and seventeen myosins. *Plant Physiol* 136: 3877–3883

Plant Systems Biology
Edited by Sacha Baginsky and Alisdair R. Fernie
© 2007 Birkhäuser Verlag/Switzerland

Case studies for transcriptional profiling

Lars Hennig[1] and Claudia Köhler[2]

[1] *Plant Biotechnology and* [2] *Plant Developmental Biology, Swiss Federal Institute of Technology (ETH) Zürich, Universitätstr. 2, 8092 Zürich, Switzerland*

Abstract

DNA microarrays are frequently used to study transcriptome regulation in a wide variety of organisms. Although they are an invaluable tool for the acquisition of large scale dataset in plant systems biology, a number of surprising results and unanticipated complications are often encountered that illustrate the limitations and potential pitfalls of this technology. In this chapter we will present examples of real world studies from two classes of microarray experiments that were designed to (i) identify target genes for transcriptional regulators and (ii) to characterize complex expression patterns to reveal unexpected dependencies within transcriptional networks.

Introduction

Since DNA microarrays have been introduced into experimental biology, scientists have used this technology to study transcriptome regulation in a wide range of organisms. Thousands of microarray studies have appeared in the literature since. In Foyer, Kiddle and Verrier's chapter several basic technical aspects concerning the design of DNA microarray experiments are discussed including sample preparation, hybridization conditions and statistical significance of the acquired data. These considerations are crucial for the successful design of microarray experiments and the acquisition of meaningful data in a biological context. As in all cases where large scale data are acquired, a number of surprising results and unanticipated complications can be expected that illustrate the limitations and potential pitfalls of a new technology. In this chapter, we will present examples of real world studies from two classes of microarray experiments, i.e., the identification of target genes for transcriptional regulators and the characterization of complex expression pattern to reveal unexpected dependencies within transcriptional networks.

Identification of target genes

To obtain a closer understanding of a particular biological process it is often helpful to search for mutants with defects in this process. The knowledge of the mutant gene that is responsible for the observed phenotype can give important insights into the process of investigation. However, to understand the molecular basis for a mutant phenotype it is essential to know which genes are deregulated in this mutant. This is of particular importance for the functional analysis of transcriptional regulators, as to understand the biological function of a transcriptional regulator itself it is often necessary to know the genes that this factor regulates. One classical approach to identify target genes regulated by a transcription factor is to compare the transcriptional profile of a mutant for that transcription factor with that of the corresponding wild type. More advanced approaches make use of an inducible complementation of the mutant phenotype, e.g., by applying the steroid inducible rat glucocorticoid receptor-binding domain fused to the protein of interest. The application of the steroid hormone dexamethasone causes the translocation of the transcription factor from the cytoplasm into the nucleus where it can activate its target genes. The challenge in both approaches is to identify the genes that are directly controlled by the transcription factor and to distinguish these primary target genes from genes that are deregulated in response to the deregulated primary targets. Subsequently, potential primary target genes are validated using Chromatin Immunoprecipitation (ChIP). The transcription factor should be directly associated with the locus of its target gene. Therefore, after immunoprecipitation with specific antibodies directed against the transcription factor the DNA of the target locus should become enriched in the precipitate. Figure 1 gives an overview about the typical steps in identifying target genes. In the following two sections we will discuss two approaches that have been successfully applied to identify primary target genes for the *Arabidopsis* Polycomb group protein MEDEA and the transcription factor LEAFY.

PHERES1 is a direct target gene of a plant Polycomb group complex

Polycomb group (PcG) genes have been initially identified in *Drosophila* by the isolation of mutations that cause strong homeotic transformations. PcG proteins form multimeric complexes that keep their target genes in a transcriptionally repressed state, which is stably transmitted over several mitotic divisions. PcG genes are evolutionary well conserved and have been identified in animals and plants (reviewed in [1, 2]). In plants, PcG proteins regulate major developmental decisions. In most flowering plants, seed development starts after the fusion of the two male gametes with the two female gametes, giving rise to the embryo and the endosperm. The maternally derived seed coat surrounds embryo and endosperm. Seed coat, embryo and endosperm together constitute the seed (reviewed in [3]). Mutants of the *fertilization independent seed* (*fis*) class bypass the strict requirement of fertilization and can start an autonomous endosperm development. If *fis* mutants are fertilized, the developing embryo and endosperm have proliferation defects and the seed aborts. Thus, the FERTILIZATION INDEPENDENT SEED (FIS) PcG proteins not only repress autonomous seed development but also coordinate the development of embryo and endosperm (reviewed in [4, 5]). To gain a

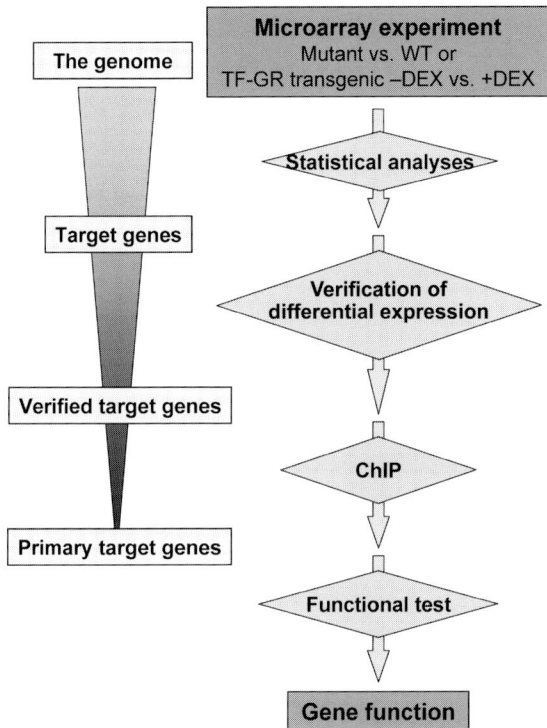

Figure 1. Scheme of the typical experimental strategy to identify target genes of transcriptional regulators. This approach establishes gene function from a microarray experiment. First, transcriptomes are measured on a genome-wide scale with microarrays. This can be a comparison of a mutant to its wild type. Alternatively, transgenic lines can be used that express a transcription factor glucocorticoid receptor hormone-binding domain fusion (TF-GR). In the absence of the steroid hormone dexamethasone (-DEX) the TF-GR protein remains in the cytosol and does not affect gene expression. Upon DEX treatment (+DEX), TF-GR migrates into the nucleus and activates target genes. If translation is not repressed with cycloheximide, both primary and secondary targets will be affected. Statistics are then used to select candidate target genes, which are verified by independent expression analysis. Chromatin immunoprecipitation (ChIP) is used to identify direct, primary target genes. Finally, the biological relevance of the finding will be addressed by functional tests. The funnel shape symbolizes number of genes analyzed at any step.

closer insight into the function of the FIS complex Köhler and colleagues aimed at the identification of direct target genes of the FIS complex [6]. The first two identified FIS genes are *MEDEA* (*MEA*) and *FERTILIZATION INDEPENDENT ENDOSPERM* (*FIE*) [7–9]. The encoded proteins MEA and FIE interact with each other and are part of a common protein complex [10–12]. Therefore, the identification of target genes of the FIS complex started with the transcriptional analysis of the *mea* and *fie* mutants assuming that in both mutants a common set of target genes would be deregulated. As the main interest of Köhler and colleagues was the identification of primary FIS target genes, the analysis focused on the identification of genes that were deregulated in *mea* and *fie* mutants at very early developmental stages, before any phenotypic aberrations were observed [6]. Mutant *mea* and *fie* plants as well as wild type plants were grown under the same environmental conditions and siliques were harvested. In the first sampling, only the *mea* mutant and wild type plants were harvested. Several weeks later a second sampling that was done including also the *fie* mutant in addition to the *mea* mutant and wild-type plants. To minimize effects of plant-to-plant transcriptional variation, material was collected and pooled from at least ten different plants for each sample. To identify commonly deregulated genes of the *mea* and *fie* mutants probe sets were selected that changed more than two-fold and were commonly affected in all three mutant RNA samples. According to these criteria, no probe set detected common down-regulation of a gene in all mutant samples. In contrast, two probe sets detected increased gene expression in all three samples. The identified deregulated genes encode for a MADS-box transcription factor and an S-phase kinase-associated protein1. The deregulated expression of both genes in *mea* and *fie* mutants was confirmed by real-time PCR of independently collected material. The gene encoding the MADS-box protein was named *PHERES1* (*PHE1*) and it was shown by ChIP that *PHE1* is a direct target gene of the FIS complex. Furthermore, the functional relevance of PHE1 could be demonstrated by introducing a knock-down construct of *PHE1* into the *mea* mutant background. The reduced *PHE1* expression in *mea* mutant seeds caused a partial complementation of seed abortion in *mea* plants indicating that enhanced *PHE1* expression in the *mea* mutant is causally related with the *mea* mutant phenotype.

Identification of direct target genes for LEAFY using inducible complementation of the leafy mutant

LEAFY (LFY) is a plant specific transcription factor that controls the switch from vegetative to reproductive development [13, 14]. Despite the biological importance of this developmental decision, *APETALA1* (*AP1*) was until recently the only known direct target gene of LEAFY [15]. However, the phenotype of *lfy* mutants was significantly stronger than the phenotype of the strongest *ap1* mutant allele. Therefore, it was assumed that *AP1* is not the only gene regulated by LFY [16]. The Wagner laboratory constructed a conditional *lfy* mutant by introducing a fusion protein of LFY with the rat glucocorticoid receptor hormone-binding domain (LFY-GR) into the *lfy* mutant background. The application of the steroid hormone dexamethasone causes the translocation of the LFY-GR fusion protein from the cytoplasm to the nucleus (Fig. 2) causing a rescue of the *lfy* mutant phenotype [15]. To find LFY dependent targets William and colleagues used 9-day-old seedlings that showed a strong LFY dependent up-regulation of *AP1* after steroid treatment [16]. *AP1* was also upregulated in the presence of cycloheximide (CHX). CHX inhibits the eukaryotic ribosomal peptidyltransferase and is used as an effective inhibitor of protein synthesis. The application of CHX allows to discriminate between primary (not

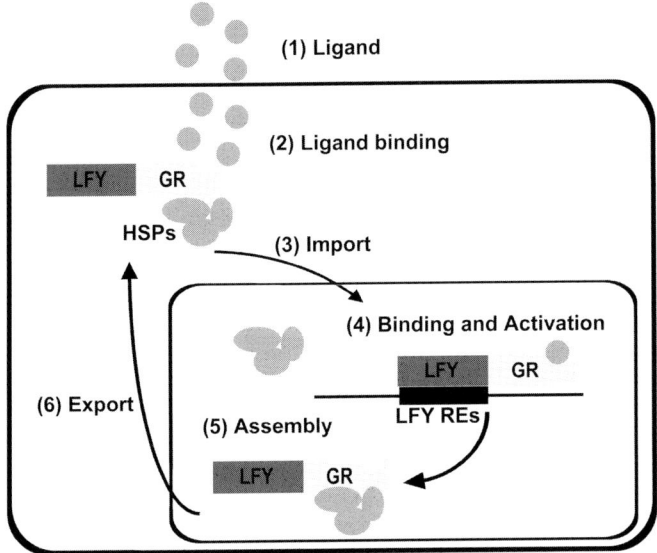

Figure 2. Nucleocytoplasmic shuttling of LEAFY-GR fusion proteins. Within the cytoplasm, heat shock proteins (HSPs) bind the LEAFY-Glucocorticoid receptor (LFY-GR) fusion protein and retain this protein in the cytoplasm. Binding of Dexamethasone (ligand) to the LEAFY-Glucocorticoid receptor fusion protein causes the translocation to the nucleus. The heat shock proteins (HSPs) dissociate from the receptor and LEAFY can bind to DNA response elements (LFY-REs) and activate transcription. Unliganded LFY-GR associates again with HSPs and is exported form the nucleus.

CHX sensitive) and secondary (CHX sensitive) target genes. *AP1* induction is independent of protein synthesis and thus probably not a secondary effect mediated by primary LFY targets. Most likely, *AP1* is a primary target of LFY. The following sample sets were generated and analyzed: (1) LFY-GR seedlings treated either with or without steroid, (2) LFY-GR seedlings treated either with or without steroid but in the presence of CHX, (3) seedlings constitutively overexpressing *LFY* (*35S::LFY*) in comparison to untreated wild-type seedlings. All samples were generated in duplicate using independently treated seedlings. The analysis concentrated on genes that were at least two-fold upregulated after steroid treatment resulting in 134 upregulated genes for sample set 1 and 152 genes for sample set 2. Because of a likely habituation of the seedlings to higher LFY expression levels, the threshold in sample set 3 was lowered to 1.4-fold upregulation, resulting in 753 upregulated genes. Out of this rather large number of deregulated genes, only 14 genes were commonly upregulated in all three sample sets. The identified genes were considered as good candidates for direct target genes of LFY as they were directly activated by LFY (without protein synthesis) and they were expressed at elevated levels in plants that ectopically express LFY. Williams and colleagues focused their further analysis on the five most highly expressed genes that encoded either potential transcription factors or signal transduction components. Those genes were confirmed to be upregulated in a LFY dependent manner but independently of protein synthesis. Finally, ChIP confirmed that LFY is indeed a direct activator of the identified genes as it can bind to the respective promoter regions. This study succeeded in the identification of five new direct target genes of LFY establishing that the inducible complementation of a mutant is an effective approach for the isolation of direct target genes of transcription factors.

Characterization of transcriptional profiles

In contrast to experiments like those described above, which aim to identify target genes of certain proteins of interest, other transcriptional profiling experiments aim to characterize expression patterns during development or in response to certain signals. Such experiments usually identify groups of genes collectively involved in certain biological processes and help to establish hypotheses about the biological functions of uncharacterized genes. Commonly these experiments involve time course designs and require different approaches for data mining than the simpler identification of target genes. Such advanced methods include, among others, regression analysis to find genes with particular expression patterns, clustering to group genes according to their expression profiles, pathway analysis and analysis of gene ontology (GO) terms to identify affected processes. Here, we will describe two examples from our own laboratories.

Cell cycle-regulated gene expression in Arabidopsis

The ability to divide is a fundamental property of cells, and multicellular organisms strictly control cell proliferation to ensure regulated development and growth. Therefore, understanding processes involved in cell division and their control is of great interest to developmental biology but also to tumor medicine. Others have studied gene expression during the cell cycle of yeast or mammalian cells [17, 18] and we used *Arabidopsis* suspension cells [19, 20]. For the experiments, we used a protocol to synchronize dividing cells in early S-phase by treatment with the DNA-polymerase inhibitor aphidicolin [21]. After washing out the drug, cells synchronously continue through one entire cell cycle, which lasts in these cells about 22 hours. Material was collected just before drug removal and subsequently at two hours intervals (Fig. 3). RNA was extracted, labeled and hybridized to Affymetrix GeneChip® microarrays. In order to enrich for relevant changes, only genes that passed a biological variation filter were selected. This filter was based on MAS5 'presence' and 'difference' calls [22], and required at least one 'P' (= present) and one 'D' or 'I' (= decreased or increased) for a gene to be considered. Transcripts that show a cell cycle modulated expression were identified using a method suggested by Shedden and Cooper [23]: This method assumes that the expression profile $Y_i(t)$ of cell cycle regulated genes can be modeled with a sine wave. The phase of the wave function relates to the expression maximum during

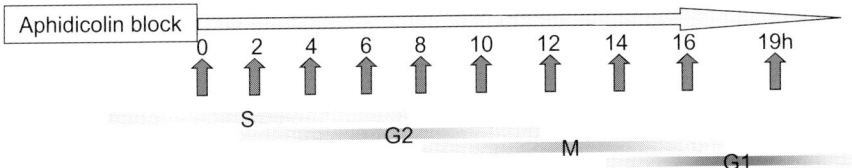

Figure 3. Scheme of experimental set-up for the transcriptional profiling of the plant cell-cycle. Asynchronously growing *Arabidopsis* suspension cells were incubated with the DNA polymerase inhibitor aphidicolin, which arrests cells in S-phase. At time zero, aphidicolin was washed out and cells synchronously re-entered the cell cycle. Samples were taken at given times during an entire cell cycle period. S, G2, M and G1 represent S-phase, G2-phase, mitosis and G1 phase, respectively.

the cell cycle. For every gene, $Y_i(t)$ can be decomposed into a periodic component $Z_i(t)$ with T = 22 h and a component $R_i(t)$ that is a-periodic or has a period substantially different from 22 h. The proportion of variance explained by the Fourier basis (Fourier proportion of variance explained (PVE)) is the ratio $m_i = var(Z_i(t))/var(Y_i(t))$, which can range from 0 to 1. Values closer to 1 indicate greater sinusoidal expression with a period of 22 h, whereas values closer to 0 indicate a lack of periodicity or periodicity with a period that is substantially different. Because among several thousand measurements some genes would display a periodic expression profiles even by chance, significance was estimated by shuffling the time points randomly and calculating a reference distribution of PVE values m based on the randomized data. Genes with a statistically significant ($p < 0.05$) greater periodic expression in the experiment than the randomized data set were selected for downstream analysis.

Out of the 22,800 probe sets on the ATH1 microarray, 9,910 passed the biological variation filter of which 1,605 had a significant periodicity. Out of these 1,605 genes, 1,016 had a fold change that was at least once larger 2 or smaller –2. Hierarchical and SOM clustering grouped these genes into several clusters with preferred expression in various phases of the cell cycle. A total of 669 genes had their expression maximum in S phase (0–4 h), 20 genes in G2 (6–8 h), 198 in mitosis (10–14 h) and 129 genes in G1 (16–19 h). In addition, a large number of signal transduction and regulatory components had strongly changing expression values but did not always fit a sine wave. These genes encode 93 receptor like kinases (RLKs), nine mitogen-activated protein kinase (MAPK) cascade members, eight protein phosphatase 2C (PP2C) and 79 annotated transcription factors (TF). Because only 18 TF genes were significantly oscillating, it is possible that the factors that regulate cell cycle oscillation will show expression during the cell cycle that is not necessarily periodic. It was also striking that there was a higher percentage of G2 genes in this set of genes than in the set of periodic genes. This analysis found back most of the known cell cycle regulators in *Arabidopsis* but identified also many other genes that were not known to be expressed cell cycle-dependent and likely include unknown regulators of the cell cycle. Thus, these results provide starting points for future targeted reverse genetic approaches.

Transcriptional programs of early reproductive stages in Arabidopsis

In addition to basic cellular functions like progression through the cell cycle, developmental programs are commonly studied using transcriptional profiling. We have characterized gene expression during plant reproduction [24]. Here, we analyzed RNA from three developmental stages of *Arabidopsis*, namely closed flower buds shortly before pollination (stage I), open pollinated flowers (stage II), and siliques 2 d after pollination (stage III). First, we compared the expression data to similar data sets from seedlings, roots or rosette leaves to identify transcripts that preferentially accumulate in flowers and developing fruits (reproductive set). Second, we selected genes that change expression upon pollination and initiation of seed and fruit development (regulated set). In the reproductive set, we found a significant overrepresentation of YABBY-, MADS-box- and MYB-type transcription factors. In the regulated set we found a significant overrepresentation YABBY-, MADS-box-, NAC-, CCAAT-HAP3- and MYB-type transcription factors. These results strongly suggest a dominating role of members of these transcription factor families in seed plant reproduction. Indeed, evolution of MADS-box transcription factors and evolution of plant reproductive organs are closely connected [25].

To identify various groups of regulated genes in the reproductive set, we used a regression approach with nine predefined patterns of interest. Assigning functional categories to genes, we observed that transcription factors were significantly overrepresented among the constantly expressed reproductive genes. By contrast, genes related to metabolism were significantly overrepresented among the upregulated, downregulated or transiently changed genes. These results show that organ and tissue specificity is to a large extent defined by specific transcription factors that remain expressed throughout the experiment, while genes for metabolic enzymes have often a highly dynamic pattern during the tested developmental stages. One metabolic pathway was analyzed in more detail, and it turned out that expression of enzymes for flavonoid metabolism is heavily regulated: Genes for flavonol synthesis were mostly downregulated, genes for anthocyanin synthesis were transiently upregulated, and genes for proanthocyanins were continuously upregulated. Intriguingly, the expression pattern of the structural genes of this pathway reflected closely the expression patterns of genes for transcription factors known to control gene expression for flavonoid synthesis. These results provide a molecular and genomic basis for existing physiological data about the importance of flavonoid biosynthesis during flower development [26]. Flavonoids, which are synthesized in several floral organs, are required for pollen function. Anthocyanins are transiently formed in *Arabidopsis* pistils after pollination, and proanthocyanins are synthesized in the developing testa to form condensed tannins of the seed coat [27].

Because reproductive development relies on intricate coordination of cell cycle activity, the data were also analyzed using the previously established information on cell cycle dependent gene expression. None of the known core-cell cycle genes in *Arabidopsis* was in the set of regulated genes demonstrating that the core cell cycle regulators fulfill basic cellular functions that are not specific to particular developmental stages. Surprisingly, when the maximal expression during the cell cycle for the reproductive genes and for all genes was compared, it was found that mitosis-specific genes are strongly overrepresented and S-phase-specific genes were largely lacking from the reproductive gene set. These results imply that S-phase relies during reproductive development on proteins that are important in other stages of the life cycle as well. By contrast, the G2 and M phases of cell proliferation during reproductive development involve often-specific proteins. Such functions could for instance involve the control of the division plane, which is essential for plant morphogenesis.

Another surprise from this dataset was the observation that genes encoding small secreted proteins were strongly overrepresented among the upregulated, downregulated and the transiently changed genes but not among the constantly expressed genes. Cell–cell signaling based on small, secreted proteins or peptides is well established in plants, e.g., the WUSCHEL CLAVATA1 (CLV1)-CLV3 system or sporophytic self-incompatibility in the Brassicaceae [28]. Only a few enzymes are smaller than 15 kDa, and therefore many of the regulated small secreted proteins could function directly as signaling molecules or as precursors for peptide hormones, similar to the ZmEA1 peptide of maize [29].

Conclusions

Microarray studies can involve very diverse experimental designs and analysis strategies. Because the biological question determines the best design and strategy, it is essential that this question is exact and precise. Nevertheless, even with a well-defined question, a well-suited experimental system and a powerful analysis strategy, verification of results with independent techniques is often essential.

After a microarray experiment, diverse reasons call for verification and follow-up experimentation. First, any statistical analysis will generate errors. Type I errors (false positives) arise when genes are called differentially expressed although in reality they are not. Most experimental researchers are aware of type I errors and try to control it with appropriate statistical measures. In transcriptomics and other highly parallel experiments, the conventional statistical confidence level α (typical set to 0.05) is commonly replaced by the false discovery rate FDR. In contrast to α which reflects the probability of any false positive occurring in the selected gene list, the FDR reflects the percentage of false positives among the selected genes. Although a certain fraction of false positives can usually be tolerated, it requires independent experiments to obtain certainty about the regulation of any particular gene. While type I errors are false positives, type II errors are false negatives that arise when true signals are missed. Often, experimental researchers are not aware of type II errors, and usually the rate of type II errors is not known. Only more highly parallel tests can efficiently reduce type II errors, and therefore it is usually of no or only limited relevance if certain genes do not appear in the final selection in a microarray data experiment.

Second, statistical significance is not necessarily equivalent with biological relevance. Tests for errors in the selected gene lists always involve transcript measurements (e.g., Northern-blots or RT-qPCR). In contrast, biological relevance will be revealed only by functional experiments. To this end, researchers typically choose reverse genetic approaches using transgenics (e.g., ectopic overexpression or RNAi) or mutants (e.g., TILLING or T-DNA insertion lines [30–32]) to modify the dosage of selected genes. One reason why differential transcript levels identified with microarrays are not always biological relevant, are other levels of regulation, like differential splicing or translation as well as posttranslational modifications of proteins and altered metabolite abundance. Technologies to measure such effects will be discussed in the following chapters.

References

1. Otte AP, Kwaks TH (2003) Gene repression by polycomb group protein complexes: a distinct complex for every occasion? *Curr Opin Genet Dev* 13: 448–454
2. Ringrose L, Paro R (2004) Epigenetic regulation of cellular memory by the Polycomb and Trithorax group proteins. *Annu Rev Genet* 38: 413–443
3. Drews GN, Yadegari R (2002) Development and function of the angiosperm female gametophyte. *Annu Rev Genet* 36: 99–124
4. Köhler C, Grossniklaus U (2002) Epigenetic inheritance of expression states in plant development: the role of polycomb group proteins. *Curr Opin Cell Biol* 14: 773–779
5. Hsieh TF, Hakim O, Ohad N, Fischer RL (2003) From flour to flower: how polycomb group proteins influence multiple aspects of plant development. *Trends Plant Sci* 8: 439–445
6. Köhler C, Hennig L, Spillane C, Pien S, Gruissem W, Grossniklaus U (2003) The Polycomb-group protein MEDEA regulates seed development by controlling expression of the MADS-box gene *PHERES1*. *Genes Dev* 17: 1540–1553
7. Grossniklaus U, Vielle-Calzada JP, Hoeppner MA, Gagliano WB (1998) Maternal control of embryogenesis by *MEDEA*, a polycomb group gene in *Arabidopsis*. *Science* 280: 446–450
8. Ohad N, Yadegari R, Margossian L, Hannon M, Michaeli D, Harada JJ, Goldberg RB, Fischer RL (1999) Mutations in *FIE*, a WD Polycomb group gene, allow endosperm development without fertilization. *Plant Cell* 11: 407–416
9. Luo M, Bilodeau P, Koltunow A, Dennis ES, Peacock WJ, Chaudhury AM (1999) Genes controlling fertilization-independent seed development in *Arabidopsis thaliana*. *Proc Natl Acad Sci USA* 96: 296–301
10. Spillane C, MacDougall C, Stock C, Köhler C, Vielle-Calzada J, Nunes SM, Grossniklaus U, Goodrich J (2000) Interaction of the Arabidopsis Polycomb group proteins FIE and MEA mediates their common phenotypes. *Curr Biol* 10: 1535–1538
11. Luo M, Bilodeau P, Dennis ES, Peacock WJ, Chaudhury A (2000) Expression and parent-of-origin effects for FIS2, MEA, and FIE in the endosperm and embryo of developing *Arabidopsis* seeds. *Proc Natl Acad Sci USA* 97: 10637–10642
12. Köhler C, Hennig L, Bouveret R, Gheyselinck J, Grossniklaus U, Gruissem W (2003) Arabidopsis MSI1 is a component of the MEA/FIE Polycomb group complex and required for seed development. *EMBO J* 22: 4804–4814
13. Weigel D, Meyerowitz EM (1994) The ABCs of floral homeotic genes. *Cell* 78: 203–209
14. Weigel D, Nilsson O (1995) A developmental switch sufficient for flower initiation in diverse plants. *Nature* 377: 495–500
15. Wagner D, Sablowski RW, Meyerowitz EM (1999) Transcriptional activation of *APETALA1* by LEAFY. *Science* 285: 582–584

16. William DA, Su Y, Smith MR, Lu M, Baldwin DA, Wagner D (2004) Genomic identification of direct target genes of LEAFY. *Proc Natl Acad Sci U S A* 101: 1775–1780
17. Spellman PT, Sherlock G, Zhang MQ, Iyer VR, Anders K, Eisen MB, Brown PO, Botstein D, Futcher B (1998) Comprehensive identification of cell cycle-regulated genes of the yeast *Saccharomyces cerevisiae* by microarray hybridization. *Mol Biol Cell* 9: 3273–3297
18. Cho RJ, Huang M, Campbell MJ, Dong H, Steinmetz L, Sapinoso L, Hampton G, Elledge SJ, Davis RW, Lockhart DJ (2001) Transcriptional regulation and function during the human cell cycle. *Nat Genet* 27: 48–54
19. Menges M, Hennig L, Gruissem W, Murray JAH (2002) Cell cycle-regulated gene expression in *Arabidopsis*. *J Biol Chem* 277: 41987–42002
20. Menges M, Hennig L, Gruissem W, Murray JA (2003) Genome-wide gene expression in an *Arabidopsis* cell suspension. *Plant Mol Biol* 53: 423–442
21. Menges M, Murray JAH (2002) Synchronous *Arabidopsis* suspension cultures for analysis of cell-cycle gene activity. *Plant J* 30: 203–212
22. Liu WM, Mei R, Di X, Ryder TB, Hubbell E, Dee S, Webster TA, Harrington CA, Ho MH, Baid J et al. (2002) Analysis of high density expression microarrays with signed-rank call algorithms. *Bioinformatics* 18: 1593–1599
23. Shedden K, Cooper S (2002) Analysis of cell-cycle-specific gene expression in human cells as determined by microarrays and double-thymidine block synchronization. *Proc Natl Acad Sci USA* 99: 4379–4384
24. Hennig L, Gruissem W, Grossniklaus U, Köhler C (2004) Transcriptional programs of early reproductive stages in *Arabidopsis*. *Plant Physiol* 135: 1765–1775
25. Becker A, Theissen G (2003) The major clades of MADS-box genes and their role in the development and evolution of flowering plants. *Mol Phylogenet Evol* 29: 464–489
26. Shirley BW (1996) Flavonoid biosynthesis – new functions for an old pathway. *Trends Plant Sci* 1: 377–382
27. Xie DY, Sharma SB, Paiva NL, Ferreira D, Dixon RA (2003) Role of anthocyanidin reductase, encoded by *BANYULS* in plant flavonoid biosynthesis. *Science* 299: 396–399
28. Matsubayashi Y (2003) Ligand-receptor pairs in plant peptide signaling. *J Cell Sci* 116: 3863–3870
29. Marton ML, Cordts S, Broadhvest J, Dresselhaus T (2005) Micropylar pollen tube guidance by EGG APPARATUS 1 of maize. *Science* 307: 573–576
30. McCallum CM, Comai L, Greene EA, Henikoff S (2000) Targeted screening for induced mutations. *Nat Biotechnol* 18: 455–457
31. Alonso JM, Stepanova AN, Leisse TJ, Kim CJ, Chen H, Shinn P, Stevenson DK, Zimmerman J, Barajas P, Cheuk R et al. (2003) Genome-wide insertional mutagenesis of *Arabidopsis thaliana*. *Science* 301: 653–657
32. Sessions A, Burke E, Presting G, Aux G, McElver J, Patton D, Dietrich B, Ho P, Bacwaden J, Ko C et al. (2002) A high-throughput *Arabidopsis* reverse genetics system. *Plant Cell* 14: 2985–2994

Regulatory small RNAs in plants

Cameron Johnson and Venkatesan Sundaresan

Plant Biology and Plant Sciences, University of California, Davis, CA 95616, USA

Abstract

The discovery of microRNAs in the last decade altered the paradigm that protein coding genes are the only significant components for the regulation of gene networks. Within a short period of time small RNA systems within regulatory networks of eukaryotic cells have been uncovered that will ultimately change the way we infer gene regulation networks from transcriptional profiling data. Small RNAs are involved in the regulation of global activities of genic regions via chromatin states, as inhibitors of 'selfish' sequences (transposons, retroviruses), in establishment or maintenance of tissue/organ identity, and as modulators of the activity of transcription factor as well as 'house keeping' genes. With this chapter we provide an overview of the central aspects of small RNA function in plants and the features that distinguish the different small RNAs. We furthermore highlight the use of computational prediction methods for identification of plant miRNAs/precursors and their targets and provide examples for the experimental validation of small RNA candidates that could represent trans-regulators of downstream genes. Lastly, the emerging concepts of small RNAs as modulators of gene expression constituting systems networks within different cells in a multicellular organism are discussed.

Introduction

Prior to the discovery of microRNAs in the last decade and the mechanisms of RNA silencing, protein coding genes were considered to be the only significant components for regulation of gene networks. Within a short period of time the discovery of small RNA systems within regulatory networks of eukaryotic cells has substantially altered this paradigm and will ultimately change the way we infer gene regulation networks from transcriptional profiling data (see Chapters by Foyer et al. and Hennig and Köhler). It is now recognized that small RNAs are involved in processes including the regulation of global activities of genic regions via chromatin states, as inhibitors of 'selfish' sequences (transposons, retroviruses), in establishment or maintenance of tissue/organ identity, and as modulators of the activity of transcription factor as well as 'house keeping' genes. Small RNAs such as miRNAs, short interfering RNAs (siRNAs), and in plants, the transacting-siRNAs, are 21–24 nt single stranded RNAs that are sequence-specific negative regulators that are produced from longer double stranded RNA (dsRNA) molecules. SiRNAs exactly match the RNA from which they are produced and result in cleavage and elimination of these source RNAs, whereas miRNAs are produced from RNA hairpin precursor molecules and act to negatively regulate unrelated target RNAs by transcript cleavage if matching exactly, or predominantly by translational inhibition if insufficient pairing occurs between the miRNA and the target transcript. These RNA negative regulators are part of a complex network of pathways for which the central component encompasses the many potential variants of the RNA-induced silencing complex (RISC), the details of which are reviewed elsewhere (reviewed in [1–4]). The RISC complexes are characterized by their ability to use a Dicer-processed small RNA for sequence specific target recognition. These Dicer or a Dicer-related proteins belong to the PAZ domain containing RNase-III class of proteins that produce double stranded RNA cleavage products with 2 nt 3' overhangs, one strand of which is loaded onto RISC. Central to each RISC is a protein of the Argonaute family, each of which contain a PAZ and a PIWI domain, and is thought to hold the single stranded small RNA (reviewed in refs. 5, 6). Target site recognition by the active RISC may lead to mRNA cleavage, translational inhibition of the mRNA or transcriptional silencing at the genomic locus, with the exact outcome dependant on the degree of complementarity between the small RNA and the target, but also probably on the particular type of RISC as determined by the specific Argonaute protein.

This chapter provides a brief overview of the central aspects of small RNA function in plants and the features that distinguish the different small RNAs, the use of computational prediction methods for identification of plant miRNAs/precursors and their targets, the experimental validation of small RNA candidates that could represent trans-regulators of downstream genes, and the emerging concepts of small RNAs as modulators of gene expression constituting systems networks within different cells in a multicellular organism.

Origin of small RNAs in plants

Dicer processed small RNAs are believed to be derived from double stranded RNA from at least four different sources: 1) Double stranded intermediates of viral or retrotransposon origin (siRNAs). 2) Annealed duplexes of sense transcripts with cis- and trans-natural anti-sense transcripts (siRNAs). 3) Double stranded products resulting from the action of RNA dependant RNA polymerase (involved in the production of siRNAs, including trans-acting siRNAs). 4) miRNA precursors consisting of locally folded RNA structures. siRNAs may also be derived from extended inverted repeats over larger stretches of RNA than in the case of miRNA precursors. In *Arabidopsis* and likely also in other plants, these different sources of dsRNA are thought to be processed by overlapping and partially redundant pathways [7] each thought to incorporate at least one dicer-like and Argonaute protein. Other factors specific to each pathway such as RNA dependent RNA polymerases for trans-acting

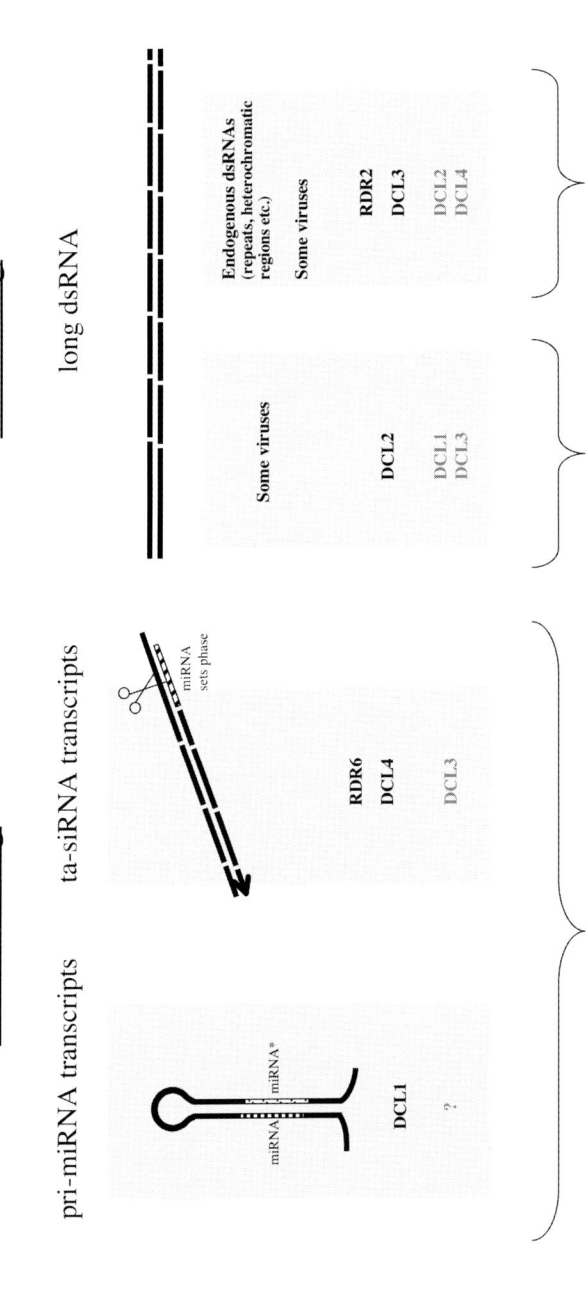

Figure 1. Biogenesis of small RNAs. Transacting small RNAs are produced from endogenous loci that are distinct from the targets they act on and these are conserved at the sequence level due to the continued requirement to match the sites within the transcripts of their targets. On the other hand, self-acting or autonomous siRNAs are general products of the RNAi pathway and are usually derived from viruses or repeated sequences within the genome. These small RNAs represent defense molecules with specificity against the sequences from which they are derived. DCL1 and DCL4 produce 21 nt small RNAs [7, 13], whereas DCL2 and DCL3 produce 22–23 nt and 24 nt, respectively [7]. In the absence of the primary DCL for each pathway, other DCLs (grey type) can partially compensate [12]. Since the size of small RNAs appears to be determined by the processing DCL, the resulting small RNAs will be of the size produced by the substituting DCL.

siRNAs, are also required (Fig. 1). In *Arabidopsis* there are four members of the Dicer-like protein family, DCL1, DCL2, DCL3 and DCL4, and the functions of these have diverged and partially specialized to process particular dsRNA substrates. DCL1 appears to be specialized for processing the imperfect base pairing that occurs in the stem region containing the miRNA/miRNA* sequences within miRNA precursors. The other dicer members do not appear to be able to substitute for this function in development, since the *dcl1* null mutants are embryo lethal [8]. Furthermore, miRNAs have been reported to be undetectable in *dcl1* weak alleles [9–11]. DCL4 is specialized for ta-siRNA production along with the RNA dependant RNA polymerase RDR6 [12], while DCL2 and DCL3 appear to be general producers of siRNAs. DCL2 is able to process viral RNA from turnip crinkle virus but not CMV or TuMV [12], while DCL3 might be primarily involved in endogenous siRNAs from silent heterochromatic regions [12]. The sizes of small RNAs appear to be determined by which dicer member processes the dsRNA. It has been shown that DCL1 and DCL4 produce 21 nt small RNAs [7, 13], DCL3 produces 24 nt siRNAs and DCL2 produces 22–23 nt siRNAs [7].

Biogenesis and distinguishing features of siRNAs and miRNAs

The general RNAi mechanism results in the production of siRNAs that are directed against invasive elements such as viruses and retro-transposons. These siRNAs are self-acting or autonomous in that they act on the same molecular sequences that they are generated from, and as a result match their targets exactly. It is thought that when siRNAs incorporated into a RISC exactly match the source RNA, that this results in cleavage and subsequent degradation of matching copies of the RNA and may result in complete suppression of these elements. The function of siRNAs as a defense mechanism is enhanced by the systemic transfer of siRNAs throughout a plant. It has been shown in plants that siRNAs can act systemically via the phloem and result in the protection of the entire plant from a virus that has initiated its infection at a local site. The mobility of the signal has been shown to depend on RDR6 [14, 15] which might contribute to signal amplification and propagation through the phloem. In addition to this, siRNAs have the ability to induce transitive RNA interference, in which primary siRNAs specific for one section of an RNA transcript can induce the production of secondary siRNAs from a different part of the same transcript enabling the spread of silencing along the nucleic acid sequence [16]. This process presumably provides inherent protection against different but related viruses from that which caused the initial siRNA induction, but also acts to amplify the signal. As well as transitive RNAi, genomic silencing of selfish nucleic acids that become integrated into the DNA genome may occur via siRNAs that are associated with a complex like the *S. pombe* RNA-induced initiation of transcriptional gene silencing (RITS) complex, which would enable localized action of siRNAs to maintain silencing epigenetic states (reviewed in [17]). Unlike siRNAs, miRNAs are derived from single RNA molecules by processing of a double stranded region of a folded RNA precursor. In animal systems the secondary structure of precursors is relatively simple and the dimensions of these precursors appear to be restricted from between 60 and 100 nt in length. In plants the precursor structure and size appear to much less constrained. Within the secondary structure, side branches and multiple end loops are frequent and the precursor sizes range from about 60 nt to over 300 nt in length. In animal systems miRNA genes have been identified within intergenic regions but also within introns (reviewed in [18]). For miRNAs and their targets within plants, there is often a mismatch between the terminal nucleotides of the miRNA and the corresponding nucleotides in the target transcript. These mismatches may be involved in preventing the production of siRNAs from other parts of the target transcript via transitive RNAi through the action of an RDR. Alternatively, or in addition, an RDR may need to be led to the target transcript by an appropriate RISC complex, and the miRNA specific DCL1-containing RISC may not be able to associate with RDR6 or RDR2 in order for such a process to occur. More recently another species of small RNA, called trans-acting siRNAs (ta-siRNAs), that also act as regulators of gene expression have been found in *Arabidopsis* and other plants [19]. ta-siRNAs found in *Arabidopsis* are thought to be derived from transcripts in which the required phasing results from a predefined dicer processing start point achieved by miRNA directed cleavage, and subsequently made double stranded by an RNA dependent RNA polymerase. In contrast with the cis-acting siRNAs, the sequences of trans-acting miRNAs and ta-siRNAs and their co-evolving but genomically distinct target sites are constrained by the functional requirement that they continue to match their targets. The resulting conservation of sequences across 18–22 nt facilitates their computational prediction within and between species.

Computational prediction of miRNAs and their targets

Cloning and sequencing small RNAs has been a central strategy for identifying miRNA sequences from within genomic sequence datasets, and has been responsible for the initial identification of many of the currently recognized miRNAs in *Arabidopsis*. Cleavage products of RNase III type enzymes, such as dicer, contain a 5' phosphate which has enabled enrichment for miRNAs and siRNAs from other small RNAs resulting from other mechanisms such as ribosomal and mRNA degradation [20]. In addition to cloning, technologies such as MPSS and 454 sequencing, which allow high throughput direct sequencing of expressed RNAs, represent more sensitive approaches to small RNA detection (see [21], http://mpss.dbi.udel.edu/ and http://www.454.com). However, experimental strategies have technical limitations. First, although highly expressed miRNAs can be relatively easily identified from among the many clones in a small RNA library, miRNAs that are expressed in a relatively small number of cells or only under specific conditions or time of development may not be represented in many small RNA libraries. Second, despite the enrichment based on the 5' phosphate, miRNAs often represent only a small proportion of the total cloned small RNAs in a library. Furthermore, the functional basis of this enrichment process has been questioned, at least for use in *Drosophila*, where an endogenous kinase activity was suggested to have added phosphate groups to the 5' end of small RNAs derived from other processes such as RNA degradation [22].

Clues to the identity of miRNA sequences from within small RNA libraries can be derived bioinformatically when a relatively complete genome sequence is available. The sequence of a miRNA should be found embedded in a genomic sequence, that if expressed would be part of a double stranded stem region of a predicted RNA secondary structure. Sometimes the miRNA* sequence is also found within the small RNA library thus revealing the two nucleotide 3' overhangs RNase III signature that in turn supports the processing of a single RNA molecule rather than a duplex of two different RNA molecules derived from the two different genomic strands. In addition the miRNA sequence, by definition, should have a matching target sequence within another region of the genome. However, without molecular evidence of the miRNA* sequence the existence of the other features do not by themselves confirm the classification as a miRNA. This is because the regulatory specificity of miRNAs is determined within such a short sequence that can occur by chance alone, and almost all genomic sequences, when represented as RNA, can be folded into a predicted secondary structure that contain double stranded helical regions. In addition to the classification of experimentally derived small RNA sequences as either miRNAs or siRNAs, computational strategies have been used to provide a means to predict new miRNA candidates from available genome sequence data. Several different strategies and algorithms have been devised, as shown in Table 1, and the principles of some approaches are discussed below.

Unlike protein coding genes, miRNA genes do not have open reading frames, codon bias or other significant internal characteristics that can help in their identification. The requirement for miRNAs to match their targets provides a constraint on both the miRNA sequence and the sequence of their target(s). The miRNA* sequence is also constrained, but to a lesser degree, due to the requirement for the miRNA to be processed from a double stranded region in the stem of the miRNA precursor (pre-miRNA). Therefore, not surprisingly, most computational strategies for identifying miRNA genes have incorporated a comparative genomics component to search for conserved sequences in related species (summarized in Tab. 1). Among the first algorithms using comparative genomics were MiRscan, miRseeker and srnaloop which were produced for analyzing animal genomes, and an algorithm MIRFINDER that was used on the *Arabidopsis* and rice genomes. All these algorithms use relatively complete sequence data available from two or more genomes and look for the existence of interspecies conservation of the precursor-embedded miRNA and miRNA* subsequences. A more recent comparative genomic algorithm, phylogenic shadowing, is best used with several closely related genomes and overcomes the problem of insufficient divergence having occurred between two closely related species. In this approach the genomes are aligned to produce a multiple sequence alignment in which less important nucleotide residues will more often vary across the species while important ones will be conserved across most if not all the species. This variation in residue conservation can be graphically represented in vista plots with the miRNA and miRNA* sequences visualized as two peaks of increasing conservation in a region of relatively low conservation [23].

Table 1. Abridged summary of computational methods for miRNA prediction in animals and plants

Algorithm/Approach	Genomes involved in analysis	Premises/Aims	Methods and considerations summary	References
MiRscan	Human, Mouse, C. elegans, fugu	o miRNA count estimate o miRNA prediction	o miRNA conservation o Precursor features o Log odds scoring	24, 25 and http://genes.mit.edu/mirscan
miRseeker	D. melanogaster; D. pseudoobscura	o miRNA prediction	o miRNA conservation o Minimal features o Arbitrary scoring	26
srnaloop	C. elegans, D. melanogaster, human	o miRNA prediction	o miRNA conservation o Minimal features o Arbitrary scoring	27
MIRFINDER	Arabidopsis, Oryza	o miRNA prediction	o miRNA conservation o Some filtering	28
—	Arabidopsis, Oryza	o miRNA prediction	o miRNA conservation o Precursor features o Arbitrary scoring	29
—	Arabidopsis, Oryza	o Target prediction	o Simple mismatches o Arbitrary scoring	30
phylogenic shadowing	primates	o Proof of concept o miRNA count estimate	o miRNA conservation	23
findMiRNA	Arabidopsis, Oryza	o Interactive database o miRNA/target candidates	o miRNA/target pairing o Arbitrary scoring o miRNA conservation	31 and http://sundarlab.ucdavis.edu/mirna
TargetScan	Human, mouse, rat, fugu	o Target prediction o Target count estimate	o miRNA 5' seed match o Markov model analysis	32
MovingTargets	D. melanogaster	o Research software o Drosophila miRNA targets	o miRNA/target pairing features o Arbitrary scoring	33

Using comparative genomics methods, estimates of the total number of miRNAs in a single species will depend on the evolutionary distance between the genome under study and the comparison genome. The greater the distance between the two species, the fewer miRNAs can be identified from the comparison, but the strength of the evidence is perhaps stronger due to the increased divergence of other neighboring sequences. Using closely related species in the comparison increases the total number of predicted miRNAs, which will approach the total number of miRNAs that actually exist in the genome of interest, except that several genomes are needed in the comparison, as in phylogenic shadowing, in order to detect the conserved miRNA sequences in an otherwise relatively un-diverged set of genome sequences. The phylogenic shadowing approach has produced results for primates that suggest that there are possibly twice as many miRNA genes in the human genome than was previously believed from earlier studies using more distantly related species [23].

In addition to sequence conservation, some of the algorithms use additional criteria to more specifically identify miRNA precursors from among the conserved sequences. In this respect, the most advanced algorithm is probably MiRscan, which takes into consideration features such as the distance of the miRNA from the end loop, extension of base pairing around the miRNA/miRNA* double stranded segment, the presence of a 5' U residue in the miRNA, localized conservation within the 5' and 3' ends of the miRNA, nucleotide bias in the first five positions, and base pairing and bulge symmetry in the miRNA/miRNA* duplex region. Other algorithms used on metazoan genomes with a more limited use of precursor/miRNA features analysis include miRseeker and srnaloop [26, 27]. Bioinformatic approaches similar to these latter methods have been used on plants (see [28, 29] and Tab. 1).

The use of comparative genomics methods in plants has enabled the discovery of many miRNA genes in *Arabidopsis* and rice. In addition, algorithms employing relatively straightforward homology searches enabled the identification of potential precursor orthologs/homologs in other plant species such as poplar, as well as lower plants [34, 35]. However, these methods cannot identify species–specific miRNAs, such as miR161, miR163 and miR173, which are specific to *Arabidopsis* and were initially identified by cloning. Interestingly these miRNAs are represented by single precursor loci unlike other *Arabidopsis* precursors that exist in families. This would mean that even intra-specific sequence comparison would not have revealed these miRNA precursors, and they may not have been identified at all if their expression levels were too low for experimental detection. Therefore, there is a need for bioinformatic strategies that can enable the identification of miRNAs without relying on sequence conservation. An alternate target-based strategy has been developed using an algorithm called findMiRNA (Tab. 1), which exploits the requirement that any miRNA must have a matching target sequence elsewhere in the genome, probably within a transcript encoding a protein [31]. This requirement enabled the mapping of almost all good miRNA-target candidate pairs existing as matches between subsequences of intergenic/intronic regions (with hairpin potential) and subsequences of protein coding transcripts. At this stage the dataset represents mostly false positive miRNA candidates in addition to the true positives. A post-processing step (that incorporated the characteristic divergence pattern of miRNA precursor sequences) was applied to the resulting large dataset which enabled identification of novel miRNAs. The large unfiltered dataset is available at <http://sundarlab.ucdavis.edu/mirna/> together with custom filters provided for various characteristic miRNA/precursor parameters, which can be deployed to reduce or eliminate the background of spurious candidates.

There is still a need for the implementation of an algorithm for use in plants with a more comprehensive set of specific features associated with miRNAs, similar to those used by MiRscan. The identification of additional miRNA specific features is continuing, and in the future it may be possible to develop algorithms that will be capable of identifying single copy miRNA genes without the use of comparative genomics.

Confirmation of candidate miRNAs and targets

As is the case for many bioinformatic problems, there is no perfect algorithm for predicting miRNA precursors. Rules that can be applied to absolutely distinguish miRNA precursors from other sequences currently do not exist. For this reason each miRNA candidate identified by an algorithm needs to be validated before it should be included as a confirmed miRNA. This validation process often seeks to obtain molecular evidence for the existence of a miRNA by detection of the miRNA itself and/or by detecting the effect of the miRNA on target transcripts. Methods to detect miRNAs include small RNA cloning, RNA blot hybridization (miRNA Northerns) and more recently PCR-based approaches. The use of *Arabidopsis* plants expressing the viral suppressor of RNA silencing P1/HC-pro, in which the levels of most miRNAs are significantly elevated, can increase the signal still further [10, 31, 36]. Early studies tended to conclude miRNA status if a strong signal was detected on a miRNA Northern. Later, as more weakly expressed miRNAs were being assessed, confusion arose between miRNAs and siRNAs. Signals arising from miRNAs should be in the range of 21–22 nt in size, as is expected for *Arabidopsis* DCL1 processed small RNAs. Such signals may also arise from DCL4 processed double stranded RNA as is the case for ta-siRNAs. If weak signals of two or more bands of similar strength in the range of 23–24 nt is observed, this is more likely the product of other dicers such as *Arabidopsis* DCL2 and DCL3.

Genetic approaches are also available to distinguish miRNAs from siRNAs. Since, unlike miRNAs, the production of most endogenous siRNAs require the action of RNA dependent RNA polymerase 2 (RDR2) (Fig. 1) while ta-siRNA production requires RDR6. Control RNA isolated from rdr2 and rdr6 mutants should resolve the issue. Unlike siRNAs, the molecular levels of *bona fide* miRNAs should be unaffected in plants that are mutant for RDR2 or RDR6. Hybridization methods, including microarrays, have the limitation that the exact sequence being detected is not known. This also means the boundary of the detected small RNA sequence remains unknown and therefore the exact miRNA sequence predicted cannot be confirmed with such a method. PCR-based methods offer dramatically increased sensitivity and the sequence data may also include sequence boundary information [37–40].

A commonly used technique for the validation of miRNA targets, and therefore also by implication the existence of the small RNA, is the detection of mRNA cleavage products using 5' RACE. These PCR amplified cleavage products can be sequenced to identify the exact nucleotide sites that are cleaved by the specific RISC complex. This technique is very sensitive and has enabled the validation of the molecular interaction between many *Arabidopsis* miRNAs and their suggested targets. The sensitivity, however, represents a problem with respect to target validation, as it can be argued that the molecular interaction detected by 5'RACE can be so infrequent as to represent an interaction that has no biological significance in the life cycle of the plant, and that all one is doing is reconfirming the generally accepted mechanism that sufficiently matching 'miRNA-target' pairs can result in cleavage of the transcript by the miRNA loaded RISC. The method could be used to determine molecular targets of a miRNA that fall within the cleavage class, but a negative result does not indicate that translation of the proposed target is not affected. More biologically oriented methods may be more appropriate. Other sources of evidence supporting the biological significance of a proposed miRNA-target pair may be achievable through genetic approaches, such as the identification of a phenotype associated with a mutation that would be expected to affect miRNA-target interaction.

It is likely that purely bioinformatic approaches can also be used to provide evidence of biological significance for a particular miRNA-target pair. One possible approach might be to detect sequence conservation of the target site within otherwise divergent but related transcripts. This could be achieved by alignment of orthologous target transcript sequences from two or more sufficiently diverged genomes or the use of phylogenic shadowing for the orthologous transcripts across several closely related species.

Future prospects for computational discovery of small RNAs in plants

The ultimate goal of computational approaches to small RNA discovery is to detect miRNAs or ta-siRNAs that would otherwise not be easily identified. Use of algorithms that rely more heavily on characteristics of miRNA genes may enable predictions of miRNAs in a single genome but the presence of a high proportion of false positives precludes this method as a way to estimate miRNA gene number within a species. Approaches based on a good statistical foundation will be valuable for estimating the number of miRNAs within an organism. Some success can be achieved through extensions of already available computational tools, as in the case of the identification of the transacting-siRNA, ta-siR-ARF (TAS3) [41]. Other non-statistical methods will use additional criteria to limit the data based on features expected to be associated with trans-acting small RNAs. The effectiveness of these methods will depend on the basis of the selective criteria and how well they are integrated into the approach as a whole. With the increasing amounts of data that relates directly to the epigenetic state of any particular site within a genome, for instance whether the region is composed of repeated sequence or perhaps revealing the predominant methylation states of regions with the use of methods such as bisulfite sequencing, biologically relevant data can be used to more thoroughly and accurately analyze the available data. This will enable the effective interrogation of the available genomic sequence to identify small RNAs that are involved at both the post transcriptional level but also the transcriptional level of gene regulation.

Genes, networks and systems: Regulation by small RNAs in plants

In plants small RNAs appear to fall into two categories, those involved in 'defense' related functions and those that represent regulators of development and homeostasis (see Fig. 2). Defense-related small RNAs are siRNAs, usually of the 24 nt class, that act to generally suppress RNA production from the invading virus or a 'selfish' nucleotide sequence in the genome such as a retrotransposon. In plants these siRNA signals are capable of being transmitted between cells as well as through the phloem to result in systemic silencing. A different set of siRNA molecules are present in complexes involved in a positive feedback loop for post transcriptional gene silencing, and these act in a localized fashion on specific loci. Also every cell will have a particular miRNA expression profile, with the various miRNAs at different concentrations depending on the state of the cell or plant. The consequences of these miRNA concentrations will depend on the type of regulatory circuit being modulated. Three different potential outcomes for regulation by any expressed miRNA have been proposed [42]. An increase in the expression of a miRNA may: 1) act to switch on or turn off a biological response, 2) act to tune a biological response, and 3) is biologically neutral despite a reduction in the level of the 'target' transcript. The differences between miRNAs of the switching category and the tuning category are shown in Figure 2.

In animal systems, the matching of miRNAs to their targets is based on a much looser interaction than which occurs in plants, and as such animal miRNAs are thought to have a large number of targets with perhaps as many as 1,000 different target transcripts for each miRNA [43]. For example miR1 and miR124 from animals are likely to represent switches that define a tissue type as they have been shown to regulate the expression of large numbers of genes specific to muscle and brain respectively [44]. In plants, most miRNAs appear to act on their target transcripts in a way that resembles the action of siRNAs in both animals and plants, and therefore the regulatory networks for miRNAs might be simpler to model computationally in plant systems than in animal systems. Probably the best example in plants of such a tissue identity network is that involving the miRNAs miR165/166 that negatively regulates the transcripts of the adaxial-specific (upper surface specific) class III HD zip transcription factors PHABULOSA, PHAVOLUTA and REVOLUTA within the abaxial tissue (lower tissue) during and after leaf development in *Arabidopsis* [45, 46]. In wild type plants, the miR165/166 family is expressed in the abaxial domain of developing leaf primordia and act to exclude PHB, PHV and REV transcripts. In plants containing target site mutant alleles of these genes, the transcripts with the mutated target site are no longer excluded from the abaxial

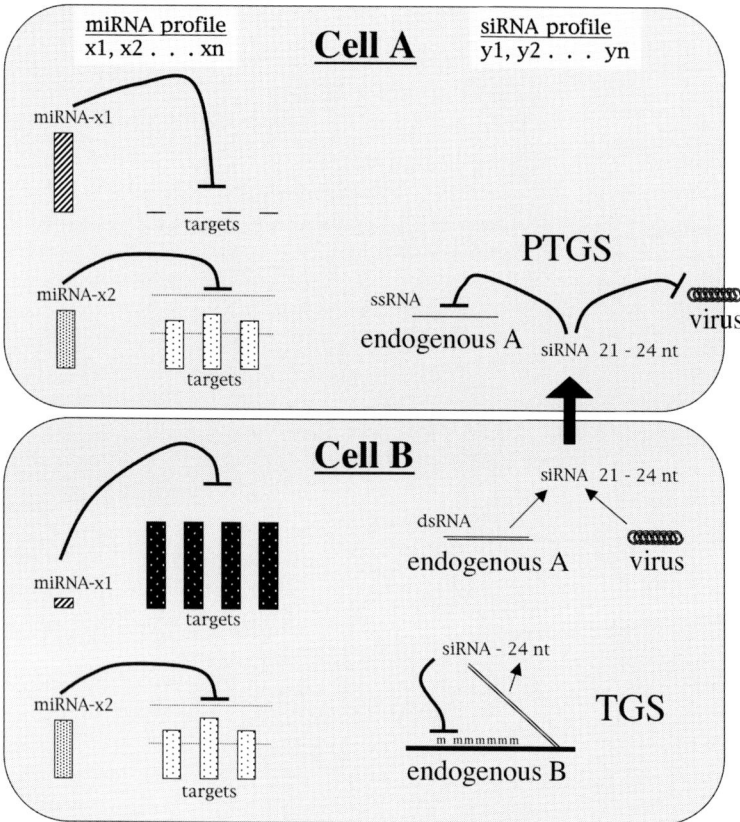

Figure 2. Cellular miRNA and siRNA profiles. Two plant cells A and B are shown with distinct miRNA profiles that result in the expression or modulation of different sets of genes. Whereas miRNA-x1 acts as a switch by reducing the protein expression of its target mRNAs below a biological threshold in cell A, miRNA-x2 acts to modulate and maintain target mRNA-product levels within appropriate upper and lower bounds indicated by fine lines. In addition, Cell B has mounted a siRNA response to a virus, as well as siRNA mediated silencing of an endogenous gene. These siRNAs constitute signals that can be transmitted to Cell A, which then alters its own siRNA profile in response.

domain and this results in a radialized leaf. This miRNA causes a change of state through the downregulation of a target transcript and thus belongs to the switching category of miRNA (category 1 in Fig. 2).

Plant miRNAs may also be involved in the control of homeostasis. An example is the targeting of two components of the sulfate assimilation pathway by miR395. This miRNA was shown to target ATP-sulfurylase [47], but a conserved target site was also identified within the 5' UTR of the sulfate transporter gene by alignment of the presumptive orthologs from *Arabidopsis* and rice [31]. These two targets represent structurally unrelated proteins that act in the same cellular process. This example could illustrate the biological utility of a tuning miRNA (like miRNA-x2 in Fig. 2), with targets that are distinct enzyme components of a nutrient assimilation pathway. In summary, it is likely that plant cells will have defining small RNA profiles that are responsive to signals from other cells, maintaining a balance of gene expression through silencing and modulation of transcripts and chromatin that will finally affect protein concentrations and metabolic and regulatory pathway activities (for details on the analysis of proteins see the following two chapters). The challenge for the future will be to incorporate these regulatory molecules and their effects into the systems biology models of plant gene expression (see also Chapters by Steinfath et al., and Schöner et al.)

References

1. Tang G (2005) siRNA and miRNA: an insight into RISCs. *TBS* 30(2): 106–114
2. Herr AJ (2005) Pathways through the small RNA world of plants. *FEBS* 579: 5879–5888
3. Preall JB, Sontheimer EJ (2005) RNAi: RISC gets loaded. *Cell* 123(4): 543–545
4. Hammond SM (2005) Dicing and slicing: the core machinery of the RNA interference pathway. *FEBS Lett* 579(26): 5822–5829
5. Sontheimer EJ (2005) Assembly and function of RNA silencing complexes. *Nat Rev Mol Cell Biol* 6(2): 127–138
6. Carmell MA, Xuan Z, Zhang MQ, Hannon GJ (2002) The Argonaute family: tentacles that reach into RNAi, developmental control, stem cell maintenance, and tumorigenesis. *Genes Dev* 16(21): 2733–2742
7. Gasciolli V, Mallory AC, Bartel DP, Vaucheret H (2005) Partially redundant functions *Arabidopsis* DICER-like enzymes and a role for DCL4 in producing trans-acting siRNAs. *Curr Biol* 15(16): 1494–1500
8. Golden TA, Schauer SE, Lang JD, Pien S, Mushegian AR, Grossniklaus U, Meinke DW, Ray A (2002) SHORT INTEGUMENTS1/SUSPENSOR1/CARPEL FACTORY, a Dicer homolog, is a maternal effect gene required for embryo development in *Arabidopsis*. *Plant Physiol* 130(2): 808–822
9. Reinhart BJ, Weinstein EG, Rhoades MW, Bartel B, Bartel DP (2002) MicroRNAs in plants. *Genes Dev* 16(13): 1616–1626
10. Kasschau KD, Xie Z, Allen E, Llave C, Chapman EJ, Krizan KA, Carrington JC (2003) P1/HC-Pro, a viral suppressor of RNA silencing, interferes with *Arabidopsis* development and miRNA unction. *Dev Cell* 4(2): 205–217
11. Finnegan EJ, Margis R, Waterhouse PM (2003) Posttranscriptional gene silencing is not compromised in the *Arabidopsis* CARPEL FACTORY (DICER-LIKE1) mutant, a homolog of Dicer-1 from *Drosophila*. *Curr Biol* 13(3): 236–240
12. Xie Z, Johansen LK, Gustafson AM, Kasschau KD, Lellis AD, Zilberman D, Jacobsen SE, Carrington JC (2004) Genetic and functional diversification of small RNA pathways in plants. *PLoS Biol* 2(5): E104
13. Qi Y, Denli AM, Hannon GJ (2005) Biochemical specialization within *Arabidopsis* RNA silencing pathways. *Mol Cell* 19(3): 421–428
14. Himber C, Dunoyer P, Moissiard G, Ritzenthaler C, Voinnet O (2003) Transitivity-dependent and -independent cell-to-cell movement of RNA silencing. *EMBO J* 22(17): 4523–4533

15. Schwach F, Vaistij FE, Jones L, Baulcombe DC (2005) An RNA-dependent RNA polymerase prevents meristem invasion by potato virus X and is required for the activity but not the production of a systemic silencing signal. *Plant Physiol* 138(4): 1842–1852
16. Sijen T, Fleenor J, Simmer F, Thijssen KL, Parrish S, Timmons L, Plasterk RH, Fire A (2001) On the role of RNA amplification in dsRNA-triggered gene silencing. *Cell* 107(4): 465–476
17. Zilberman D, Henikoff S (2005) Epigenetic inheritance in *Arabidopsis*: selective silence. *Curr Opin Genet Dev* 15(5): 557–562
18. Ying SY, Lin SL (2004) Intron-derived microRNAs – fine tuning of gene functions. *Gene* 342: 25–28
19. Allen E, Xie Z, Gustafson AM, Carrington JC (2005) microRNA-directed phasing during trans-acting siRNA biogenesis in plants. *Cell* 121(2): 207–221
20. Lau NC, Lim LP, Weinstein EG, Bartel DP (2001) An abundant class of tiny RNAs with probable regulatory roles in *Caenorhabditis elegans*. *Science* 294(5543): 858–862
21. Lu C, Tej SS, Luo S, Haudenschild CD, Meyers BC, Green PJ (2005) Elucidation of the small RNA component of the transcriptome. *Science* 309(5740): 1567–1569
22. Aravin AA, Lagos-Quintana M, Yalcin A, Zavolan M, Marks D, Snyder B, Gaasterland T, Meyer J, Tuschl T (2003) The small RNA profile during *Drosophila melanogaster* development. *Dev Cell* 5(2): 337–350
23. Berezikov E, Guryev V, van de Belt J, Wienholds E, Plasterk RH, Cuppen E (2005) Phylogenetic shadowing and computational identification of human microRNA genes. *Cell* 120(1): 21–24
24. Lim LP, Glasner ME, Yekta S, Burge CB, Bartel DP (2003) Vertebrate microRNA genes. *Science* 299(5612): 1540
25. Lim LP, Lau NC, Weinstein EG, Abdelhakim A, Yekta S, Rhoades MW, Burge CB, Bartel DP (2003) The microRNAs of *Caenorhabditis elegans*. *Genes Dev* 17(8): 991–1008
26. Lai EC, Tomancak P, Williams RW, Rubin GM (2003) Computational identification of *Drosophila* microRNA genes. *Genome Biol* 4(7): R42
27. Grad Y, Aach J, Hayes GD, Reinhart BJ, Church GM, Ruvkun G, Kim J (2003) Computational and experimental identification of *C. elegans* microRNAs. *Mol Cell* 11(5): 1253–1263
28. Bonnet E, Wuyts J, Rouze P, Van de Peer Y (2004) Detection of 91 potential conserved plant microRNAs in *Arabidopsis thaliana* and *Oryza sativa* identifies important target genes. *Proc Natl Acad Sci USA* 101(31): 11511–11516
29. Wang XJ, Reyes JL, Chua NH, Gaasterland T (2004) Prediction and identification of *Arabidopsis thaliana* microRNAs and their mRNA targets. *Genome Biol* 5(9): R65
30. Rhoades MW, Reinhart BJ, Lim LP, Burge CB, Bartel B, Bartel DP (2002) Prediction of plant microRNA targets. *Cell* 110(4): 513–520
31. Adai A, Johnson C, Mlotshwa S, Archer-Evans S, Manocha V, Vance V, Sundaresan V (2005) Computational prediction of miRNAs in *Arabidopsis thaliana*. *Genome Research* 15: 78–91
32. Lewis BP, Shih IH, Jones-Rhoades MW, Bartel DP, Burge CB (2003) Prediction of mammalian microRNA targets. *Cell* 115(7): 787–798
33. Burgler C, Macdonald PM (2005) Prediction and verification of microRNA targets by MovingTargets, a highly adaptable prediction method. *BMC Genomics* 6(1): 88
34. Floyd SK, Bowman JL (2004) Gene regulation: ancient microRNA target sequences in plants. *Nature* 428(6982): 485–486
35. Axtell MJ, Bartel DP (2005) Antiquity of microRNAs and their targets in land plants. *Plant Cell* 17(6): 1658–1673

36. Mallory AC, Reinhart BJ, Bartel D, Vance VB, Bowman LH (2002) A viral suppressor of RNA silencing differentially regulates the accumulation of short interfering RNAs and micro-RNAs in tobacco. *Proc Natl Acad Sci USA* 99(23): 15228–15233
37. Chen C, Ridzon DA, Broomer AJ, Zhou Z, Lee DH, Nguyen JT, Barbisin M, Xu NL, Mahuvakar VR, Andersen MR et al. (2005) Real-time quantification of microRNAs by stem-loop RT-PCR. *Nucleic Acids Res* 33(20): e179
38. Raymond CK, Roberts BS, Garret-Engele P, Lim LP, Johnson JM (2005) Simple, quantitative primer-extension PCR assay for direct monitoring of microRNAs and short-interfering RNAs. *RNA* 11: 1737–1744
39. Shi R, Chiang VL (2005) Facile means for quantifying microRNA expression by real-time PCR. *Biotechniques* 39(4): 519–525
40. Lu DPP, Read RLL, Humphreys DTT, Battah FMM, Martin DIK, Rasko JEJ (2005) PCR-based expression analysis and identification of microRNAs. *J RNAi Gene Silencing* 1(1): 44–49
41. Williams L, Carles CC, Osmont KS, Fletcher JC (2005) A database analysis method identifies an endogenous trans-acting short-interfering RNA that targets the *Arabidopsis* ARF2, ARF3, and ARF4 genes. *Proc Natl Acad Sci USA* 102(27): 9703–9708
42. Bartel DP, Chen CZ (2004) Micromanagers of gene expression: the potentially widespread influence of metazoan microRNAs. *Nature* 5: 396–400
43. Lewis BP, Burge CB, Bartel DP (2005) Conserved seed pairing, often flanked by adenosines, indicates that thousands of human genes are microRNA targets. *Cell* 120(1): 15–20
44. Lim LP, Lau NC, Garret-Engele P, Grimson A, Schelter JM, Castle J, Bartel DP, Linsley PS, Johnson JM (2005) Microarray analysis shows that some microRNAs downregulate large numbers of target mRNAs. *Nature* 433: 769–773
45. Emery JF, Floyd SK, Alvarez J, Eshed Y, Hawker NP, Izhaki A, Baum SF, Bowman JL (2003) Radial patterning of *Arabidopsis* shoots by class III HD-ZIP and KANADI genes. *Curr Biol* 13(20): 1768–1774
46. Williams L, Grigg SP, Xie M, Christensen S, Fletcher JC (2005) Regulation of *Arabidopsis* shoot apical meristem and lateral organ formation by microRNA miR166gand its AtHD-ZIP target genes. *Dev* 132(16): 3657–3668
47. Jones-Rhoades MW, Bartel DP (2004) Computational identification of plant microRNAs and their targets, including a stress-induced miRNA. *Mol Cell* 14(6): 787–799

Plant Systems Biology
Edited by Sacha Baginsky and Alisdair R. Fernie
© 2007 Birkhäuser Verlag/Switzerland

Differential display and protein quantification

Erich Brunner, Bertran Gerrits, Mike Scott and Bernd Roschitzki

Functional Genomics Center Zürich, Winterthurerstr. 190, 8057 Zürich, Switzerland

Abstract

High-throughput quantitation of proteins is of essential importance for all systems biology approaches and provides complementary information on steady-state gene expression and perturbation-induced systems responses. This information is necessary because it is, e.g., difficult to predict protein concentrations from the level of mRNAs, since regulatory processes at the posttranscriptional level adjust protein concentrations to prevailing conditions. Despite its importance, quantitative proteomics is still a challenging task because of the high dynamic range of protein concentrations in the cell and the variation in the physical properties of proteins. In this chapter we review the current status of, and options for, protein quantification in high-throughput experiments and discuss the suitability and limitations of different existing methods.

Introduction

Quantitative proteome analysis, the global analysis of protein expression, is a complementary method to study steady-state gene expression and perturbation-induced changes. In comparison to gene expression analysis at the mRNA level, proteome analysis provides more accurate information about biological systems and pathways since the measurement directly focuses on the actual biological effector molecules. It is, e.g., difficult to predict protein concentrations from the level of mRNAs, since regulatory processes at the posttranscriptional level adjust protein concentrations to prevailing conditions. Quantitative information on proteins is necessary to infer regulatory events that take place between the expression of a gene and the metabolite that is synthesized by the gene product (Fig. 1). Recent analyses with different biological systems revealed that in many cases no apparent correlations between transcript, protein and metabolite levels exist, suggesting that regulation occurs at different nodes in the network. These cases particularly comprise conditions where rapid responses of the system towards, e.g., stress conditions are required.

Quantitative analysis of protein expression is therefore an important tool for the examination of complex biological systems. Albeit its importance, quantitative proteomics is still a challenging task because of the high dynamic range of protein

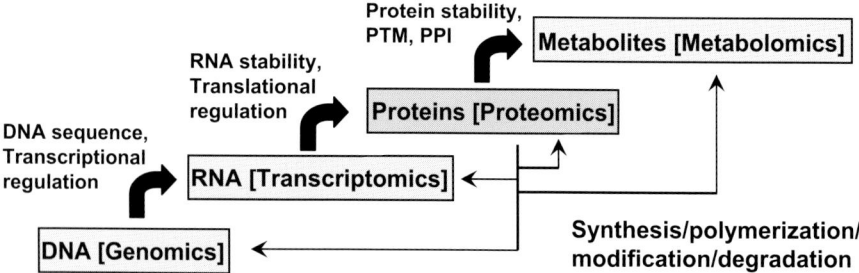

Figure 1. Regulatory network of gene expression. Regulation occurs at different nodes in the network.

amounts in the cell and the variation in the physical properties of proteins. The current methods to determine protein expression levels are applicable to most biological systems or any model organism and therefore are described here from a very general point of view. As a general rule, the applicability of a certain quantification strategy is mainly determined by the method that is used to separate and analyse the proteins: Gel-based proteomics provoked and generated different quantitation strategies than gel free approaches. For each of the quantitative approaches described below, the general features, a range of possible applications as well as their advantages or limitations are outlined. By means of a candidate experiment the reader is guided step by step through the experimental set up thereby receiving a comprehensive overview over the prevailing tools and techniques in quantitative proteomics.

Quantitative two-dimensional gel electrophoresis

Introduction

Two-dimensional gel electrophoresis (2-DE) is a well-established electrophoretic method for separating proteins in a gel matrix [1]. In the most common approach, proteins are extracted and non-protein substances are removed. The proteins are then dissolved in a buffer for isoelectric focusing. The proteins are then electrophoretically separated in an immobilized pH gradient (IPG) gel strip; each protein migrates to its isoelectric point. This process is called isoelectric focusing (IEF). The focused proteins on the strip are then loaded onto a sodium dodecyl sulfate (SDS) polyacrylamide gel. The SDS-denatured proteins are then migrated in the presence of an electrical field across the length of the gel: SDS-PAGE [2]. Over the course of this electrophoresis small proteins will migrate further than large proteins. At the conclusion of this stage, the proteins have been resolved in the first dimension according to isoelectric point (pI) and the second dimension according to molecular weight (MW). The proteins are then fixed in the gel, stained and scanned. The resulting images can be analyzed and compared. After image analysis, spots of interest can be picked. The proteins are then digested with trypsin, de-salted, spotted to a MALDI target, and analyzed by MALDI-MS [3].

Gel-based quantitation versus *LC approaches*

Using 2-DE as a fractionation technique has distinct differences from LC-based quantitative proteomics. The most obvious is that whole proteins are separated, and the quantitation of integrated optical spot density is done before the mass spectrometry. Since the gel can be calibrated, MS identification of spot digests can be validated with respect to pI and MW.

Another advantage of gel-based proteomics is the orientation of spot patterns indicating post-translational modifications (PTMs) (Fig. 2). A variety of PTM-specific stains exist [4]. Using a PTM-specific stain prior to a general protein stain can serve as a useful approach for both quantitation and MS data validation [5].

One should never assume that one spot (even a nicely symmetrical spot) on a gel corresponds to a single protein [6]. However, the MALDI analyses are quantified with respect to the position on the MALDI target, and the digest from each gel spot goes to a single MALDI target location. Thus, the number of coincident proteins is never great. Usage of narrow range ('zoom') IPG strips reduces coincident proteins even more. Usage of zoom IPG strips (approximately 1.5 pI unit range) is necessary to perform quantitative gel-based proteomics, as a wide range strip will generally have many spots with greater than one protein per spot.

Gel-based proteomics involves many transfer steps, and some protein is lost at each transfer [7]. Such losses necessitate consistent technique for all gels processed in any comparative study. For more precise quantitation protein samples from two different conditions can be covalently labeled at lysine residues with different

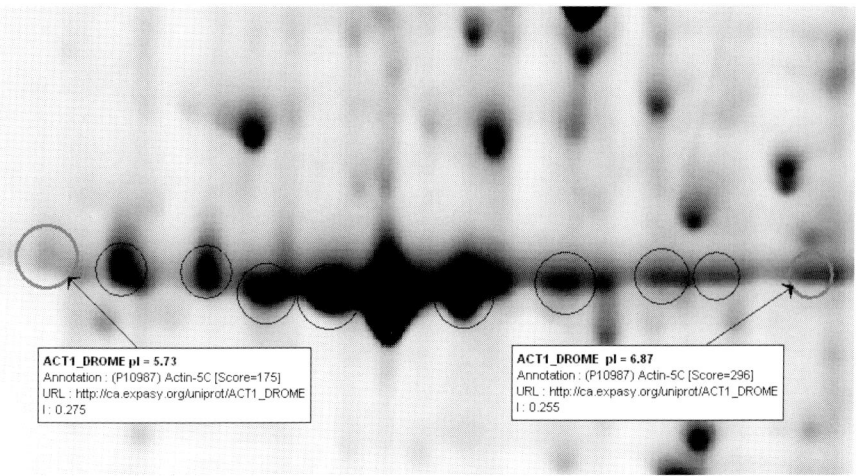

Figure 2. Example for a 2-D-PAGE gel showing spot tailing as a result of urea-induced carbamylation.

fluorescent cyanine dyes. To facilitate an internal standard, a third pooled sample is labeled with a third cyanine dye. All three labeled extracts are pooled and run in a single gel [8]. This approach to 2-DE is called 2-D Fluorescence Difference Gel Electrophoresis (DIGE). A DIGE approach significantly increases precision of measurement of protein expression ratios for two reasons: elimination of gel-to-gel variability, and the use of an internal standard for quantitation of spot density ratios.

Gel-based techniques can only resolve the proteins within the pI range of the IPG strip. A LC-based approach will yield a mix of peptides irrespective of pI. Pre-fractionation methods based on pI do exist: free flow electrophoresis (FFE) and liquid-phase isoelectric focusing (e.g., Rotofor) [9]. These approaches are important for the use of zoom IPG strips, unless one can tolerate overloading the strip and sacrificing the proteome beyond the pI range of the strip.

Protein sample preparation and fractionation strategies

Protein samples for 2-DE must be of sufficient purity for IEF. Lipids, carbohydrates, salts, surfactants, and insoluble residues can all cause difficulties in IEF. Thus, samples must have interfering substances removed before IEF. A universal problem with sample purification is alteration of the proteomic composition of the sample: any purification step will cause losses, and the losses will not be proportionate to the composition of the sample. For example, not all proteins have the same (in)solubility in cold acetone. Thus, for quantitative 2-DE proteomics, the general approach should be to clean the sample just enough to allow for efficient IEF. Some traces of salts and other interferents can be tolerated, especially if absorbent pads are used in IEF [10].

Given the wide range of differences between different organisms, there is no single approach that is appropriate to go from tissue to protein isolate. For example, some tissues present problems from high fat content, other tissues may have high levels of insoluble material. The experimentalist must consult the literature or Internet resources to help locate relevant protocols. Protease inhibitors are almost always required to be included in the initial preparation step [1].

Acetone/TCA precipitation has been shown to be an effective approach with proteomic studies [11]. Many vendors offer clean-up kits based on this approach. For very difficult samples, one may use a phenol extraction approach [12, 13]. Phenol extraction will result in a very clean sample, but it is unknown how the proteome is biased using this approach. Lengthy dialysis steps may be avoided by the use of spin filters [14].

Fractionation of the sample is nearly always a good idea. The proteome of organisms and whole cell lysates is far too complex to resolve using a 2-DE approach. For an expert technician, over 5,000 proteins may be resolved on a 24 cm × 20 cm 2-D gel; 2,000 protein spots may be routinely resolved by less experienced individuals [1]. Sequential extraction [15] results in multiple fractions based on aqueous solubility. FFE and Rotofor techniques [9] are useful for the fractionation of proteins based on pI; this approach assures the experimentalist that high levels of proteins will not migrate off the IPG strip. Subcellular fractionation techniques [16] should be used when feasible.

The first dimension: isoelectric focusing

The quantity of protein to apply to a gel is dependent on the size of the gel, the staining approach, and the sensitivity of the mass spectrometer to be used. For a ruthenium tris-bathophenanthrolate stained 24 cm × 20 cm × 0.1 cm 2-D gel, 150 µg is generally sufficient for identification of the top 80–90% of the spots in the gel [17]. For a Coomassie gel, 300 µg is generally sufficient, but one can go lower.

IEF buffer composition should be varied depending on sample type [1, 18–20]. To avoid streaking in the alkaline range, dithiothreitol (DTT) should never be used with IPG strips with pIs above 7. Instead, use a nonionizable reducing agent [21] such as tributylphosphine (TBP), or the thiol-protecting agent hydroxyethyl disulfide (HED). Also, IEF buffers should contain 10% isopropanol and 5% glycerol to prevent streaking due to electroendoosmotic flow [19]. Streaking and loading efficiency are also affected by loading style and IEF voltage programming [1, 22, 23]. The surfactant of choice is usually CHAPS, but ASB-14 is showing increasing promise as a surfactant to increase representation of membrane proteins in 2-D gels [24–26].

After IEF, the strips need to be (double) equilibrated in reducing agent and alkylated to prevent disulfide formation at cysteine thiols. Some choose to reduce and alkylate prior to IEF, but this is not generally recommended due to shifting the pI before IEF. Alternatively, one can equilibrate the IPG strips in HED in a single step [18]. The resulting mass spectra must be searched with consideration of the cysteine S-mercaptoethanol modification. Other compounds such as tris(2-carboxyethyl)-phosphine and vinylpyridine have been used for preventing IEF streaking [27].

IEF is the stage of 2-DE which is most in flux. There exist a great variety of approaches in buffer composition, IEF voltage programming, and strip equilibration techniques. The experimentalist is encouraged to choose wisely then stick to one's experimental design. Gels are difficult enough to compare without adding extra variability from 'tinkering' from experiment to experiment.

SDS-PAGE and gel stains

SDS-PAGE for 2-DE is generally performed in the discontinuous buffer system of Laemmli [28] and modifications thereof. Due mainly to insolubility problems, proteins heavier than 150 kDa are not suitable for traditional Laemmli SDS-PAGE. Low MW proteins can be resolved in a Tris-tricine buffer system [29]. The second dimension of 2-DE is much more established than IEF: The IEF strip is loaded to the top edge of a gel, and sealed in place with a warm agarose solution colored with bromophenol blue for tracking the electrophoretic migration. For a 24 cm × 20 cm × 0.1 cm gel, a two-stage program is recommended: 2 Watts/gel for 45 min for loading proteins, and 17 Watts/gel for electrophoresis at 25°C. The migration time is variable; 4 h for 20 cm is typical. If one prefers to run SDS-PAGE overnight, a suitable protocol is a 45 min loading step at 7 mA per gel, and then increasing to 15 mA per gel for 18 h at 20°C. The proteins in a gel need to be fixed after SDS-PAGE. Diffusion of lower molecular weight proteins in PA gels becomes apparent after 6 h. Excessively low pH will cause esterification [30] at protein carboxyl groups. Thus, TCA fixing is to be avoided when possible.

A variety of stains are available for staining 2-D gels [4, 5, 31–33]. Silver staining is sensitive, but has a number of disadvantages for quantitative proteomics. Silver-stained gels have poor linear response [32] with concentration. Silver-stained gels also tend to form crater spots, which complicate quantitation. While most staining techniques have the greatest intensity at the center of the protein spot, a crater spot has reduced signal intensity at the center. In a three-dimensional view, most non-silver stained spots appear as a conical peak. In a three-dimensional view of many silver spots, a profiled crater spot appears as a volcanic caldera. Relative to other staining techniques, silver staining can reduce signal intensity for MALDI-MS, even when using the Shevchenko method [34].

Coomassie staining has numerous advantages. Coomassie staining is relatively inexpensive and compatible with mass spectrometry. Newer formulations of colloidal Coomassie Brilliant Blue (CBB) along with improved protocols [31] have increased the sensitivity of CBB to near silver levels. CBB spots are visible, and thus do not require a fluorescent scanner for imaging.

For a high-sensitivity stain with long-term stability, MS-compatibility, and good linear response, the best approach is using ruthenium (II) tris-(bathophenanthroline disulfonate), [RuBP]. RuBP can be easily used as the commercial formulation SYPRO Ruby [Invitrogen Corporation] [35]. The main disadvantage of SYPRO Ruby is the expense. RuBP staining can be done without the expense of SYPRO Ruby by the use of 1 µM aqueous RuBP solution according the Lamanda protocol [32]. The expense is 100-fold less. The synthesis of RuBP concentrate is relatively simple, and the 20 mM concentrate is stable for years at 4°C (personal communication) [36]. Aliquot the concentrate into 1.5 mL tubes, and freeze them at –20°C for long-term storage.

Staining with epicocconone (Deep Purple) is sensitive and MS-compatible. However, Deep Purple is not as photostable as RuBP [37]. It is also quite expensive. Deep Purple has been reported to have a linear response to four orders of concentration [33].

For the highest levels of accuracy and precision in quantitative gel proteomics, one must approach 2-DE using a pre-stained internal standard in the gel along with two other stains for the two conditions to be studied. This approach is usually referred to as DIGE (see, *Gel-Based Quantitation versus LC Approaches,* discussed previously). With the DIGE technique, one can see finer changes in up- or down-regulation between different conditions. The 'staining' is a reaction which adds a charged cyanine at a similarly charged lysine residue. The reaction adds 0.5 kDa per lysine, and is staining is minimal (no more than one cyanine per molecule) [38]. DIGE gels can be fluorescently scanned immediately after SDS-PAGE. They are scanned three times, once for each fluorophore, and the images can be combined for a visual comparison. The separate images are analyzed to see how the intensity ratios vary between the individual conditions, and the internal standard, which contains all the proteins. Gel spot match quality is generally excellent due to co-electrophoresis of control and treated sample within the same gel. DIGE experiments are quite expensive, with cyanine dye expenses in the hundreds of dollars per gel. However, the DIGE approach is certainly the gold standard for quantitative 2-DE.

Spot analysis software and experimental design

Several 2-DE pattern software packages are available. Since the author only has extensive experience with one software package, no review will be offered. All software allows for comparison of groups of gels where each group is a specific biological condition (e.g., control vs. treated). The coefficient of variation (CV) of spot intensity within a group is a key factor to use to determine if between-group differences are significant. Of course, biological replicates must be considered when generating groups for expression analysis. A rule of thumb for one-color comparison of groups is four gels per group. Given the complexity of a 2-DE experiment, it is recommended that five gels be run for each condition. If one of the five gels is of poor quality, four gels will remain for generating CVs.

For the 'typical' experiment where one is searching for proteomic changes, the following approach is recommended:

1. Run two control gels and two treated gels though the 2-D workflow. Analyze and pick spots of interest. See if you can identify some interesting proteins in the gel. If you can separate and identify the proteins, move on to the next step.
2. Run four (or five) gels per condition. Given careful one-color staining, proteins up- or downregulated by 60% or greater can be identified.
3. Run DIGE to refine your findings. Quantitative 2-DE experiments can yield striking results. However, the required level of technical lab bench skill for 2-DE is high, and it can take weeks or months to generate high quality data. When one has the option of using an LC-MS approach as opposed to 2-DE, it should be carefully considered (see below).

Quantitative proteomics by metabolic labeling

Isotope-based quantitative analysis by mass spectrometry has long been used in the small molecule field [39] and later on in structural biology where researchers applied this technology to detect phase shifts in NMR studies by replacing all ^{14}N atoms using ^{15}N media. In 1999 this substitution technology was applied to bacteria and yeast for simultaneous identification and quantitation of individual proteins by mass spectrometry and for determining changes in the levels of modifications at specific sites on individual proteins [40, 41]. Since ^{15}N-substituted media are difficult and expensive to make for mammalian systems, the particular method employed was restricted to microorganisms. Additionally, the degree of incorporation is not necessarily 100%. Because there are varying numbers of nitrogen atoms in the different amino acids, automated interpretation of the resulting spectra has proven difficult.

The principle of metabolic isotope-coded labeling of all proteins in mammalian cell culture was first reported by the laboratory of Matthias Mann (stable isotope labeling by amino acids in cell culture (SILAC) [42]). With this technology cell lines are grown in media in which a standard essential amino acid (which is not synthesized *de novo* by these cells) is substituted by an isotopically labeled isoform, most often used is deuterated leucine (Leu-d3) (Fig. 3A). The substituted amino acids are incorporated normally into all proteins as they are synthesized and as a result all the proteins in the cell are completely tagged after a few generation cycles. No chemical labeling or affinity purification steps are necessary and the method is compatible with virtually any cell culture system, including primary cells. Even the autotrophic plant cells that can synthesize all amino acids from inorganic nitrogen were shown to be compatible with the SILAC technology [43].

Recently, metabolic labeling of two multicellular organisms such as the nematode *Caenorhabditis elegans* or the fruit fly *Drosophila melanogaster* has been demonstrated [44]. This was achieved by feeding these model organisms with ^{15}N-labeled *E. coli* or yeast, respectively. 98% of the nematode's proteins were labeled in the second generation, whereas for the fly a single live-cycle was sufficient to generate almost complete N-labeled offspring.

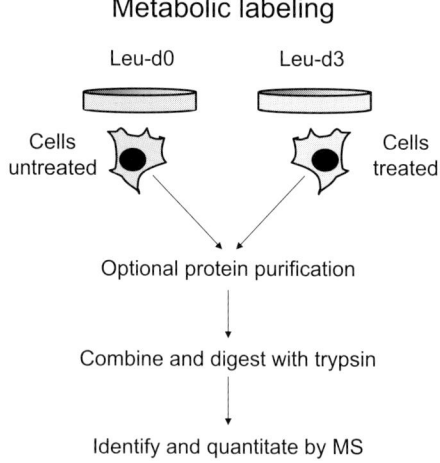

Figure 3 (A). Workflow of a typical SILAC experiment: Protein populations from both control and treated samples are then harvested, and because the label is encoded directly into the amino acid sequence of every protein, the extracts can be mixed directly. Purified proteins or peptides will preserve the exact ratio of the labeled to unlabeled protein, as no more synthesis is taking place, and therefore no scrambling can take place at the amino acid level. The proteins and peptides can then be analyzed in any of the ways in which they are analyzed in non-quantitative proteomics. Quantitation takes place at the level of the peptide mass spectrum or peptide fragment mass spectrum, exactly the same as in any other stable isotope method (such as ICAT); after [42].

It seems just a matter of time until this technology will be applied to other model organisms.

In Figure 3A the general set up of SILAC experiment is illustrated. In brief the two cell populations to be compared (e.g., induced vs. non-induced cells) are grown in either standard cell culture medium or medium supplied with an essential isotope-bearing amino acid. The proteins from both samples are then extracted. Since the label is included directly into the amino acid sequence of every protein, the extracts can be mixed directly. The purified proteins or peptides will preserve the exact ratio of the labeled to unlabeled protein, as no more synthesis is taking place and the proteins or peptides can be analyzed by mass spectrometry. Quantitation takes place at the level of the peptide mass spectrum or peptide fragment mass spectrum, identical to any other stable isotope method (see below). It is important to note that the absence of chemical steps implies the same sensitivity and throughput for SILAC as for non-quantitative methods.

Being a simple and rather cheap technology the SILAC method has become widely used in many laboratories. Furthermore, different protocols for cell fractionation and protein separation such as 2-DE or strong cation exchange chromatography can be used in combination with SILAC making it the method of choice for many applications.

Isotope coded affinity tags (ICAT^{TM})

In the previous paragraph we described the quantitation of proteins through metabolic labeling. This technology, however, is limited to unicellular organisms or cell culture systems. Complete proteome labeling by SILAC in multicellular organisms remains, with a few exceptions [44] utterly impossible. In 1999 Aebersold and colleagues developed another technique for quantitative proteome profiling that is also based on stable isotope incorporation into the proteins allowing to perform a quantitative proteome analysis of two samples irrespective of the protein source [45]. The crucial difference to SILAC, however, is that the protein-tagging takes place by chemical means after the proteins have been extracted. Protein labeling is based on a class of reagents termed isotope-coded affinity tags (ICAT, Fig. 3B). The reagent consists of three elements: an affinity tag (biotin), which is used to isolate ICAT-labeled peptides; a linker that can incorporate stable isotopes; and a reactive group with specificity toward thiol groups (cysteine residues). Since the ICAT reagents are available in two flavors (a so-called isotopic light and an isotopic heavy label) they allow to compare protein expression levels in two different samples. ICAT-labeled peptides elute as pairs from a reverse-phase column. By calculating the ratio of the areas under the elution profile curve for identical peptide peaks labeled with the light and heavy ICAT reagent, the relative abundance of that peptide in each sample can be determined, which is directly related to the abundance of the corresponding protein (Fig. 3B). Originally the ICAT reagents featured either eight hydrogen or deuterium atoms in the linker [45] in the isotope coding linker region. However, ^{2}H and ^{1}H labeled peptides show slightly different elution profiles during reversed-phase separation (RP), which makes it difficult to quantify at a single moment in time [46]. In addition, the relatively hydrophobic biotin tag causes peptides to elute in a relatively narrow time window during RP-chromatography. To circumvent these shortcomings and to minimize the effects of the label, a novel set of ICAT reagents, called cleavable ICAT (cICAT) has been developed [47]. First the polyethylene glycol linker has been replaced by an acid cleavable linker that enables clipping of the biotin tag after affinity purification. Second, the isotope coding by eight deuterium atoms has been replaced by nine ^{13}C atoms in the heavy version of the new cICAT reagents. Li and colleagues [48] demonstrated the improved performance and identical behavior of differentially labeled peptides on a RP-column.

In order to determine the absolute amount of a target protein or proteins in a complex biological sample using this technology further development of the ICAT strategy lead to the generation of the so-called VICAT reagents [49]. The principle was to generate three distinct isotope-coded tags of which one is used to label an internal reference peptide of known concentration. The technology however has never become widely accepted. It has rather become substituted by the iTRAQ technology.

The ICAT approach is based on two fundamental principles. First, pairs of peptides tagged with the light and heavy ICAT reagents, respectively, are chemically identical and therefore serve as ideal mutual internal standards for accurate quantification. Second, a short sequence of contiguous amino acids from a protein (5–25 residues) contains sufficient information to identify that unique protein. This principle is corroborated by that fact that every quantifiable peptides contains cystein, which is a rare amino acid that is frequently a component of novel tryptic peptides – peptides whose sequence is found only once in an organism's proteome.

The ICAT technology is illustrated in Figure 3B and the processing of the probes includes the following sequential steps: First proteins from the two samples (tissues, cells, whole organisms) to be compared are separately isolated and resolubilized under strong denaturing conditions using urea and SDS. The extracted proteins in one sample, representing for instance a tissue in a normal state, are then reduced before the cysteinyl residues are derivatized with the isotopically light form of the ICAT reagent. The equivalent groups in the second sample derived for instance from a tissue in a diseased state are derivatized with the isotopically heavy reagent. After the labeling is complete, the two samples are combined. This is a crucial step, because both samples undergo the same treatment thus conserving the appropriate abundance ratios of the proteins. In the subsequent step, the protein mixture is subjected to protease treatment generating two different tryptic peptide populations: a) a minor fraction (roughly 10%) consisting of (light or heavy) tagged cysteine-containing peptides, and b) a major fraction (90%) consisting of untagged non-cysteine-containing peptides. By selectively isolating the protein-tagged cysteine-containing peptides on an avidin affinity column through the biotin tag, one achieves a major reduction in peptide complexity before subjecting the mixture to mass spectrometric analysis and thus allows the analysis of quantifiable peptides under less crowded analytical conditions. Finally, the isolated peptides are separated and analyzed by LC-MS/MS (a detailed description of the underlying principles can be found in the

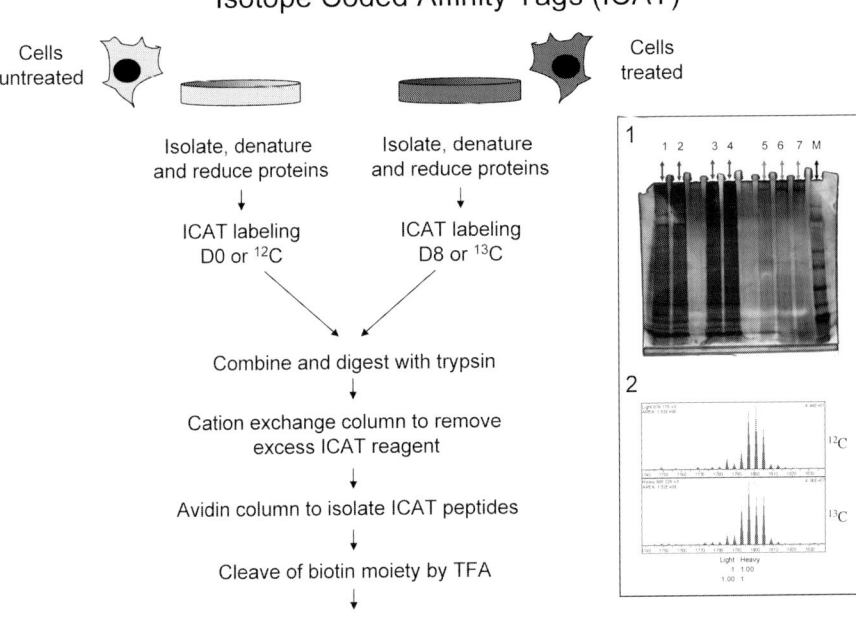

Figure 3 (B). Workflow of a typical ICAT experiment: Proteins isolated from a control sample (untreated cells) are treated with the light reagent, while proteins from the test sample are treated with the heavy reagent. The samples are mixed and the protein pool digested with trypsin. Following tryptic digestion of the pooled proteins, the peptides are separated from the byproducts of the labeling and digestion reactions on cation exchange chromatography. The ICAT-reagent-labeled peptides are then separated from the other peptides by avidin affinity chromatography. Following the avidin elution step, the ICAT-reagent-labeled peptides are evaporated to dryness and reconstituted in concentrated trifluoroacetic acid (TFA) to cleave the biotin portion of the tag from the labeled peptides. The reaction mix is kept at 37°C for 2 h and is followed by a second evaporation step to remove the acid. The peptides are then placed in an autosampler for reversed-phase capillary LC/MS/MS analysis. Inset 1: To assess whether the labeling and protease treatment processes were successful, small aliquots of the initial samples (lane 1 (sample 1) and lane 2 (sample 2)), each labeled fraction (after labeling, lane 3 (sample 1 + light ICAT) and lane 4 (sample 2 + heavy ICAT)), and the trypsinized mixture (combined samples incubated with trypsin for 4 h (lane 5), 8 h (lane 6), 16 h (lane 7), are collected after each step, run on a polyacrylamide gel and examined after the gel has been fixed and silver stained. Proper labeling of the samples can be monitored if bands show a decreased mobility. The mobility shift may, however, be subtle and hard to detect on gels with a high poly acryl amide concentration. More important is that the bands show the same strength before and after the labeling procedure indicating that no degradation of the proteins occurred. The tryptic digest is considered to be complete if distinct protein bands are no longer visible (inset 1). Inset 2: Quantitation of an ICAT experiment. Quantitation of two coeluting, differentially labeled peptides (^{12}C designates cysteine labeled with the light form of ICAT reagent, while ^{13}C designates cysteine labeled with the heavy form of ICAT reagent), the peptide elution profiles indicating the relative abundance, and the calculated ^{12}C: ^{13}C ratio obtained using XPRESS software [75].

following chapter). In this last step, both the quantity and sequence identity of the proteins from which the tagged peptides originated are determined by automated multistage MS: When peptides from the two sources are analyzed concurrently, two distinct peaks representing the differentially labeled species are detected by MS. Relative quantitation is done by comparing the areas of the related peaks of the identical, yet isotopically distinct, peptides.

To assess whether the labeling and protease treatment processes were successful, small aliquots of the initial samples, each labeled fraction (before combining them) and the trypsinized mixture are collected after each step, run on a polyacrylamide gel and examined after the gel has been fixed and silver stained. Proper labeling of the samples can be monitored if bands show a decreased mobility. The mobility shift may, however, be subtle and hard to detect on gels with a high poly acryl amide concentration. More important is that the bands show the same strength before and after the labeling procedure indicating that no degradation of the proteins occurred. The tryptic digest is considered to be complete if distinct protein bands are no longer visible (Fig. 3B).

The original ICAT protocol uses ion exchange chromatography after the ICAT labeling and mixing of the two samples to remove excess derived reagents. Another option was developed by Li [48]. By running the labeled ICAT proteins (prior to digestion) on a 1D SDS PAGE, excess ICAT reagents, salts, and detergents, can easily be removed and allows easy buffer changes for the following digestion step. Moreover, proteins are pre-fractionated according to molecular weight which can be used as an additional criterion for the evaluation of protein identifications.

This basic ICAT protocol can not only be applied to whole proteome comparisons of whole tissues, sorted cells, subcellular fractions or perturbed cell culture populations but can also be used to determine candidate interaction partners of specific proteins (bait) by immuno precipitation (IP). This is achieved by labeling the proteins that co-immunoprecipitate with the bait with one ICAT label and to tag the appropriate control IP (lacking the bait) with the corresponding tag and processing and analyzing the two samples as described. Proteins that show a 1:1 ratio are equally present in either of the samples indicating an unspecific binding of this protein to the beads or affinity column. A specific interaction of a protein with the bait is represented by an increased relative intensity signal in the specific IP. The feasibility of this approach has been demonstrated by Ranish and colleagues [50].

Alternatively, it has been demonstrated that the 2-DE and the ICAT labeling technology can be combined into a single differential display platform [51]. Proteins from two different samples are labeled with heavy and light ICAT reagents, combined and then separated by 2-D gel electrophoresis. The gel-separated proteins are detected with a sensitive protein stain, excised, cleaved with trypsin and analyzed by MS.

This method closely parallels the DIGE methodology with some important improvements – both, the DIGE and the ICAT technology decrease the electrophoretic mobility of proteins. Since the cysteine residues are modified with a pH-neutral ICAT group, the isoelectric point is preserved for all but the most basic proteins. While DIGE requires controlled labeling with the hydrophobic cyanine dyes, ICAT labeling is done to completion which is readily accomplished using excess ICAT reagent. This makes the labeling and quantification by ICAT more robust and reproducible; the labeling of proteins using cyanine dyes is more prone to generate molecular mass ladders of spots with varying degrees of dye incorporation. Moreover, since the ICAT reagent is relatively hydrophilic, migration problems do not arise during electrophoresis.

One important application of ICAT in combination with 2-D gels (instead of a separation of peptides in liquid phase) is for the assessment of the relative abundances of protein isoforms that may arise from posttranslational modification.

The ICAT technology has a number of advantages but also limitations which shall be discussed in more detail. First and foremost is its ability to reduce peptide complexity by 90% at the slight expense of being unable to identify, on theoretical grounds, some 10–15% of a cell's proteins. Second, the chemical reaction in the ICAT alkylation can be performed in the presence of urea, sodium dodecyl sulfate (SDS), salts, and other chemicals that do not contain a reactive thiol group. Therefore, proteins are kept in solution with powerful stabilizing agents until they are enzymatically digested. Third, the sensitivity of the LC-MS/MS system is critically dependent on the sample quality. In particular, commonly used protein-solubilizing agents are poorly compatible with MS. Avidin affinity purification of the tagged peptides completely eliminates contaminants incompatible with MS. Fourth, the quantification and identification of low-abundance proteins requires large amounts (milligrams) of starting protein lysate. Isotope-coded affinity tag analysis is compatible with any biochemical, immunological, or cell biological fractionation methods that reduce the mixture complexity and enrich for proteins of low abundance while quantification is maintained. It should be noted that accurate quantification is only maintained over the course of protein enrichment procedures if all manipulations preceding combination of the differentially labeled samples are strictly conserved. Fifth, unlike the $^{14}N/^{15}N$ labeling scheme, the ICAT method is a post-isolation isotopic labeling approach that does not require cells to be cultured in specialized media. Finally, the ICAT approach can be extended to include reactivity towards other functional groups. One weakness of the current ICAT method is that it requires proteins to contain cysteine residues flanked by appropriately spaced protease cleavage sites. In *Arabidopsis* approximately 5% contain no cysteinyl residues and are therefore missed by using thiol-specific ICAT reagents. Moreover, the quantitative information on posttranslational modifications of proteins is rarely available since the modified amino acid residue needs to coincide in a quantifiable cystein-containing peptide. Recently, an improved approach analogous to ICAT called iTRAQ has been developed that renders the cysteine-free proteins as well as any PTM susceptible to quantitative analysis.

Isobaric peptide tagging using iTRAQ™

iTRAQ is a primary amine specific (N-terminus) stable isobaric labeling method well suited for relative and absolute protein quantitation using mass spectrometry

[52]. A set of four labels are available adding flexibility to the experimental approach including time course analyses, biological replicates and accurate quantitation using internal standards. In general, all the steps for sample handling and post label-processing as described for the ICAT approach can be applied. As a primary difference to the ICAT technology, peptides and non-intact proteins are subjected to labeling with iTRAQ. Due to the large number of tagged peptides produced, biochemical fractionation on iTRAQ samples, for instance by SCX chromatography, are indispensable prior to MS analysis.

As a major advantage, quantitative information is not restricted to cystein-containing peptides as in the ICAT methodology, but is in effect available for any peptide class including those that underwent posttranslational modification. As a consequence, higher quantitative peptide coverage is achieved than with the ICAT method. In addition, the labeled peptides are isobaric, i.e., they do not differ in mass and hence also identical in the single MS mode (Fig. 3C). The differentially labeled isobaric peptides sum up to an increased precursor signal, improved MS/MS fragmentation and eventually result in better confidence identifications. Quantitation is elegantly and easily achieved during MS/MS fragmentation where each of the four labels generates distinct diagnostic signature ions in the low mass range with a Δ-mass of 1 Dalton (114–117 Daltons). Finally, iTRAQ is well suited to perform absolute quantitation [53] of individual proteins in complex mixtures by spiking the sample with one or more iTRAQ-tagged synthetic protein-specific peptides in known concentrations.

These tremendous improvements are achieved at the expense of an increase in sample complexity as well as an analysis being restricted to the use of mass spectrometers that cover the low mass range. However, the tremendous sample complexity demands for high throughput instruments such as ion-traps, which unfortunately still have a restricted dynamic range and in most cases cannot detect the diagnostic fragment ions. In addition, it has recently been reported that in a direct comparison of the two methods, the ICAT technology has the potential to detect a higher proportion of lower-abundance proteins than the iTRAQ methodology [54].

For both, the ICAT and the iTRAQ technology, companies offer fully-fledged solutions including the necessary reagents, MS instruments, and application software.

In a similar study, Choe and co-workers compared the reproducibility and variation in quantitation of proteins in a mixture analyzed by 2-DE and the iTRAQ technology [55]. Whereas the analysis of the 2-DE resulted in a total 68 proteins, the shotgun iTRAQ approach quantified 527 proteins. For a direct comparison of the protein expression ratio consistency, only the 55 proteins quantified with both methods (shared proteins) were included in the analysis. The variability was determined by calculating the so-called coefficent of variation (CV) and was determined to be between $CV = 0.31$ and 0.81 for 2-DE and $CV = 0.24$ to 0.53 for the isobaric tagging method. Taken together, not only could more proteins be identified but also quantification was more accurate using the isobaric iTRAQ labeling method. Moreover, spots of lower staining intensity (which correspond in most cases to lower abundance proteins) were shown to offer less consistency in quantitation by 2-DE

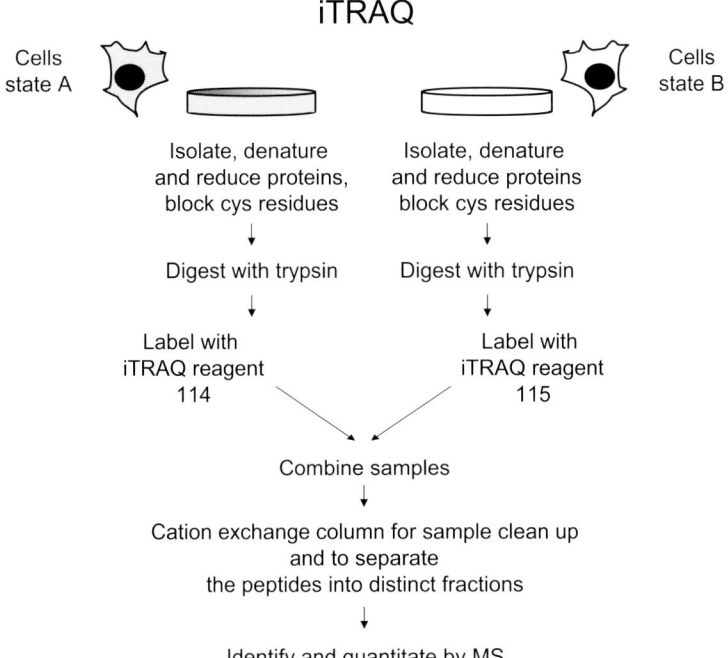

Figure 3 (C). Workflow of a typical iTRAQ experiment: Although up to four different samples can be analyzed in any given experimental procedure, for simplicity, Figure 3 shows an experiment using only two. Protein isolates are reduced, alkylated and digested with trypsin in an amine free buffer system, in parallel. The resulting peptides are then labeled with the iTRAQ reagents. Upon completion of labeling the samples are then combined. Depending on sample complexity, samples are either directly analyzed via LC-MS/MS after a one-step elution from a cation exchange column to remove reagent byproducts or, in the case of complex samples, cation exchange chromatographic fractionation to reduce overall peptide complexity.

whereas isobaric tags are capable of providing more consistent quantitation for lower intensity proteins.

Quantitation of protein levels using protease incorporated ^{18}O

This paragraph deals with another post expression labeling method, namely the incorporation of ^{18}O by proteases. One of the first applications of this method was to facilitate the interpretation of *de novo* sequencing of mass spectrometric derived peptide fragments [56] and for creating peptide internal standards [57]. However, the increased interest over the last couple of years in protein quantitation, both relative and absolute, shed new light into this particular technology.

Proteases, proteinases, or the more modern name peptidases, describe the same group of enzymes that catalyze the hydrolysis of the peptide bond in the peptide backbone of a protein. Per definition, all peptidases that incorporate oxygen from the surrounding matrix during the protein/peptide hydrolysis can be used. But for clarity, this paragraph will only deal with one specific protease, namely the most commonly used protease in proteomic experiments, trypsin.

Trypsin, a serine protease, uses a mechanism that is based on nucleophilic attack of the targeted peptidic bond by a serine. Figure 4 shows a schematic overview of the mechanism of the hydrolysis of a peptide bond. The mechanism consists essentially of six steps (see also Fig. 5) [58]:

1. Substrate binds.
2. Nucleophilic attack of the side chain oxygen of serine 195 in the active site of trypsin, on the carbonyl carbon of the readily cleavable bond, forming a tetrahedral intermediate.
3. Breakage of the peptide bond with assistance from histidine 57 (proton transfer to the new amino terminus).
4. Release of the first product.
5. Nucleophilic attack of water on the acyl-enzyme intermediate with assistance of histidine 57 and formation of the tetrahedral intermediate.
6. Decomposition of acyl intermediate and release of the second product.

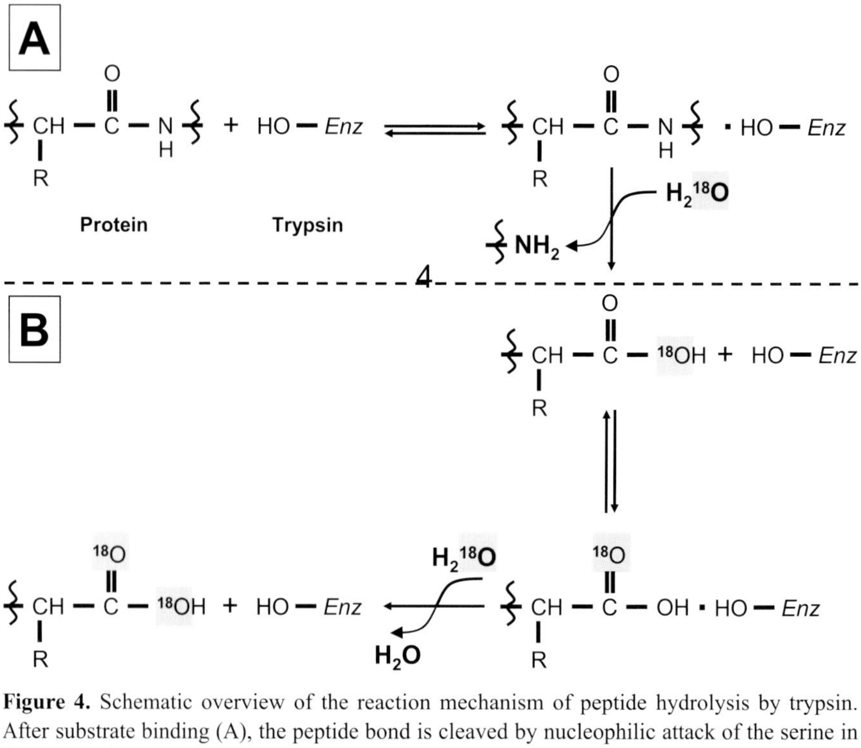

Figure 4. Schematic overview of the reaction mechanism of peptide hydrolysis by trypsin. After substrate binding (A), the peptide bond is cleaved by nucleophilic attack of the serine in the active site of trypsin. After releasing the first intermediate product, there is a carboxyl oxygen exchange (B). There is double oxygen incorporation after complete cleavage of the peptide bond. Figure adapted from [58].

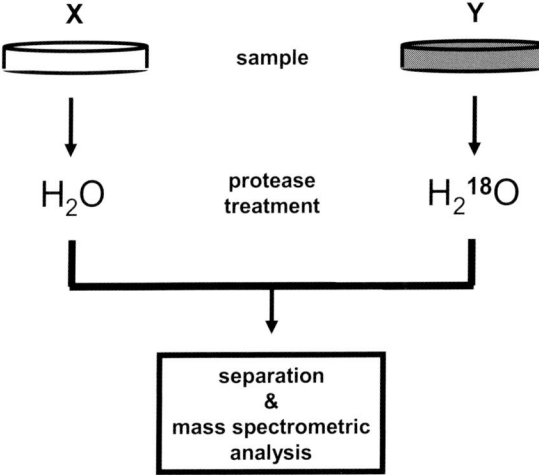

Figure 5. Experiment design for protein level quantitation using ^{18}O labeled peptides. For a two way comparison of relative protein amount, equal amounts of sample X and Y are digested independently using ordinary water and ^{18}O, respectively. Samples are then combined and subjected to subsequent peptide separation and mass spectrometric analysis.

During the hydrolysis of the peptide backbone bond by trypsin, two oxygen atoms from the surrounding matrix are incorporated into the product on the c-terminus side of either arginine or lysine. It is exactly this fact that is being made use of. By using ^{18}O enriched water ($H_2^{18}O$), ^{18}O is incorporated instead of the 'usual' ^{16}O isotope from 'normal' water ($H_2^{16}O$). Normal water does naturally contain $H_2^{18}O$, but at negligible amounts.

The actual experimental set up is straightforward and is being represented by the schematic in Figure 6. Samples are compared in a pair wise manner, e.g., sample X *versus* sample Y. Approximate equal amounts of protein from the two samples are important to the data analysis. To this, typically, a simple protein determination is performed. However, small offset differences can be corrected by using a so-called set factor in the data analysis.

Sample X is then digested in the presence of normal water, while sample Y is digested in the presence of $H_2^{18}O$. The samples are combined in a one to one ratio and subjected to subsequent peptide separation and mass spectrometric analysis.

Protein identification and quantification can then be performed using one typical LC-MS/MS run. Where the fragmentation data functions for the identification, the MS scan functions as the quantitative information. An example of a real measurement by high accuracy ion cyclotron resonance Fourier transform mass spectrometry (ICR-FT MS) is shown in Figure 6. The zoom-in shows the single charged peptides from sample X and sample Y. The double incorporation of oxygen gives rise to the distinct 4 Da difference between the mono-isotopic peaks at m/z 804.3908 and 808.3994 for sample X and Y, respectively. The ratio of the relative intensity is then a measure for the relative protein/peptide quantification.

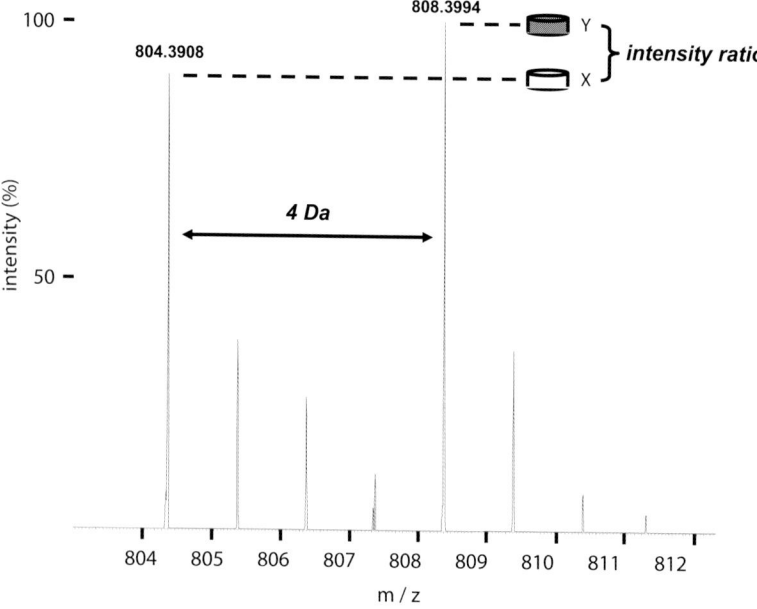

Figure 6. Example of a MS-survey scan of ^{18}O labeled and non-labeled peptide. A zoom in from a MS-survey scan of a singly charged peptide is shown which has been digested in the presence of normal water (804.3908 Da) and ^{18}O labeled water (808.3994). The double incorporation of oxygen reveals the distinct difference of 4 Da. The ratio of the relative intensity of the different peptides is used for the relative protein/peptide quantification.

A number of groups have developed software for analyzing this type of data. Mann and co-workers have developed a neat tool called MSQuant [59], which is designed to analyze isotopic labeled samples, not only ^{18}O but for instance also SILAC [42] derived samples. The software can be downloaded from http://msquant.sourceforge.net/. The software has a standard Mascot search and one or more raw files. Raw files from all major instrument vendors are supported.

A number of interesting applications using ^{18}O incorporation by different enzymes have been published [60–64]. The strong point of this particular method is that it is easy. There is no need for complex lengthy chemical labeling protocols or expensive labor intensive tissue culture work. However, $H_2^{18}O$ is rather expensive and is also less suited for complex sample analysis without further complexity reduction. An example of such an approach was demonstrated by Bonenfant and co-workers [65], where they analyzed a complex sample to quantify changes in protein phosphorylation using ^{18}O incorporation by trypsin followed by IMAC [66] enrichment.

Ion intensity-based quantitative approach

In the last few paragraphs we have described various techniques that allow the identification and quantification of proteins in complex mixtures – all of them involve the stable modifications of proteins in one way or another. As a matter of fact it would be nice to have reliable and reproducible quantitative methods for absolute protein quantification using mass spectrometry based on signal intensity only; however, comprehensive quantitative proteomics remains technically challenging due to the issues associated with sample complexity, sample preparation, and the wide dynamic range of protein abundance. Generally, signal intensity in mass spectrometry increases with the amount of analyte. A number of reports account for linear correlations between signal intensity and the amount of analyte in special applications [67, 68] but there are also concerns regarding nonlinearity of signal intensity and ion suppression effects for complex proteomic samples [69].

A very rough idea about protein concentration in complex mixtures can be gained using protein abundance indices (PAI) introduced by Rappsilber and colleagues (2002) [70]. The basis of the PAIs describes the number of identified peptides divided by the number of observable peptides per protein. This approach has been used to analyze the human spliceosome complex. This approach could only describe relative ratios of proteins within a given sample. The next step towards absolute quantification was the finding that the protein amount has a logarithmic dependency to the PAI. With this exponentially modified PAI they investigated known amounts of 46 proteins in a complex cell lysate with an average deviation factor of 1.74 ± 0.79 [71]. Despite the still strong variation of this method it has the great advantage that quantitative results can be obtained from already measured samples simply by reanalyzing them with the emPAI approach (Equation 1). With the knowledge of the total amount of protein you have applied you can recalculate the amount of your protein of interest.

$$\text{protein content } (mol\ \%) = \frac{emPAI}{\Sigma(emPAI)} \times 100 \qquad \text{(Eq. 1)}$$

Typically, absolute quantification of proteins requires the use of one or more external reference peptides to generate a calibration-response curve for specific polypeptides from that protein (i.e., synthetic tryptic polypeptide product). The absolute quantity of the protein under investigation is determined from the observed signal response for its polypeptides in the sample compared to the signal response from the calibration curve. In cases where absolute quantities of a number of different proteins are required, separate calibration curves are necessary. Absolute quantification would allow not only to determine changes between two conditions but also to perform quantitative protein comparisons within the same sample.

Gerber and co-workers describe a conventional technique for absolute quantification (called AQUA) of proteins and their corresponding modified states in complex mixtures using a synthesized peptide as a reference standard [72]. The reference peptide is chemically identical to the naturally occurring tryptic peptides of a given protein but one residue contains stable isotopes (^{13}C and/or ^{15}N). The reference standard is introduced to a complex mixture and the mixture is analyzed using LC/MS to measure the corresponding signal intensity for the spiked peptide along with the endogenous peptide. This intensity signal response is compared with an intensity calibration curve created using the introduced synthetic molecule to determine the amount of the endogenous protein in the mixture. A disadvantage with using synthetic peptides is that extra steps are required to synthesize an authentic sample, and to later 'spike' the synthetic standard prior to being able to determine the absolute quantity of the protein itself. To perform an absolute quantification for a number of proteins within a mixture requires a synthetic standard for each protein of interest (see above) [72].

Another method for absolute quantification of proteins requires that a known quantity of intact protein of a different species is spiked into the protein mixture of interest prior to digestion with trypsin or that a known quantity of pre-digested peptide is spiked into the mixture after it has been digested. The average MS signal response for the three most intense tryptic peptides is calculated for each well-characterized protein in the mixture, including those to the internal standard protein(s). The average MS signal response from the internal standard protein(s) is used to determine a universal signal response factor (counts/mol of protein), which is then applied to the other identified proteins in the mixture to determine their corresponding absolute concentration. The absolute quantity of each well-characterized protein in the mixture is determined by dividing the average MS signal response of the three most intense tryptic peptides of each well-characterized protein by the universal signal response factor described above.

Silva and co-workers observed a linear response of MS signal intensity from digested peptides correlating with protein concentration. Six proteins were analyzed in various dilutions from 6 fmol to 900 fmol total protein. All detected monoisotopic components were extracted with their accurate mass and retention time, to compare chemically identical components by using the Expression Informatics Software from Waters®. Upon decreasing protein concentrations the number of measurable peptides and their corresponding signal intensity responses decreased in a linear fashion but the relative signal intensity pattern between different proteins was constant. An average signal response of around 26,000 counts per pmol of each protein on column was observed with a CV of 4.9%. Because the response curve was independent of the protein that has been used the response factor of the spiked protein can be used to obtain absolute quantification of other well-characterized proteins in this sample. The standard protein mixture was spiked in a complex protein sample (human serum) and re-analyzed. Although there was a ~20% decrease of signal response in the signal response factor (counts/pmol) the signal intensity ratios are internally consistent. With this signal suppression effect the CV increased from 4.9% to 8.4% in the more complex sample. With this response factor it was possible to determine the absolute amount of 11 serum proteins. The results obtained from the replicate analysis were better than 15% variability [73].

Wang and co-workers reported a quantification method without labeled or spiked standards. This method relies on a number of data manipulations, e.g., baseline subtraction, data smoothing, de-isotoping, charge state normalization, and appropriate peak detection in order to identify peaks that are valid for quantitation. The authors used a test sample of five proteins where the amount of three proteins was kept constant and the amount of two proteins was varied. The relative intensity of these proteins was close to linear in a range of one order of magnitude with a CV of 33% ± 4. The quantitation method was used to analyze 105 human serum samples with spiked non-human proteins. 80 samples were tested on a Thermo Finnigan LCQ Deca ESI-Ion Trap and 25 samples were measured on a Micromass LCT ESI-ToF mass spectrometer (a detailed explanation can be found in the following chapter). The higher resolution power of the ToF instrument provides a 20 times lower detection limit compared to the LCQ-Deca instrument. One of the serum samples was arbitrarily chosen for reference (e.g., house keeping proteins) and used to adjust all LC-MS retention times. MS signal intensities were normalized with one normalization constant for the entire sample. This procedure showed the smallest variations between the samples. The result showed a linear MS response for the test proteins between 100 fmol and 100 pmol on column [74].

All ion intensity-based quantification methods were performed on samples with limited complexity. It is therefore still an open question as to whether these methods are also applicable to more complex tissue samples. Once more the studies discussed above illustrate that mass resolution, ionization efficiency, reproducibility, and sufficient pre-fractionation are crucial for MS-based quantification methods.

Summary and conclusions

Over the last 20 years several elegant techniques have been established that allow quantifying protein levels in complex biological samples. Each of these methods has advantages and none of them are without flaws. All of the technologies cover a wide range of experimental designs and for each of them there is a scientific question for which a particular approach is best suited. However, none of the techniques has won the race making the others obsolete. There are rather several important considerations to be made in the design of quantitative proteomics experiments in order to avoid dissatisfactory results, and thus, before subjecting precious biological samples to labor intensive and costly quantitative proteomic analyses. There is the urgent need to formulate the scientific questions to be answered, delineate the expected results, but also to consider the own resources and to calculate the costs of any envisaged approach. Where applicable, a reasonable solution may be to subject the same probe to more than one quantitative measurement. In any case, it is important to note that any quantitative measurement and especially any conclusion drawn thereof needs to be confirmed in the context of the corresponding biological system by other means. Emerging technologies, such as the ion-intensity-based quantitation in conjunction with the rapid improvements in MS technology, will bring along more accurate and more comprehensive measurements and carry a promise for the future.

References

1. Gorg A, Weiss W, Dunn MJ (2004) Current two-dimensional electrophoresis technology for proteomics. *Proteomics* 4(12): 3665–3685
2. Cleveland DW, Fischer SG, Kirschner MW, Laemmli UK (1977) Peptide mapping by limited proteolysis in sodium dodecyl sulfate and analysis by gel electrophoresis. *J Biol Chem* 252(3): 1102–1106
3. Westermeier RN, Naven T (2002) Proteomics in Practice. Wiley-VCH, Freiburg
4. Steinberg TH, Haugland RP, Singer VL (1996) Applications of SYPRO orange and SYPRO red protein gel stains. *Anal Biochem* 239(2): 238–245
5. Schulenberg B, Goodman TN, Aggeler R, Capaldi RA, Patton WF (2004) Characterization of dynamic and steady-state protein phosphorylation using a fluorescent phosphoprotein gel stain and mass spectrometry. *Electrophoresis* 25(15): 2526–2532
6. Gygi SP, Corthals GL, Zhang Y, Rochon Y, Aebersold R (2000) Evaluation of two-dimensional gel electrophoresis-based proteome analysis technology. *Proc Natl Acad Sci USA* 97(17): 9390–9395
7. Zhou S, Bailey MJ, Dunn MJ, Preedy VR, Emery PW (2005) A quantitative investigation into the losses of proteins at different stages of a two-dimensional gel electrophoresis procedure. *Proteomics* 5(11): 2739–2747
8. Marouga R, David S, Hawkins E (2005) The development of the DIGE system: 2D fluorescence difference gel analysis technology. *Anal Bioanal Chem* 382(3): 669–678
9. Righetti PG, Castagna A, Herbert B, Reymond F, Rossier JS (2003) Prefractionation techniques in proteome analysis. *Proteomics* 3(8): 1397–1407
10. Berkelman TS, Stenstedt T (2002) 2-D electrophoresis using immobilized pH gradients: Principles and methods, AB edn: Amersham Biosciences
11. Jiang L, He L, Fountoulakis M (2004) Comparison of protein precipitation methods for sample preparation prior to proteomic analysis. *J Chromatogr A* 1023(2): 317–320
12. Hancock RE, Nikaido H (1978) Outer membranes of gram-negative bacteria. XIX. Isolation from *Pseudomonas aeruginosa* PAO1 and use in reconstitution and definition of the permeability barrier. *J Bacteriol* 136(1): 381–390
13. Riedel K, Arevalo-Ferro C, Reil G, Gorg A, Lottspeich F, Eberl L (2003) Analysis of the quorum-sensing regulon of the opportunistic pathogen *Burkholderia cepacia* H111 by proteomics. *Electrophoresis* 24(4): 740–750
14. Manza LL, Stamer SL, Ham AJ, Codreanu SG, Liebler DC (2005) Sample preparation and digestion for proteomic analyses using spin filters. *Proteomics* 5(7): 1742–1745
15. Yao R, Li J (2003) Towards global analysis of mosquito chorion proteins through sequential extraction, two-dimensional electrophoresis and mass spectrometry. *Proteomics* 3(10): 2036–2043
16. Peters TJ (1977) Application of analytical subcellular fractionation techniques and tissue enzymic analysis to the study of human pathology. *Clin Sci Mol Med* 53(6): 505–511
17. Scott TM (2005) Success rate of spot IDs in a 2D dros gel. In: FGCZ. Unpublished Results
18. Hedberg JJ, Bjerneld EJ, Cetinkaya S, Goscinski J, Grigorescu I, Haid D, Laurin Y, Bjellqvist B (2005) A simplified 2-D electrophoresis protocol with the aid of an organic disulfide. *Proteomics* 5(12): 3088–3096
19. Hoving S, Gerrits B, Voshol H, Muller D, Roberts RC, van Oostrum J (2002) Preparative two-dimensional gel electrophoresis at alkaline pH using narrow range immobilized pH gradients. *Proteomics* 2(2): 127–134

20. Pennington K, McGregor E, Beasley CL, Everall I, Cotter D, Dunn MJ (2004) Optimization of the first dimension for separation by two-dimensional gel electrophoresis of basic proteins from human brain tissue. *Proteomics* 4(1): 27–30
21. Herbert BR, Molloy MP, Gooley AA, Walsh BJ, Bryson WG, Williams KL (1998) Improved protein solubility in two-dimensional electrophoresis using tributyl phosphine as reducing agent. *Electrophoresis* 19(5): 845–851
22. Barry RC, Alsaker BL, Robison-Cox JF, Dratz EA (2003) Quantitative evaluation of sample application methods for semipreparative separations of basic proteins by two-dimensional gel electrophoresis. *Electrophoresis* 24(19–20): 3390–3404
23. Gorg A, Boguth G, Obermaier C, Weiss W (1998) Two-dimensional electrophoresis of proteins in an immobilized pH 4-12 gradient. *Electrophoresis* 19(8–9): 1516–1519
24. Chevallet M, Santoni V, Poinas A, Rouquie D, Fuchs A, Kieffer S, Rossignol M, Lunardi J, Garin J, Rabilloud T (1998) New zwitterionic detergents improve the analysis of membrane proteins by two-dimensional electrophoresis. *Electrophoresis* 19(11): 1901–1909
25. Luche S, Santoni V, Rabilloud T (2003) Evaluation of nonionic and zwitterionic detergents as membrane protein solubilizers in two-dimensional electrophoresis. *Proteomics* 3(3): 249–253
26. Twine SM, Mykytczuk NC, Petit M, Tremblay TL, Conlan JW, Kelly JF (2005) *Francisella tularensis* proteome: low levels of ASB-14 facilitate the visualization of membrane proteins in total protein extracts. *J Proteome Res* 4(5): 1848–1854
27. Bai F, Liu S, Witzmann FA (2005) A 'de-streaking' method for two-dimensional electrophoresis using the reducing agent tris(2-carboxyethyl)-phosphine hydrochloride and alkylating agent vinylpyridine. *Proteomics* 5(8): 2043–2047
28. Laemmli UK (1970) Cleavage of structural proteins during the assembly of the head of bacteriophage T4. *Nature* 227(5259): 680–685
29. Fountoulakis M, Juranville JF, Roder D, Evers S, Berndt P, Langen H (1998) Reference map of the low molecular mass proteins of *Haemophilus influenzae*. *Electrophoresis* 19(10): 1819–1827
30. Haebel S, Albrecht T, Sparbier K, Walden P, Korner R, Steup M (1998) Electrophoresis-related protein modification: alkylation of carboxy residues revealed by mass spectrometry. *Electrophoresis* 19(5): 679–686
31. Candiano G, Bruschi M, Musante L, Santucci L, Ghiggeri GM, Carnemolla B, Orecchia P, Zardi L, Righetti PG (2004) Blue silver: a very sensitive colloidal Coomassie G-250 staining for proteome analysis. *Electrophoresis* 25(9): 1327–1333
32. Lamanda A, Zahn A, Roder D, Langen H (2004) Improved Ruthenium II tris (bathophenantroline disulfonate) staining and destaining protocol for a better signal-to-background ratio and improved baseline resolution. *Proteomics* 4(3): 599–608
33. Mackintosh JA, Choi HY, Bae SH, Veal DA, Bell PJ, Ferrari BC, Van Dyk DD, Verrills NM, Paik YK, Karuso P (2003) A fluorescent natural product for ultra sensitive detection of proteins in one-dimensional and two-dimensional gel electrophoresis. *Proteomics* 3(12): 2273–2288
34. Shevchenko A, Wilm M, Vorm O, Mann M (1996) Mass spectrometric sequencing of proteins silver-stained polyacrylamide gels. *Anal Chem* 68(5): 850–858
35. Berggren K, Chernokalskaya E, Steinberg TH, Kemper C, Lopez MF, Diwu Z, Haugland RP, Patton WF (2000) Background-free, high sensitivity staining of proteins in one- and two-dimensional sodium dodecyl sulfate-polyacrylamide gels using a luminescent ruthenium complex. *Electrophoresis* 21(12): 2509–2521
36. Scott TM (2004) RuBP Lamanda Stain Optimization. 2004: Personal Communication. Unpublished Results

37. Smejkal GB, Robinson MH, Lazarev A (2004) Comparison of fluorescent stains: relative photostability and differential staining of proteins in two-dimensional gels. *Electrophoresis* 25(15): 2511–2519
38. Tonge R, Shaw J, Middleton B, Rowlinson R, Rayner S, Young J, Pognan F, Hawkins E, Currie I, Davison M (2001) Validation and development of fluorescence two-dimensional differential gel electrophoresis proteomics technology. *Proteomics* 1(3): 377–396
39. Browne TR, Van Langenhove A, Costello CE, Biemann K, Greenblatt DJ (1981) Kinetic equivalence of stable-isotope-labeled and unlabeled phenytoin. *Clin Pharmacol Ther* 29(4): 511–515
40. Oda Y, Huang K, Cross FR, Cowburn D, Chait BT (1999) Accurate quantitation of protein expression and site-specific phosphorylation. *Proc Natl Acad Sci USA* 96(12): 6591–6596
41. Lahm HW, Langen H (2000) Mass spectrometry: a tool for the identification of proteins separated by gels. *Electrophoresis* 21(11): 2105–2114
42. Ong SE, Blagoev B, Kratchmarova I, Kristensen DB, Steen H, Pandey A, Mann M (2002) Stable isotope labeling by amino acids in cell culture, SILAC, as a simple and accurate approach to expression proteomics. *Mol Cell Proteomics* 1(5): 376–386
43. Gruhler A, Schulze WX, Matthiesen R, Mann M, Jensen ON (2005) Stable isotope labeling of *Arabidopsis thaliana* cells and quantitative proteomics by mass spectrometry. *Mol Cell Proteomics* 4(11): 1697–1709
44. Krijgsveld J, Ketting RF, Mahmoudi T, Johansen J, Artal-Sanz M, Verrijzer CP, Plasterk RH, Heck AJ (2003) Metabolic labeling of *C. elegans* and *D. melanogaster* for quantitative proteomics. *Nat Biotechnol* 21(8): 927–931
45. Gygi SP, Rist B, Gerber SA, Turecek F, Gelb MH, Aebersold R (1999) Quantitative analysis of complex protein mixtures using isotope-coded affinity tags. *Nat Biotechnol* 17(10): 994–999
46. Regnier FE, Riggs L, Zhang R, Xiong L, Liu P, Chakraborty A, Seeley E, Sioma C, Thompson RA (2002) Comparative proteomics based on stable isotope labeling and affinity selection. *J Mass Spectrom* 37(2): 133–145
47. Hansen KC, Schmitt-Ulms G, Chalkley RJ, Hirsch J, Baldwin MA, Burlingame AL (2003) Mass spectrometric analysis of protein mixtures at low levels using cleavable 13C-isotope-coded affinity tag and multidimensional chromatography. *Mol Cell Proteomics* 2: 299–314
48. Li J, Steen H, Gygi SP (2003) Protein profiling with cleavable isotope-coded affinity tag (cICAT) reagents: the yeast salinity stress response. *Mol Cell Proteomics* 2(11): 1198–1204
49. Lu Y, Bottari P, Turecek F, Aebersold R, Gelb MH (2004) Absolute quantification of specific proteins in complex mixtures using visible isotope-coded affinity tags. *Anal Chem* 76(14): 4104–4111
50. Ranish JA, Yi EC, Leslie DM, Purvine SO, Goodlett DR, Eng J, Aebersold R (2003) The study of macromolecular complexes by quantitative proteomics. *Nat Genet* 33(3): 349–355
51. Smolka M, Zhou H, Aebersold R (2002) Quantitative protein profiling using two-dimensional gel electrophoresis, isotope-coded affinity tag labeling, and mass spectrometry. *Mol Cell Proteomics* 1(1): 19–29
52. Ross PL, Huang YN, Marchese JN, Williamson B, Parker K, Hattan S, Khainovski N, Pillai S, Dey S, Daniels S et al. (2004) Multiplexed protein quantitation in *Saccharomyces cerevisiae* using amine-reactive isobaric tagging reagents. *Mol Cell Proteomics* 3(12): 1154–1169
53. Unwin RD, Pierce A, Watson RB, Sternberg DW, Whetton AD (2005) Quantitative proteomic analysis using isobaric protein tags enables rapid comparison of changes in transcript and protein levels in transformed cells. *Mol Cell Proteomics* 4(7): 924–935

54. DeSouza L, Diehl G, Rodrigues MJ, Guo J, Romaschin AD, Colgan TJ, Siu KW (2005) Search for cancer markers from endometrial tissues using differentially labeled tags iTRAQ and cICAT with multidimensional liquid chromatography and tandem mass spectrometry. *J Proteome Res* 4(2): 377–386
55. Choe LH, Aggarwal K, Franck Z, Lee KH (2005) A comparison of the consistency of proteome quantitation using two-dimensional electrophoresis and shotgun isobaric tagging in *Escherichia coli* cells. *Electrophoresis* 26(12): 2437–2449
56. Gaskell SJ, Haroldsen PE, Reilly MH (1988) Collisionally activated decomposition of modified peptides using a tandem hybrid instrument. *Biomed Environ Mass Spectrom* 16(1–12): 31–33
57. Desiderio DM, Kai M (1983) Preparation of stable isotope-incorporated peptide internal standards for field desorption mass spectrometry quantification of peptides in biologic tissue. *Biomed Mass Spectrom* 10(8): 471–479
58. Kraut J (1977) Serine proteases: structure and mechanism of catalysis. *Annu Rev Biochem* 46: 331–358
59. Schulze WX, Mann M (2004) A novel proteomic screen for peptide–protein interactions. *J Biol Chem* 279(11): 10756–10764
60. Heller M, Mattou H, Menzel C, Yao X (2003) Trypsin catalyzed 16O-to-18O exchange for comparative proteomics: tandem mass spectrometry comparison using MALDI-TOF, ESI-QTOF, and ESI-ion trap mass spectrometers. *J Am Soc Mass Spectrom* 14(7): 704–718
61. Hicks WA, Halligan BD, Slyper RY, Twigger SN, Greene AS, Olivier M (2005) Simultaneous quantification and identification using 18O labeling with an ion trap mass spectrometer and the analysis software application 'ZoomQuant'. *J Am Soc Mass Spectrom* 16(6): 916–925
62. Rao KC, Carruth RT, Miyagi M (2005) Proteolytic 18O labeling by peptidyl-Lys metalloendopeptidase for comparative proteomics. *J Proteome Res* 4(2): 507–514
63. Sun G, Anderson VE (2005) A strategy for distinguishing modified peptides based on post-digestion 18O labeling and mass spectrometry. *Rapid Commun Mass Spectrom* 19(19): 2849–2856
64. Hood BL, Lucas DA, Kim G, Chan KC, Blonder J, Issaq HJ, Veenstra TD, Conrads TP, Pollet I, Karsan A (2005) Quantitative analysis of the low molecular weight serum proteome using 18O stable isotope labeling in a lung tumor xenograft mouse model. *J Am Soc Mass Spectrom* 16(8): 1221–1230
65. Bonenfant D, Schmelzle T, Jacinto E, Crespo JL, Mini T, Hall MN, Jenoe P (2003) Quantitation of changes in protein phosphorylation: a simple method based on stable isotope labeling and mass spectrometry. *Proc Natl Acad Sci USA* 100(3): 880–885
66. Andersson L, Porath J (1986) Isolation of phosphoproteins by immobilized metal (Fe^{3+}) affinity chromatography. *Anal Biochem* 154(1): 250–254
67. Purves RW, Gabryelski W, Li L (1998) Investigation of the quantitative capabilities of an electrospray ionization ion trap linear time-of-flight mass spectrometer. *Rapid Commun Mass Spectrom* 12(11): 695–700
68. Voyksner RD, Lee H (1999) Investigating the use of an octupole ion guide for ion storage and high-pass mass filtering to improve the quantitative performance of electrospray ion trap mass spectrometry. *Rapid Commun Mass Spectrom* 13(14): 1427–1437
69. Muller C, Schafer P, Stortzel M, Vogt S, Weinmann W (2002) Ion suppression effects in liquid chromatography-electrospray-ionisation transport-region collision induced dissociation mass spectrometry with different serum extraction methods for systematic toxicological analysis with mass spectra libraries. *J Chromatogr B Analyt Technol Biomed Life Sci* 773(1): 47–52

70. Rappsilber J, Ryder U, Lamond AI, Mann M (2002) Large-scale proteomic analysis of the human splicesome. *Genome Res* 12(8): 1231–1245
71. Ishihama Y, Oda Y, Tabata T, Sato T, Nagasu T, Rappsilber J, Mann M (2005) Exponentially modified protein abundance index (emPAI) for estimation of absolute protein amount in proteomics by the number of sequenced peptides per protein. *Mol Cell Proteomics* 4(9): 1265–1272
72. Gerber SA, Rush J, Stemman O, Kirschner MW, Gygi SP (2003) Absolute quantification of proteins and phosphoproteins from cell lysates by tandem MS. *Proc Natl Acad Sci USA* 100(12): 6940–6945
73. Silva JC, Gorenstein MV, Li G-Z, Vissers JPC, Geromanos SJ (2006) Absolute quantification of proteins by LCMSE: A virtue of parallel ms acquisition. *Mol Cell Proteomics* 5(1): 144–156
74. Wang WX, Zhou HH, Lin H, Roy S, Shaler TA, Hill LR, Norton S, Kumar P, Anderle M, Becker CH (2003) Quantification of proteins and metabolites by mass spectrometry without isotopic labeling or spiked standards. *Anal Chemistry* 75(18): 4818–4826
75. Han DK, Eng J, Zhou H, Aebersold R (2001) Quantitative profiling of differentiation-induced microsomal proteins using isotope-coded affinity tags and mass spectrometry. *Nat Biotechnol* 19(10): 946–951

Plant Systems Biology
Edited by Sacha Baginsky and Alisdair R. Fernie
© 2007 Birkhäuser Verlag/Switzerland

Protein identification using mass spectrometry: A method overview

Sven Schuchardt[1] and Albert Sickmann[2]

[1] *Fraunhofer Institute of Toxicology and Experimental Medicine, Drug Research and Medical Biotechnology, Nikolai-Fuchs-Strasse 1, 30625 Hannover, Germany*
[2] *Rudolf-Virchow-Center, DFG-Research Center for Experimental Biomedicine, University of Wurzburg, Versbacherstr. 9, 97078 Würzburg, Germany*

Abstract

With the introduction of soft ionization techniques such as Matrix Assisted Laser Desorption Ionization (MALDI), and Electrospray Ionization (ESI), proteins have become accessible to mass spectrometric analyses. Since then, mass spectrometry has become the method of choice for sensitive, reliable and inexpensive protein and peptide identification. With the increasing number of full genome sequences for a variety of organisms and the numerous protein databases constructed thereof, all the tools necessary for the high-throughput protein identification with mass spectrometry are in place. This chapter highlights the different mass spectrometric techniques currently applied in proteome research by giving a brief overview of methods for identification of posttranslational modifications and discussing their suitability of strategies for automated data analysis.

Introduction

Since its invention in 1905, mass spectrometry (MS) has become a widely established technique for analyzing chemical structures in quantities down to trace levels. Due to a lack of suitable ionization techniques for high mass biomolecules, proteins remained inaccessible to MS analysis for decades. Since the introduction of soft ionization techniques such as Matrix Assisted Laser Desorption Ionization (MALDI) and Electrospray Ionization (ESI), MS at the end of the 1980s [1, 2] protein analysis by mass spectrometry underwent a rapid phase of development. In parallel, an increasing number of full genome sequences for a variety of organisms are now available and numerous protein databases were constructed from this information. Well-annotated, high-quality protein databases built the ground on which high-throughput protein identification with mass spectrometry can be performed.

The modular arrangement of different types of mass analyzers in combination with MALDI- or ESI has resulted in a wide variety of different mass spectrometric instrumentation (e.g., MALDI-TOF, ESI-Q-TOF, ESI-ion trap, MALDI-TOF/TOF, ESI-FT-ICR, etc.). All of these MS techniques allowed the determination of the primary structure of a protein, though they always required additional sample preparation techniques. Furthermore, the analysis of posttranslational modifications such as phosphorylation or glycosylation has become possible. Modern mass spectrometers now combine attributes like high sensitivity, mass accuracy, mass resolution, and rapid analysis as well as sophisticated data handling in a system-dependent manner. In addition to these technical aspects in mass spectrometry, greatly improved sample separation and preparation techniques have also lead to enhanced sensitivity. The quantification of chemically or metabolically labeled proteins is yet another focus of interest in mass spectrometry (see previous chapter). Despite these advances current MS approaches still have limitations and are therefore subjected to further development. The aim of this paper is therefore to highlight the different mass spectrometric techniques currently applied in proteome research by giving a brief overview of methods for identification of posttranslational modifications and discussing their suitability of strategies for protein quantification.

General technical considerations

Mass spectrometry is a highly sensitive and accurate method for the determination of molecular masses of different types of molecules. All common mass spectrometers consist of three functional units: the ion source which ionizes the analyte, the mass analyzer which separates the resulting ions according to their mass-to-charge ratio (m/z), and the ion detector, whose signals can be recorded and processed by a computer. The order, which is given here, reflects the direction of the ion's path through a standard mass spectrometer. For every unit of a mass spectrometer, different designs are available, all of which can be arranged in a multitude of ways. For mass analyzers in particular, different arrangements of units can be incorporated into a single mass spectrometer. For example, the coupling of two mass selective devices for tandem mass spectrometry (MS/MS) has expanded the field's application enormously, resulting in a profusion of experimental set ups and designs in modern protein analyzing mass spectrometers. For a better understanding of the variety of instrumentation, a brief introduction to the functional principles of the most common designs is essential.

Ion sources

The ion source is designed to generate analyte ions and transfer them into the gasphase, where they can enter the vacuum of the mass spectrometer. The ions are generated by loss or gain of charge (e.g., electron capture, electron ejection, protonation, deprotonation or cationization). Electron ionization (EI) was the most common ionization technique for mass analysis until the development of MALDI and ESI ionization. The electron ionization technology is limited to compounds with masses well below the range of peptides and proteins, due to the involatility of large biomolecules in a vacuum by thermal desorption. Nevertheless, electron ionization still plays an important role in the routine analysis of small molecules. The first satisfactory biomolecule ionization was achieved with techniques such as plasma desorption [3] and fast atom bombardment (FAB) [4], which still have several limitations. With the introduction of 'soft' ionization techniques (e.g., MALDI and ESI) in mass spectrometry, problems like thermal decomposition and excessive fragmentation of large biomolecules such as peptides could be overcome. In both cases, the ionization is primarily accomplished by protonation of the analyte in a liquid phase which is supplemented with a proton donor (e.g., an organic acid).

MALDI – Source and sample introduction

For this ionization technique, the purified analyte is generally dissolved in a matrix solution, spotted onto a solid target and co-crystallized with the matrix. The matrix, which typically contains a UV sensitive aromatic compound, is used to facilitate UV-laser energy-absorption and energy-transfer. The irradiated area of the crystals and the analyte embedded therein are vaporized by the laser energy uptake (Fig. 1). Although the mechanism of ion formation during the MALDI process is still a matter of some controversy [5], the efficiency of ionization and the initial ion velocity can be controlled by the choice of matrix or the composition of the analyte sample. Typical matrix compounds include 2,5-dihydroxybenzoic acid (DHB), 3,5-dimethoxy-4-hydroxy-cinnamic acid (sinapinic acid), and α-cyano-4-hydroxy-cinnamic acid (HCCA). The analyte molecules are normally ionized by simple protonation, leading to the formation of the typical singly charged $[M+H]^+$ type species (where M is the mass of the analyte molecule). Trace contaminations of earth alkali metals in the matrix will especially generate $[M+X]^+$ ions (where X = Li, Na, K, etc.). Once the ions are vaporized, they are accelerated in an electric field and different mass analyzers can be used to measure their *m/z*. The most commonly used instrument type is the MALDI-TOF-MS design whose performance has dramatically improved due to the introduction of delayed ion extraction [6, 7] and reflectron technology. The MALDI evaporation process generates ions with an initial velocity distribution, which normally causes low resolution due to start-time errors. This effect is compensated with delayed ion extraction by the use of a two-stage acceleration field in combination with a delay time resulting from appropriate acceleration voltages following the laser pulse.

MALDI-TOF instruments are capable of analyzing intact proteins and complex peptide mixtures since they have an almost unlimited mass range that can be analyzed within their flight tube. The MALDI technique generates singly-charged molecules [8] with a typical detection limit in the low femtomol range. MALDI has long been considered a 'soft' ionization technique that apparently generates almost exclusively intact ions. In fact, a significant degree of metastable decay occurs after ion acceleration which is used in reflectron TOF or in modern TOF-TOF analyzers for simple post-source decay (PSD) analysis. Such an analysis provides some structural information about an analyte ion, which can be used for the interpretation of the mass spectrum and the identification of the analyte molecule.

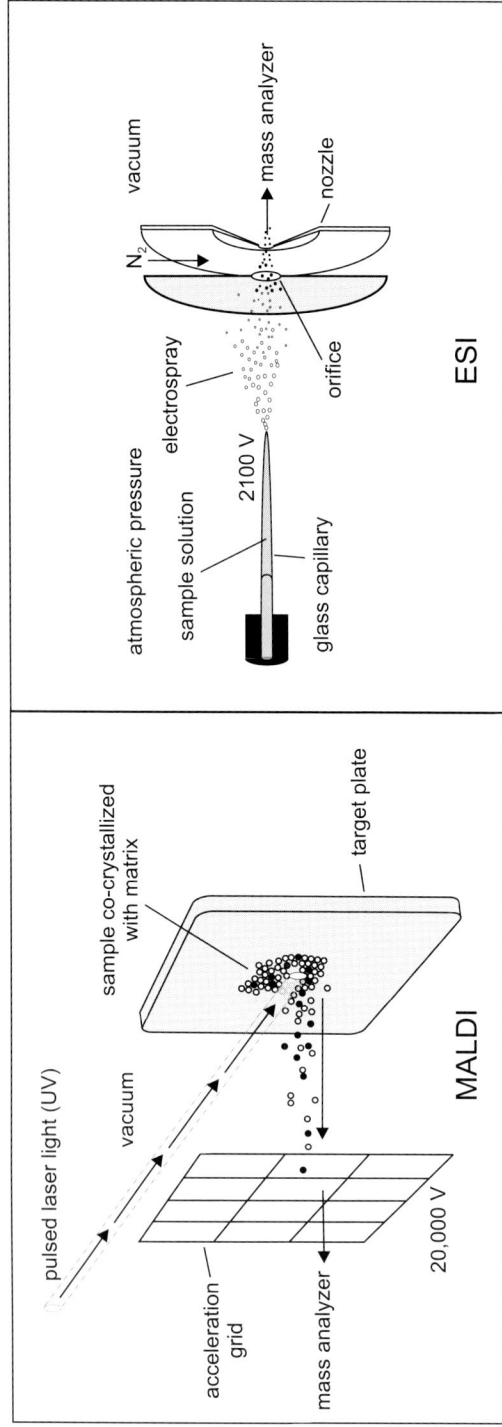

Figure 1. Schematic illustration of the ionization methods MALDI and ESI. For MALDI the sample co-crystallized with matrix is dried on the target plate and is placed in the vacuum of the mass spectrometer. After irradiation with pulses of UV laser light the sample and matrix molecules desorb from the condensed state. Once in the vapor phase the ions are accelerated out of the source by application of a high potential (approx. 20 kV). The ESI process is carried out under atmospheric pressure with a capillary containing the sample solution. The strong electric field attracts the ions to the orifice of the mass spectrometer. Solvent evaporation can be facilitated with a dry gas stream (N_2), with the low-flow nanospray setup (NSI), the evaporation process occurs also in the absence of gas to completion. The ions are further accelerated and focused under vacuum conditions by a series of extraction electrodes and lenses.

ESI – Source and sample introduction

The introduction of charged molecules into the mass spectrometer with ESI sources is carried out using different quantities of aqueous sample under atmospheric pressure conditions [2]. In nanoelectrospray (nanoES) technology [9], for example, only a few microliters of sample are needed for spraying from the highly charged (up to 3,000 V) tip of a metal coated glass needle to the inlet of the mass spectrometer (Fig. 1). The finely pointed nozzle generates a strong electric field, which helps to accelerate the charged droplets and to form a constant spray of 20–200 nL/min. Evaporation of the solvent, which is normally supported by a dry gas, decreases the droplet size and thus increases the surface charge density, finally releasing solvent-free ionized analyte molecules. Here, organic solvents, e.g., 2-propanol or acetonitrile, facilitate the evaporation process and enhance the formation of a stable spray. The resulting ions are directed into an orifice and focused stepwise under increasing vacuum conditions by electrostatic lenses to form an ion beam. The ESI technique generates primarily multiply charged molecules. It has been demonstrated that the maximum charge states and charge state distributions of ions generated by electrospray ionization are influenced by solvents that are more volatile than water [10, 11].

Mass analyzers

Time of Flight (TOF) mass analyzer

An attractive feature of the TOF mass spectrometer is its graspable design. Mass analysis simply involves measuring the flight time of the ions on their way through the field-free-drift region in a flight tube after acceleration. The velocity of the ions in the analyzer tube is dependent on their m/z values. The greater the m/z, the lower the speed and the longer the time needed to travel the distance to the detector. Unfortunately, for a simple linear tube design, the mass resolution is relatively poor due to the inevitable initial energy spread from the evaporation process. This disadvantage was eliminated by the introduction of the reflectron [1], which is located at the end of the flight tube and compensates the fuzziness in flight times by focusing ions with the same m/z in space and time before they hit the detector (Fig. 2). Thus, with a reflectron TOF mass analyzer design high resolution up to 25,000 can be effortlessly accomplished.

Another feature of MALDI-TOF instruments is the post-source decay (PSD) technique that makes use of the fact that some of the MALDI generated ions undergo metastable decay during flight through the mass analyzer. For simple reflectron MALDI-TOF devices a composite PSD mass spectrum is generated stepwise due to the kinetic energy range dependent focusing potential. However, modern MALDI-TOF-TOF-MS devices provide a faster and more precise MS/MS-spectrum generation comparable with other common tandem-MS devices.

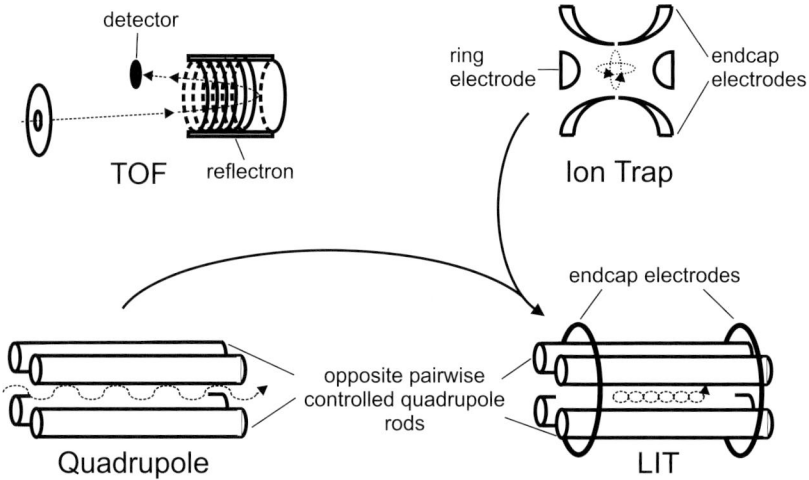
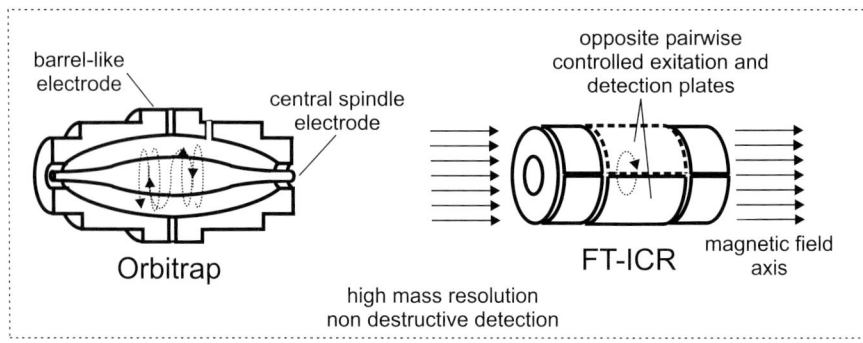

Figure 2. Schematized mass analyzer types. **TOF**: Some time of flight reflectron analyzer are capable of PSD- or LIFT-tandem MS and provide generally high mass resolution. **Ion Trap**: The Paul ion trap can usually perform fast MS^n experiments but suffers normally from low mass resolution and accuracy. **Quad**: Multiple quadrupol mass filters in combination with a collision cell are suitable for tandem MS with good mass accuracy. **LIT**: Linear ion trap are simplified a synthesis of a Quad and an ion trap analyzer (connecting arrows) with over all improved performance. Within the end caps it can trap strings of ions. **Orbitrap**: It can be considered as a highly modified ion trap with an exceptional resolution and mass accuracy. **FT-ICR**: This mass analyzer provides the highest resolution power and the best mass accuracy of all currently known devices. All these analyzers can be combined with each other and with ion sources and detectors in various ways.

Quadrupole (Q or Quad) mass analyzer

The principle of a quadrupole mass filter is based on the fact, that ions have an m/z-dependent trajectory in an alternating radio-frequency field [94]. The oscillating field is generated by two pairs of rod electrodes which focuses ions in two dimensions (i.e., two axes). The ions are alternately accelerated to the active attracting electrode. At any given field oscillation of the amplitude and the frequency a number of ions with a specific *m/z* value are stabilized in between the electrodes, while the majority of ions are discarded. For this reason, quadrupole mass analyzers are described as mass filters. With different electrode designs, the ions can be trapped in a defined volume (ion trap), or drift through a third dimension as in quadrupole mass filters (ion path). The range of the scanning mass gate is highly field modulation-dependent. If the mass window is increased, more selected ions pass in stable trajectories through the analyzer, increasing the signal but reducing the resolution. Triple quadrupole (triple quad) and the Q-TOF mass spectrometers are commonly used set ups to perform tandem MS with quadrupole mass analyzers.

Ion trap mass analyzer

In principle, the ion trap functionality is similar to the quadrupole analyzer [94], the difference being that the ions are trapped in three dimensions due to the specific assembly of the electrodes. The trapping volume for selected ions is defined by a ring electrode and two end-cap electrodes in a compact shape. The operation of ion trap analyzers is more sophisticated, since several gate drives can be applied for demanding mass analyses. The operation of an ion trap instrument is, in many ways, similar to that of a triple quadrupole mass spectrometer. The triple quad performs ion selection, collisional dissociation and mass analysis in three aligned mass analyzers separated in time and space, whereas the ion trap performs each operation sequentially in a single device only separated in time. A major drawback of the ion trap design is the limitation in the number of ions that can be trapped. The more ions are located in the limited volume of the ion trap, the more they interact with each other, e.g., repulsion by identical charges, and the more deviation from their predicted behavior can be observed. A significant loss of resolution and mass accuracy are direct consequences of excessively high ion density. This 'space charge' phenomenon requires additional scanning and control procedures to ensure that a suitable number of ions are trapped during every scan. Normally 0.5 amu can easily be resolved if 'space charging' is minimized. Following collision induced dissociation (CID), fragment ions can be scanned out of the trap to generate an MS/MS spectrum. If required, more MS stages can usually be performed with ion trap instruments (MS^n). However, n is usually less than 7 depending on the ion yields from former experiments. Fast scanning rate, sensitivity, flexibility, robustness and relatively low cost are the considerable advantages of the ion trap mass analyzers.

Orbitrap

Despite the fact that the Orbitrap uses constant electrostatic fields while the ion trap uses an oscillating electric field, the Orbitrap can be regarded as a highly modified ion trap (Fig. 2). The electrode geometry of the Orbitrap is a completely new design and resembles an elongated circular outer barrel with a central spindle-like electrode [12]. These axially symmetric electrodes create a combined quadro-logarithmic electrostatic potential, leading to stable ion trajectories around the central electrode and a simultaneous oscillation in the axial direction. The Orbitrap design provides high resolution (up to 150,000), high mass accuracy (2–5 ppm), and an appropriate dynamic range [13] and can be operated with MALDI and ESI sources [12, 14]. Although the applicability of the Orbitrap in tandem mass spectrometry is currently being scrutinized in different laboratories, this new type of high resolution mass analyzer has the potential to become a cost-effective alternative to FT-ICR-MS instruments (next section). However, to date, insufficient practical data are available to evaluate the future impact of Orbitrap instruments in mass spectrometric protein analysis.

Fourier Transform-Ion Cyclotron Resonance (FT-ICR) mass analyzer

This smart type of mass analyzer is having a great impact on MS derived protein and peptide analysis. FT-ICR-MS offers a higher resolution and mass accuracy than any other currently available mass spectrometer designs. The analyte ions are trapped in a combination of electric and strong magnetic fields, which give rise to the high performance of the FT-ICR analyzer (Fig. 2). Ions trapped by a static electric field are constrained to move in circular orbits in the presence of a uniform static magnetic field. The frequency of the circular motion (cyclotron frequency) is a function of the *m/z* of the ion and the magnetic field strength. The radius of this circular motion is dependent on the momentum of the ions in the plane perpendicular to the magnetic field. Thus, under high vacuum conditions, ions can be contained for a long period of time and ion excitation and detection of their cyclotron frequencies can be performed repeatedly. This technique allows nondestructive detection of the ions and subsequent acquisition of the spectra with a broadband amplifier for all ions simultaneously. Fourier transformation of the induced image current signals provides a complete mass spectrum with very high mass accuracy. Unfortunately, every aspect of FT-ICR-MS performance improves at higher magnetic fields which normally originate from superconducting magnets. Currently available superconducting magnetic materials must be operated at extremely low temperature (typically <10 K). Using superconducting magnets in FT-ICR analyzers constrains the design of these instruments and requires a balance for the analysis-space (a large space is desirable since 'space charge' phenomena can be avoided) and the limited size of the homogeneous magnetic fields that are technically achievable with superconducting magnets. These challenging technical demands make FT-ICR-MS technology very cost-intensive, rendering this design economically less attractive. However, coupled to a MALDI or an ESI source, FT-ICR is the most effective and promising mass spectrometric technology and has undoubtedly become an important research tool in protein analysis.

Ion detectors

With exception of the Orbitrap and the FT-ICR instruments a destructive ion detection is the general approach to register incoming ions from the different mass analyzers. Ions are generally detected by secondary electron multipliers (SEM) or by microchannel plate (MCP) detectors. Usually, the detector enables the mass spectrometer to generate an analog signal, by producing secondary electrons, which are further amplified. The analog signal from the detector is finally digitized and processed by a computer. Several additional designs and applications for ion detection are in use, e.g., photon-sensitive detectors [15] but are beyond the scope of this review.

Analysis of proteins and peptides by mass spectrometry

In this section, the most widely used modern mass spectrometry techniques for identification of proteins and peptides will be described. At present, the typical approach for analyzing proteins is to gather protein spots from 2-D gels, to convert them into peptides, obtain sequence tags of the peptides, and then identify the corresponding proteins from matching sequences in a database. The procedure for a successful protein identification is thereby arranged in a hierarchy of methods depending on the degree of protein sample complexity.

General analytical considerations

Peptide mass fingerprinting (PMF) is the fastest method for identifying proteins recovered from 2D-PAGE or other samples containing only one or two proteins making sophisticated upstream protein fractionation workflows necessary. A detailed description of the sample treatment prior to mass spectrometric analysis is given in the next section. The MALDI-MS analysis and the appropriate database search can easily be done within a few minutes per sample (Fig. 3). More time consuming is the tandem-MS approach, which is often required in case of an unsuccessful PMF analysis since it provides information about the peptide structure which can be used to infer the amino acid sequence. These types of analyses are normally performed with mass spectrometers coupled to nano-HPLC and takes up to 2 h per sample although MS/MS analyses with a static spray are possible. This approach has currently become the standard protein identification method and yields a much higher identification rate compared to the PMF-approach. A brief comparison of these two mass spectrometric methods is given in Figure 4. For completely unknown proteins more labor- and time-intensive procedures are applied, e.g., *de novo* sequencing which can take between several hours and one day and Edman degradation with its high sample consumption and long analysis times. Among the approaches mentioned above, the classical Edman degradation approach is the slowest but the only fully database-independent method.

Peptide Mass Fingerprinting (PMF)-identification

The mass spectrometric analysis of in-gel digested proteins can be done easily by the peptide mass fingerprinting (PMF) approach [16–18]. The general strategy of

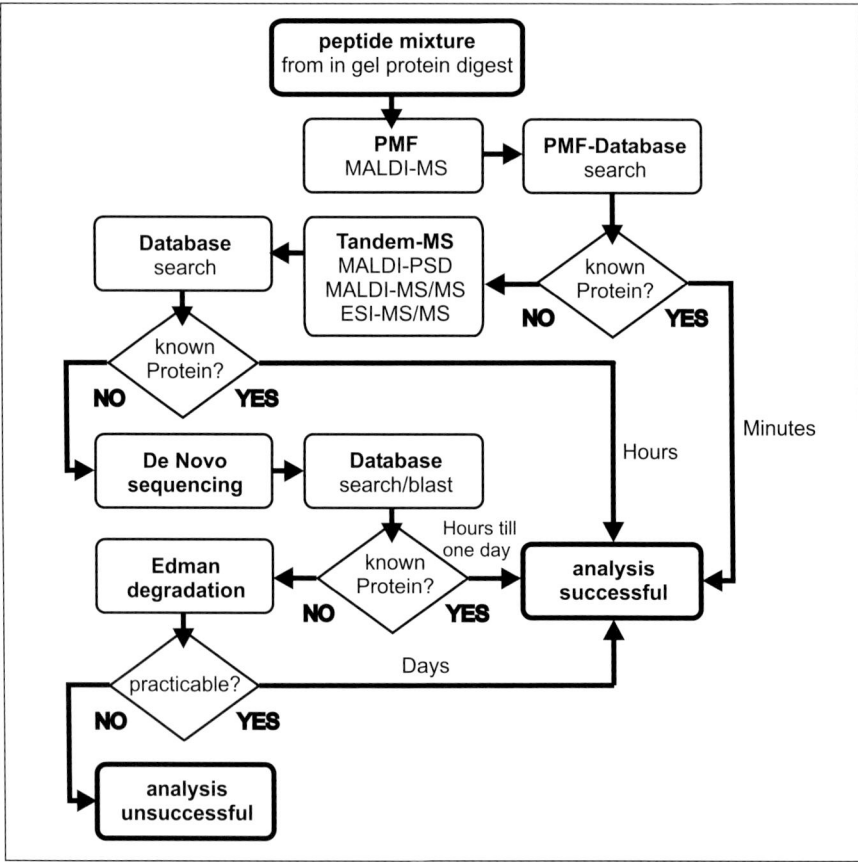

Figure 3. Workflow for a protein analysis strategy after in gel digestion. The fastest method for protein identification is the peptide mass fingerprinting (PMF) approach. Mass spectrum recording and database search can be done in less than 2 min. The analysis of tandem mass spectra takes between a few minutes and few hours. More time consuming is the full de novo sequencing approach, which can take between several hours and some days, which is comparable with the database-independent but sample and time consuming Edman degradation method.

PMF comprises the digestion of a protein by a protease with high selectivity for specific residues and a high reactivity for cleavage to give a maximum peptide yield. Trypsin, which cleaves proteins selectively at lysine and arginine residues except those adjacent to proline, meets this requirement and is therefore the most widely used protease in protein mass spectrometry analysis. After gel electrophoresis and mass spectrometry-compatible staining (such as all Coomassie-based methods, SyproRuby (see also the previous chapter) and some silver stain protocols that do not use crosslinking reagents), the protein spots are excised, washed, and digested. Since every protein digest gives rise to a unique set of peptides after cleavage with a specific protease, the identification can be performed by the comparison of the

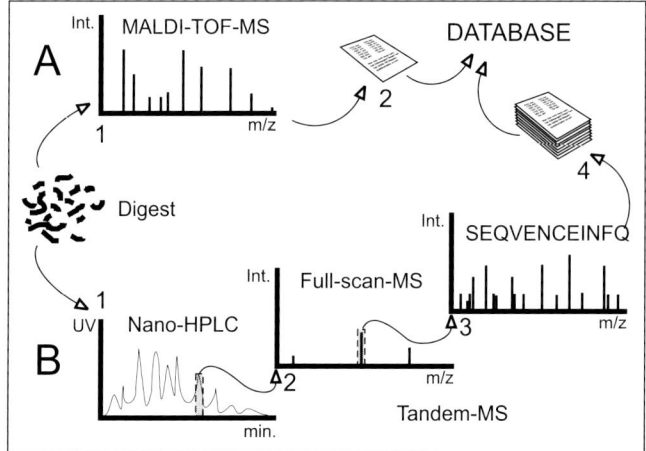

Figure 4. Comparison of data generation between simple PMF-approach (A) and the nano-HPLC-tandem-MS-approach (B). Starting point for each approach is a tryptic protein digest. In case of (A), the digest is directly analyzed by MALDI-TOF-MS (1) generating a simple peak list of the detected peptides (2). The peptide masses are then aligned with theoretical digest lists from each known protein in the database (*in silico* digestion). In case of tandem-MS combined with HPLC separation (B 1), much more data with higher peptide detection sensitivity can be generated from the digest. A tandem-MS capable instrument scans continuously through the HPLC-run (offline MALDI or online ESI) generating first a full scan spectrum for each duty circle (2). From here, all applicable peptides can be automatically selected and subsequently fragmented by user-defined routine. The MS/MS-spectrum reveals in many cases sequence information from the selected peptide (3). Thus, thousands of MS-spectra can be generated from a single HPLC-run (4) and yield a much higher reliability in database identification than the PMF approach (A).

measured peptide masses with calculated (and predicted considering the known protease cleavage site) peptide masses from database entries. In principle, any mass spectrometer can be used for determining the peptide masses. However, highly accurate mass measurements significantly increase the reliability of the database matches. Most of the MALDI-TOF instruments equipped with delayed extraction and reflectron analyzers are capable of this type of approach.

Unfortunately, several factors complicate the peptide mass fingerprinting approach. Important limiting factors are sample losses by inappropriate handling, incomplete digestion of low-abundance or hydrophobic proteins, multiple proteins in one gel spot, and the presence of contaminants (e.g., detergents, salts, human keratin). These factors are critical when analyzing protein amounts in the lower fmol range. All protein modifications such as, e.g., glycosylation or phosphorylation also complicate the PMF-approach. In such cases, the best strategy is the chemical or enzymatic removal of these modifications provided that they can be predicted. In the course of automation for high throughput proteomics, the PMF approach is very applicable, since hundreds of protein identifications can be performed per day. Unfortunately, protein digestion and the handling of low fmol amounts of protein are not routinely possible, despite the fact that modern MS instruments are reaching attomol sensitivities. Nevertheless, peptide mass fingerprinting remains primarily a protein identification technique based on the comparison of measured and calculated peptide masses. Even with up-to-date databases and improved search algorithms, scoring dependent identification will remain unsatisfactory for highly homologous proteins and for the investigation of PTMs. Other more sophisticated mass spectrometry technologies such as tandem MS are therefore required. One such promising method is the Accurate Mass and Time (AMT)-tag approach for whole protein characterization based on the analysis of low level tryptic peptides by LC-FT-ICR-MS [19, 20].

Peptide fragmentation identification

In contrast to PMF, the peptide fragment identification approach yields direct sequence information. This technique not only measures the mass of the tryptic peptides, but it also provides sequence information of the peptide fragments generated by CID and measured by tandem MS. This analytical step provides amino acid sequence tags that dramatically enhance the success rate of protein identification by database searches. However, using sequence tags for the identification of peptides and the respective proteins is frequently confused with *de novo* sequencing (next section). The identification of proteins is usually performed by searching within protein or expressed sequence tag (EST) databases using various search algorithms such as SEQUEST^{TM} [21], Mascot^{TM} [22], ProFound^{TM} [23], Phenyx [24, 25], etc.

The implementation of tandem mass spectrometry (MS/MS) has pushed the boundary of mass spectrometric peptide analysis considerably both in terms of sensitivity and information content. The coupling of two mass analyzers in combination with a collision cell has enabled the direct determination of sequence information from peptides. The first mass analyzer serves to select the target peptide for introduction into the collision cell. Here, the ions undergo multiple collisions with inert gas atoms (such as nitrogen or argon), whose kinetic energy is converted into vibrational energy, which is sufficient to cleave a single amide-backbone bond within a peptide. The second mass analyzer simply records the resulting fragment ions. Figure 5 shows ions produced by low energy collision induced dissociation (CID); this is the mode generally used by triple quadrupole, Q-TOF or ion trap instruments. The different types of positively charged fragment ions are assigned according to a generally accepted nomenclature [26, 27]. The resulting ions are called b-type fragment ions when the *N*-terminus is included (fragmentation from the C-terminus) and y-type fragment ions when the *C*-terminus is included (fragmentation form the N-terminus). In fragment ion spectra, b- and y-type fragment ion signals are commonly but not necessarily the most intense signals. Corresponding b- and y-type fragment ions can be obtained by calculating the mass differences between distinct fragment ions and the precursor ion signal. Furthermore, the amino acid sequence can be calculated by the mass differences within each ion series. The fragmentation pattern is highly dependent on the amino acid sequence of the peptide, the collisional energy and the number of charges carried by the peptide. Furthermore, the information content of a fragment spectrum depends, to a certain degree, on the instrument set up used to obtain the spectra. ESI and MALDI, quadrupole, ion trap and TOF analyzers are all complementary techniques. Easy to interpret spectra are produced by ESI-Q-TOF MALDI-TOF/TOF instruments, since they are capable of generating high resolution and high mass accuracy full range fragment ion spectra allowing, e.g., for de-isotoping. The mass accuracy of Q-TOF instruments allows the differentiation of glutamine and lysine solely on the basis of the mass difference (Q = 128.06 amu and K = 128.09 amu), whereas the isobaric amino acids isoleucine and leucine (both 113.08 amu) cannot be distinguished under low collision energy regimes. Another advantage of both types of instruments are the accessibility of the low mass range (m/z < 200), which allows the acquisition of immonium ion data for amino acid composition analysis of peptides (e.g., the marker ions 110 amu – His, 120 amu – Phe, 136 amu – Tyr, 159 amu – Trp, 175 amu – C-terminal Arg). The best fragment ion spectra can be obtained from doubly or triply charged peptides since all resulting fragment ions also remain charged and can be detected. The generated fragment ion spectra may increase in complexity because doubly charged ions are able to form singly and doubly charged fragment ions (depending where the charged residue is localized in the peptide). Triply charged ions are even capable of forming triply charged fragment ions. However, only one fragment ion can be detected from singly charged peptides and the remaining fragment is uncharged (neutral) and therefore does not respond to electric fields and is not detected (neutral loss). Since MALDI in most cases generates singly-charged ions [8], its fragmentation spectra only exist of singly charged ions which reduce the complexity of product ion spectra and yield sufficient information to determine peptide sequences. The distribution and transfer of charge plays an essential role in the peptide dissociation process and thus the relative amounts of detectable fragment ion types. Histidine, tryptophan, arginine, and lysine, which have a high basicity in the gas phase, can easily attract protons, yielding relatively high fragment ion intensities. Spectra of abnormally fragmenting peptides like proline- or histidine-rich sequences are more difficult to interpret because of an increased number of internal fragment (neither containing the N- nor the C-terminus of the precursor) ion signals or the partial deletion of serial ion signals. However, for the data interpretation it is mandatory that all intense ions are accounted for; otherwise the result can be deceptive. Another method of peptide identification is the fragmentation of isolated peptide ions by post source decay (PSD) using a common MALDI-TOF or MALDI-TOF/TOF instrument [28]. Generally the PSD fragmentation pattern favors backbone cleavages producing predominantly a-, b-, and y-type fragment ions with very little side-chain specific cleavages. The a-ions are formed by the loss of CO from corresponding b-ions explaining why x-ions cannot be detected in CID spectra (Fig. 5). A further development of MALDI-PSD is the combination with MALDI LIFT-TOF-TOF technology [29] and an additional collision cell that can also induce high energy collisions in the kiloelectron volt range. Compared to ESI instruments, the higher-energy fragments that result from side-chain cleavages, such as d-, w-

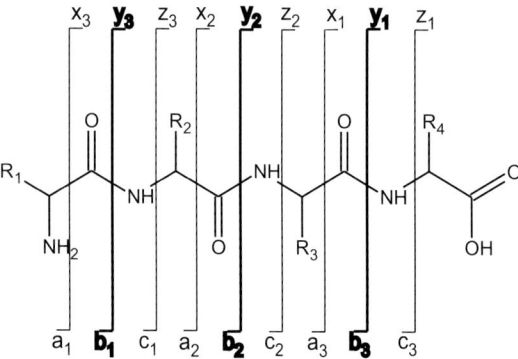

Figure 5. The generally accepted nomenclature for peptide fragment ions. N-terminal fragment ions are classified as either a, b or c and the C-terminal ions are labeled either x, y or z. A subscript represents the number of residues in the respective fragment. The y- and b- type series (bold) are the most prominent signals in low-energy collision induced dissociation of tryptic peptides.

and v-type fragment ions [95] are also observable, along with strong signals from immonium ions when gas is added to the collision cell. High-energy product ions such as w- and d-ions can be used to differentiate between the isobaric amino acids leucine and isoleucine.

The three-dimensional ion trap, with its high sensitivity, rapid duty cycles, ability to perform MS^n and excellent fragmentation efficiency has become a standard instrument in peptide fragment analysis. The greatest advantage of the ion trap instrument lies in its ability to retain the fragment ions after MS/MS so that a fragment ion can be selected for further MS/MS analysis (MS^n). In contrast, 3D ion trap mass spectrometers are limited to the low mass range region and exhibit lower mass accuracy and resolution due to the 'space charging' phenomena, even though improvements such as linear ion trap (LIT) are in implementation [30, 31]. Current linear ion trap instruments will typically produce fragment-ion mass accuracies of better than ± 0.3 amu, and the fragment ion range is presently no more limited. More highly developed and modified mass spectrometers such as FT-ICR or Orbitrap instruments guarantee high resolution mass spectra and the implementation of modern fragmentation techniques. Unfortunately, FT-ICR instruments only operate at very high vacuum which is in conflict with the commonly used CID fragmentation technique that uses gas. Consequently, alternative fragmentation techniques have been employed, such as infrared laser multiphoton dissociation (IRMPD) or electron capture dissociation (ECD) [32, 33]. The use of ECD with FT-ICR-MS instruments not only results in different fragmentation patterns but is also advantageous for analyzing protein modifications. Another advantage of FT-ICR-MS is that it enables 'top-down' protein characterization, in which the intact protein is fragmented directly in the mass spectrometer. MS/MS of intact proteins electrosprayed into the mass spectrometer has already been demonstrated [34]. High mass accuracy, resolution and the ability of FT-ICR-MS instruments to perform MS^2

experiments, allows us to make sense of the complex fragmentation patterns generated from intact proteins. Although there are continuous developments such as coupling TOF-TOF and Q-TOF technology with MALDI, most of the MS/MS approaches in proteomics are still performed using ESI in combination with ion trap or Q-TOF instruments. However, MS/MS spectra of peptides can only provide partial sequence information of a protein. Another shortcoming of sequence tag -protein identification by database search is the existence of protein modifications that are unknown or not included in the search algorithm used. Furthermore, the n search parameters (e.g., mass tolerance, size of the database) have a major impact on the search results and must be carefully adjusted to the experimental requirements.

De novo sequencing

De novo sequencing [35] is often presented as an alternative to the methods described above. However, *de novo* sequencing requires almost full sequence coverage of a peptide and is based mainly on the manual or computer aided interpretation of a-, b- and y-type fragment ion series from peptide tandem mass spectra. After the sequencing of individual peptides it is necessary to assemble the sequence information and reconstruct the whole protein sequence. Therefore, three or more different proteases, e.g., trypsin, chymotrypsin or Glu-C (see Tab. 1 for specificity of proteases), are often used independently for digestion to generate overlapping peptides. The overlapping peptide sequences may be aligned and thereby combined into longer sequences or even the entire protein sequence.

Table 1. Overview: Proteases used in protein analysis

Endopeptidase	Type	Specificity	pH range	Inhibitors
Chymotrypsin	Serine	Y, F, W	1.5–8.5	Aprotinin, DFP, PMSF
Trypsin	Serine	R, K	7.5–9.0	TLCK, DFP, PMSF
Glu C	Serine	D, E	7.5–8.5	DFP
Lys C	Serine	R	7.5–8.5	DFP, Aprotinin, Leupeptin
Arg C	Cysteine	R	7.5–8.5	EDTA, Citrate
Asp N	Metallo	D (N-terminal)	6.0–8.0	EDTA
Elastase	Serine	A, V, I, L, G	8.0–9.0	DFP, α1-Antitrypsin, PMFS
Pepsin	Acidic	F, M, L, W	2.0–4.0	Pepstatin
Papaine	Cysteine	R, K, G, H, Y	7.0–9.0	IAA, TLCK, TPCK
Proteinase K	Serine	Hydrophobic AA	7.0	IAA
Thrombin	Serine	R	7.5	DFP, TLCK, TPCK

DFP = Diisopropylfluorophosphate; PMSF Phenylmethylsulfonylfluoride; IAA = Iodoacetamide; TPCK = L-1-chloro-3-(4-tosylamido)-4-phenyl-2-butanone; TLCK = L-1-chloro-3-(4-tosylamido)-7-amino-2-heptanonhydrochloride

To make *de novo* MS/MS spectra interpretation easier, peptides can be modified by chemical means. This will help distinguishing the different ion types (mainly y- and b-ions), since they have different chemical properties. Most of the methods influence the intensity of b- or y-type fragment ions by adding negative (knock out intensity) or positive (increased intensity) charges to the various functional groups of the peptides [36–40]. Another method of simplifying fragment spectra interpretation involves the labeling of a particular ion series by introduction of stable isotopes into the peptide [41, 42]. A very easy and reliable isotopic technique uses tryptic digestion in 50% ^{18}O-labeled water to identify y-type fragment ions by modifying the C-terminus with a 2 amu shift which results in mass spectral doublets [43]. Occasionally, highly sophisticated technologies may increase the success of *de novo* sequencing. For some proteins 100% sequence coverage was achieved by using, e.g., infrared-multiphoton dissociation (IRMPD) in combination with FT-ICR-MS [44]. Moreover, the amount of mass spectrometric data generated by such experiments is constrained by the manual interpretation and validation which is necessary to infer an amino acid sequence from an MS/MS spectrum *de novo*.

Current state of instrumentation in proteome analysis by mass spectrometry

Even though the mass spectrometric instrumentation for protein analysis is improving at an amazing pace, no single instrument presently fulfils all the requirements for high-throughput proteome research in a systems biology context. In fact a formidable number of specialized instruments exist. The combination of MALDI or ESI sources with the different types of mass analyzer described above increases the total number of mass spectrometers available to date. Since different mass spectrometers have different strength and weaknesses, deep understanding of their functional principles is necessary to decide which technique to use for a specific biological question. We provide here a broad overview of the commonly used mass spectrometer types in proteome research to help elucidating optimal solutions.

Commonly used mass spectrometers in proteome analysis

The easiest and most effective approach is the direct analysis of the proteolytic digest with a MALDI-TOF mass spectrometer. However, the tandem or multiple stage mass spectrometric analysis combined with CID is steadily replacing PMF, though the quality of the generated fragment spectra varies considerably with the various modular instruments available. A typical tandem MS instrument is the triple stage quadrupole, which can perform fragmentation analysis of sufficient quality. Due to the demand for more mass accuracy and resolution the Q-TOF was developed, in which second mass analyzer is an orthogonal acceleration TOF-analyzer [45]. The relatively slow scanning rate of the Q-TOF instruments remains a problem especially when running precursor ion scans (PIS) or neutral loss scans (NLS). The low-resolution, mostly ESI coupled, ion trap mass spectrometers are very popular

Table 2. Survey of commonly used MS-configurations in current proteome analysis

Ionization Source	Mass Analyzer	Tandem MS/Fragmentation type	Resolution*	Mass* Accuracy	Field of Application
MALDI	TOF	restricted/PSD	+	+	PMF, (MS^2) intact proteins
MALDI	TOF	yes/PSD, LID-LIFT	+	+	PMF, MS^2 intact proteins
MALDI	TOF, TOF	yes/LID, CID	+	+	PMF, MS^2
ESI/MALDI	Q, Q, Q (triple quad)	yes/CID	o	+/o	MS^2, LC-MS, PIS, NLS
ESI/AP MALDI	Q, Q, LIT	yes/RE, CID	+/o	+	MS^3, LC-MS, PIS, NLS
MALDI	ion trap, TOF	yes/RE, CID	+	+/o	MS^2, LC-MS
ESI/MALDI	Q, Q, TOF	yes/CID	+	++	MS^2, PIS, NLS
ESI/AP MALDI	ion trap	yes/RE, CID	o	+/o	MS^3; LC-MS
ESI/MALDI	LIT	yes/RE, CID	+	+	MS^n; LC-MS
ESI/MALDI	Orbitrap	yes/CID	++	++	MS^3; LC-MS, non destructive detection
ESI/MALDI	FT-ICR	yes/CID, ECD, IRMPD	+++	+++	MS^3; LC-MS, intact proteins, non destructive detection

AP = atmospheric pressure; TOF = time of flight; Q = quad = quadrupole; LIT = linear ion trap; FT-ICR = Fourier Transform-Ion Cyclotron Resonance; PSD = post source decay; LID = laser induced dissociation; CID = collision induced dissociation; RE = resonance excitation; ECD = electron capture dissociation; IRMPD = infrared-multiphoton dissociation; LC-MS = MS coupled chromatography; n = 3 to 7 in most of cases, for MS^2 = tandem MS;
* based on manufacturer information.

for a number of reasons detailed above. These 'tandem-in-time' instruments normally have extremely fast scan rates which is important, since fast duty circles are critical for high sensitivity especially when combined with increasingly fats and efficient chromatographic separation techniques [46]. During the last few years various hybrid instruments (Tab. 2) have evolved rapidly to meet these changing needs. The ubiquitous space charging problem of standard ion trap cells has recently been solved by the design of the linear ion trap (LIT), which provides much higher resolution and has prepared the ground for a new generation of powerful tandem MS instruments. LIT coupled with FT-ICR is undoubtedly the most powerful mass spectrometer type currently available for protein analysis [47]. Thus, the different mass spectrometer designs employed in protein analysis vary widely in their operation and performance characteristics. A short overview of the most common mass spectrometers is given in Table 2.

Mass spectrometric coupled techniques for increasing sensitivity and specificity

Although the sensitivity of mass analyzers and detectors has reached an impressive level, additional improvements in analytical performance have come from new approaches in sample preparation and separation techniques. High sensitivity is not only a question of sophisticated mass spectrometer design and assembly; it is also affected by the selected combination of high end mass spectrometer devices and progressive sample introduction. Irrespective of which protein separation technique has been applied (e.g., gel electrophoresis or chromatographic separation) in most of the cases a complex mixture of proteins will be analyzed and a complex mixture of peptide introduced into the mass spectrometer. The analysis of a highly purified single protein is usually the exception in a proteome study. From unseparated peptide mixtures, only the most abundant peptides are usually detected since they suppress the detection of low abundance species. Pre-fractionation of complex mixtures, the removal of interfering impurities and sample preconcentration are widely-used techniques for enhancing mass spectrometric sensitivity (e.g., ZipTip^{TM} procedure for manual MALDI-MS sample preparation or trap columns for nano-HPLC separations). Hydrophilic 'anchor' surfaces, positioned onto a MALDI target plate, are also used to obtain higher mass spectrometric sensitivity improved mass accuracy and easier instrument automation [48]. Multiple peptides can be detected more successfully when MS measurements are coupled with HPLC separation. Nowadays, miniaturization of liquid chromatography (nano-HPLC) allows handling sample volumes from a few microliters up to a hundred microliters. In combination with online chromatographic pre-concentration and desalting methods, more than a 100-fold increase in sensitivity can be achieved [49]. It has been demonstrated that nano-HPLC sensitivity increases linearly with decreasing flow rate in the range of 20–400 nL/min [50]. Current nano-HPLC technology can be coupled, either online or offline, to any mass spectrometer. Consequently, nano-HPLC combined with ESI-tandem-MS instruments is nowadays the standard method for peptide identification. For the analogous MALDI-tandem-MS approach offline nano-HPLC must be applied [51]. In this case the eluate and the matrix solution are applied directly on the MALDI target by a spotting robot. The separated sample is thus 'stored' on the target and multiple MS analyses are possible over a longer period of time. Furthermore, modern MALDI-TOF-TOF mass spectrometers are capable of fully automated repeatable data acquisition. This is advantageous, since the nano-HPLC-ESI-tandem-MS method, in contrast to the offline-technique, only allows one MS-experiment per sample. The high level of automation for both processes greatly improves the overall speed and the accuracy of proteome analyses. Automation is inevitable, because the amount of data recorded by such continuous scanning mass spectrometric analysis techniques is beyond the scope of a manual data interpretation. Noteworthy is the observation that MALDI and ESI MS analysis coupled with nano-HPLC from identical samples, yield largely complementary results for protein identification [52]. This is a generally recognized problem for the use of fundamentally different analytical techniques and must therefore be taken into consideration for data interpretation.

Analysis of Posttranslational Modifications (PTMs) by mass spectrometry

The deduction of the primary amino acid sequence from a protein is a completely different task compared to mapping posttranslational protein modifications. The latter ideally requires high sequence coverage, if no modification is to be missed. As mentioned above, this cannot be performed routinely with existing mass spectrometric techniques and for the most part has to be done manually. In the following, we will discuss the analysis of protein phosphorylation, which is the most frequently occurring posttranslational modification in cellular signaling. At the same time, phosphoprotein analyses illustrate that a combination of different technical adaptations at different levels (upstream sample fractionation and mass analyzer set up) must be employed for an optimal solution to a specialized biological question.

Phosphorylation analysis

The variety of functions, in which phosphoproteins are involved, necessitates a huge diversity of phosphorylated protein species [53]. Fortunately, only a limited number of amino acids can be phosphorylated by protein kinases, O-phosphates attached to serine-, threonine-, and tyrosine-residues being the most common class [54]. Additionally, the unusual amino acid hydroxy-proline can also be O-phosphorylated. Further, relatively rare phosphorylation sites can be found on histidine and lysine (N-phosphates), cysteine (S-phosphates) as well as apartic and glutamic acid residues (acyl-phosphates). Presently, phosphorylation sites of phosphoproteins are usually identified by mass spectrometry.

Detection and enrichment of phosphoproteins and phosophopeptides

The mass spectrometric phosphoproteins and phosphopeptides signal intensities are often suppressed in comparison to the unphosphorylated species since they have unfavorable chemical characteristics for ionization. Thus, an appropriate method for the isolation or enrichment of phosphopeptide samples is advantageous before performing mass spectrometry analysis. After lysis of cells, all phosphatases and proteases are released and may cause a loss of phosphorylation sites. This can be suppressed by the addition of phosphatase inhibitors to all buffer solutions and by working at low temperatures (mostly at 4°C) during sample preparation. The stabilities of different phosphorylation sites in distinct buffer systems are well characterized and the analysis procedure should be adapted to this [54]. N-phosphates are labile at low pH-values while O-phosphates are stable to acidic conditions. Phosphorylations that are unstable in all buffer systems must be analyzed indirectly, generally using less sensitive techniques [55–57]. The most frequently applied technique for phosphopeptide and phosphoprotein enrichment is immobilized metal-ion chromatography (IMAC) [58] which was originally introduced by Porath et al. for the purification of His-tagged proteins [59]. The sustained success of the IMAC-technology in phosphoproteomics is based on its compatibility with further separation and detection techniques such as capillary electrophoresis [60], LC-MS/MS [61, 62] and target bonded MALDI-MS. Another method for enrichment is the specific binding of organic phosphates to TiO_2-columns under acidic conditions [63, 64], after which elution is accomplished at an alkaline pH. Further methods for the enrichment of phosphoproteins and phosphopeptides use chemical modification by labeling or derivatization in combination with respective HPLC purification techniques [65–69]. Despite these effective enrichment technologies, the problem of mass spectrometric identification remains, due to the frequently observed low level of a particular phosphoprotein compared to the unphosphorylated species [70, 71].

Identification and localization of phosphorylated amino acid residues

The frequently applied 'bottom-up' phosphorylation analysis of proteolytically digested samples generally yields a peptide mixture containing both phosphorylated and unphosphorylated peptides. A widely used method for the identification of phosphorylated peptides is the comparison of ESI MS spectra before and after alkaline phosphatase treatment, which gives rise to a -80 amu (-HPO_3) shift of the phosphopeptides. The phosphorylated peptides can be further analyzed by MS/MS experiments for localization of the phosphorylated residue. Under MALDI-TOF-MS conditions, serine- and threonine-phosphorylated peptides tend to lose phosphorous acid (H_2PO_3) and phosphoric acid (H_3PO_4) due to metastable decay, while phosphotyrosine residues remain intact. For identification of the phosphorylated peptides, the sample is first measured in linear-mode, where only the intact phosphopeptides can be observed. In reflector-mode, a decrease in the intensity of these phosphopeptide signals occurs and the usually low resolution signals due to the loss of -80 amu and 98 amu appear. Generally, a major drawback of mass spectrometric phosphopeptide analysis is the decreased ionization rates due to suppression effects. Phosphorylated residues only maintain a negative charge if the pH is not less than 1.5, which is not favorable while operating in the standard positive ion mode used for detecting peptides. The best way to reduce phosphopeptide suppression effects is to operate in the less sensitive negative ion mode for a full MS spectrum or to reduce the sample complexity by HPLC techniques, as mentioned above for generating MS/MS spectra. In the case of MALDI-MS, suppression effects can be partly circumvented by the use of 2',4',6'-trihydroxyacetophenone with di-ammonium citrate, a UV-sensitive matrix, resulting in a higher signal intensity for most of the phosphopeptides [72]. A similar effect can be achieved by the use of phosphoric acid as a DHB matrix additive for MALDI-MS-derived phosphorylation analysis [73, 74]. With these methods, the fragmentation patterns of peptides remain unaffected under MALDI conditions and the phosphorylation sites can be determined by a conventional PSD-experiment or by a TOF/TOF-analyzer capable of tandem-MS. Triple quadrupole and Q-TOF instruments offer the opportunity to perform precursor ion scanning (PIS) and neutral loss scanning (NLS), which, though time-consuming, are useful mass spectrometric tools for the identification and localization of phosphorylated residues. PIS is particularly useful, when stable tyrosine phosphorylation is being investigated. The first quadrupole analyzer is therefore used as a mass filter, scanning repeatedly through the entire mass range. The second quadrupole commonly serves as a collision cell in which the passing peptides are fragmented. The latter mass analyzer (quadrupole or TOF) is used for monitoring the specific fragment ion that is characteristic for the residue of interest. In the case of phosphotyrosine, the immonium ion at 216.043 amu is indicative for this type of phosphorylation. Sufficient resolution and mass accuracy is indispensable for correct determination of this immonium ion, since several dipeptides with almost identical masses exist [75]. PIS for a phosphate residue (79 amu) can be conducted in full MS negative ion mode [76], whereas for MS/MS-spectra the polarity has to be switched in order to obtain better fragmentation signals in positive ion mode. Switching the polarity during the experiment exceeds the capabilities (if actually applicable) of currently available MS instruments, resulting in decreased scanning rates. For serine- and threonine-phosphorylations a simple selective derivatization technique permits the use of PIS in positive ion mode, abolishing the necessity for polarity switching. This can be done by alkaline β-elimination of the phosphate moieties and subsequent Michael-type addition of 2-dimethylaminoethanethiol, which is followed by oxidation. Low energy CID reveals 2-dimethylaminoethanesulfoxide at 122.06 amu, which can be selected for PIS in positive ion mode [77]. Unfortunately, alkaline racemization of peptide bonds leads to strongly reduced trypsin cleavage rates and peak broadening during RP-HPLC due to the separability of the obtained diastereomers and incomplete derivatization.

In a neutral-loss scan (NLS), the first quadrupole and the third mass analyzer are scanned at the same rate with an offset of 98 amu for loss of the phosphoric acid (H_3PO_4). Both PIS and NLS are highly sensitive phosphorylation detection methods, but are only applicable on instruments with sufficient resolution and a fast scanning rate as is the case in LC-MS/MS coupling. The same can be done with a regular ion trap instrument using ion/ion-reactions for reduction of the charge state and subsequent MS^n-experiments [78]. However, FT-ICR is presently the only mass spectrometric technique capable of electron capture dissociation (ECD) [79], although this technique could be applied to any ion trap analyzer [80]. ECD fragmentation is suitable for the analysis of protein modifications that are usually labile in MS/MS-experiments. Thus, modifications such as phosphorylation or glycosylation remain intact in ECD experiments, while the peptide backbone is cleaved upon electron capture yielding c-, and z-type fragment ions [33] rather than the b- and y-type fragment ions produced by CID and PSD. Recently, electron transfer dissociation (ETD) was established as an alternative to the ECD-fragmentation by using a modified linear ion trap yielding fragmentation patterns similar to ECD [81]. In this process, electrons are transferred to the protein- or peptide-ions from anions generated by a chemical ionization source containing methane buffer gas. The possibility of using all these new mass spectrometric techniques in modified ion trap analyzers will certainly improve the analysis of all posttranslational modifications in the near future.

Mass spectrometric data handling and interpretation

The amount of data generated by mass spectrometric analysis depends on the analytical method and the objectives of the study. For single protein analysis the acquired data are generally manageable and in most cases can be evaluated manually, even when posttranslational modifications are taken into account. The manual approach is normally supported by software that is usually provided with the MS instrument. Although the final data interpretation is user dependent, the results are mostly comprehensible. This is obviously not true for the analysis of complex protein or peptide mixtures in a high throughput environment. The acquisition of tens of thousands of spectra per day makes manual methods inadequate for analysis. In the case of continuous data uptake and for automated data interpretation, specialized software tools frequently with complicated algorithms come into use. However, before database search engines are involved, the raw data must normally be converted in a MS-device independent data format (e.g., dta, mgf, xml, etc.). This is a difficult task, since the quality and the complexity of the mass spectrometric data vary considerably with the used instrument type. The following database search can alternatively be performed using a free access web-based or in-house licensed database, offered by several providers. For complex proteome studies there is a hierarchy of protein identification techniques based on peptide analysis (see sections on peptide mass fingerprinting, peptide fragment identification, and *de novo* sequencing). The sophistication of these mass spectrometric techniques and the respective data handling complexity increases in the order presented above.

The first PMF approach developed relies upon a comparison of the experimentally determined mass values with the predicted molecular mass values of the peptides, generated by a theoretical digestion of each protein in a database. Since the protein databases have grown steadily larger with an inevitable increase in redundancy, each dataset must be compared with a growing number of candidates. Consequently, the criteria for PMF analysis have become more stringent and a more precise mass assignment is necessary. Furthermore, an increasing number of matched peptides for higher sequence coverage is advantageous, which can be achieved by better sample preparation, by higher performance MS, and by more sophisticated MS interpretation algorithms [82, 83]. However, the larger the databases, the greater the likelihood of false positive results. Particularly for the PMF approach the limitations at this stage are apparent, explaining why this method is steadily being replaced by more reliable protein identification techniques. Peptide sequence determination using tandem MS is now becoming the accepted standard for protein identification. Data obtained this way are much more complex and require highly developed software for handling, especially when multidimensional peptide separation techniques are coupled with tandem mass spectrometry [84, 85]. All these considerations also apply to the subsequent database search. Currently such database searches are based on comparisons between the experimentally recorded fragment ions and all predicted fragments for all potential peptides in the database with the corresponding molecular weight. The computation of these potential fragment ions is based on known fragmentation rules [86]. The matching of multiple peptide sequences for higher sequence coverage is the goal of these calculations. High sequence coverage of the matched proteins provides greater statistical confidence in the result obtained. Error tolerant and remote sequence homology searching are additional parameters included in more powerful search algorithms, although these are time-consuming and computationally intensive [87]. Moreover, the multitude of different types of mass spectrometers available complicates the analysis of the results considerably. The algorithms must differentiate between the different charge states of the fragmented precursor ions of MALDI or ESI generated spectra. Furthermore, the different types of mass analyzer influence the data quality as well. Higher performance triple quad, Q-TOF or TOF-TOF instruments provide more accurate tandem MS data than the low-resolution but very sensitive ion trap instruments. However, the data extraction algorithm and the search engines must be accompanied by steady optimization of the data processing. Bioinformatics has therefore already taken a key position in mass spectrometric protein identification techniques.

The most sophisticated MS interpretation algorithm is needed, if no matches are found in the protein database. This indicates that the protein being sought is possibly not present in the database and therefore *de novo* peptide sequencing based on known rules for peptide fragmentation must be applied. This approach requires good quality MS/MS spectra, an accessible genome database, and a *de novo* sequencing algorithm for the interpretation of MALDI- [88] and ESI-generated [89–92] PSD- and CID-spectra. Without genome databases the *de novo* sequencing approach works adequately on a peptide level. For a completely unknown protein, manual intervention for data interpretation or even Edman degradation for sequence determination may still be required. Also, for the database search approach there may be uncertainty as to the choice of search engine and search parameters [93].

Data computation as a whole is a turbulent and rapidly developing area in mass spectrometry, which makes it difficult to establish generally accepted standards. Despite these developments, the widely accepted truth still remains: For any computer-generated protein match returned from a database, the probability of a false positive result cannot be excluded with certainty.

Concluding remarks

Protein identification by mass spectrometry is presently the most powerful tool in proteome research in a systems biology context. The variety and constant development of mass spectrometric techniques guarantees further improvements in protein identification performance and broadens the scope of proteomics analyses in general. Currently, mass spectrometric protein analysis is in a very dynamic state, making it difficult to establish long-needed standards for generally accepted procedures. Consequently, a direct comparison of different instruments is not meaningful. An understanding of the basic function of the different mass spectrometric designs discussed above is essential to design efficient strategies for protein detection or quantification. However, it is presently not foreseeable which design will become widely accepted. Although, FT-ICR-MS with its recent refinements is now the most promising of the current MS platforms, other developments should be kept in view. These include, among others, the new ion trap designs, including linear ion trap and Orbitrap, which now form the basis of a new generation of powerful tandem mass spectrometers with unsurpassed sensitivity. Such mass analyzers may provide a space-saving and less costly alternative to FT-ICR-MS systems. It is therefore more likely that several different MS technologies will continue to coexist. The single mass analyzer design, incapable of 'real' tandem MS, alone is threatened with extinction in proteome analysis.

Perhaps most challenging of all is the need for increased sample throughput connected with high performance MS. The use of automated multidimensional peptide separation techniques together with isotope tagging methods should provide mass spectrometry with a high throughput platform that promises sufficient analytical depth for proteome analyses. Furthermore, the insertion of HPLC-based peptide fractionation prior to the tandem MS techniques has made it possible to detect low abundance proteins and to compare changes in protein expression. As has already happened in genomics, increased automation of sample handling, mass spectrometric analysis, and the interpretation of MS spectra are generating a flood of qualitative and quantitative proteome data. It is becoming more and more apparent, that the high-performance computation of recorded MS data is the main bottleneck in mass spectrometric protein identification (see also Chapter by Ahrens et al.).

Although the mass spectrometric interpretation algorithms currently in use can clearly produce good results, nearly all MS protein information is based on the characterization of short peptide sequences. The demand for higher sample throughput in proteomics makes a manual and time-consuming user intervention more and more impractical, leading inevitably to an unknown number of false positive results. Thus, the elaboration of generally accepted minimum requirements for the publishing of mass spectrometric protein identification has become indispensable.

References

1. Karas M, Hillenkamp F (1988) Laser desorption ionization of proteins with molecular masses exceeding 10,000 daltons. *Anal Chem* 60: 2299–2301
2. Fenn JB, Mann M, Meng CK, Wong SF, Whitehouse CM (1989) Electrospray ionization for mass spectrometry of large biomolecules. *Science* 246: 64–71
3. Sundqvist B, Kamensky I, Hakansson P, Kjellberg J, Salehpour M, Widdiyasekera S, Fohlman J, Peterson PA, Roepstorff P (1984) Californium-252 plasma desorption time of flight mass spectroscopy of proteins. *Biomed Mass Spectrom* 11: 242–257
4. Barber M, Green BN (1987) The analysis of small proteins in the molecular weight range 10–24 kDa by magnetic sector mass spectrometry. *Rapid Commun Mass Spectrom* 1: 80–83
5. Karas M, Kruger R (2003) Ion formation in MALDI: the cluster ionization mechanism. *Chem Rev* 103: 427–440
6. Takach EJ, Hines WM, Patterson DH, Juhasz P, Falick AM, Vestal ML, Martin SA (1997) Accurate mass measurements using MALDI-TOF with delayed extraction. *J Protein Chem* 16: 363–369
7. Bahr U, Stahl-Zeng J, Gleitsmann E, Karas M (1997) Delayed extraction time-of-flight MALDI mass spectrometry of proteins above 25,000 Da. *J Mass Spectrom* 32: 1111–1116
8. Karas M, Gluckmann M, Schafer J (2000) Ionization in matrix-assisted laser desorption/ionization: singly charged molecular ions are the lucky survivors. *J Mass Spectrom* 35: 1–12
9. Wilm M, Mann M (1996) Analytical properties of the nanoelectrospray ion source. *Anal Chem* 68: 1–8
10. Iavarone AT, Jurchen JC, Williams ER (2000) Effects of solvent on the maximum charge state and charge state distribution of protein ions produced by electrospray ionization. *J Am Soc Mass Spectrom* 11: 976–985
11. Iavarone AT, Jurchen JC, Williams ER (2001) Supercharged protein and peptide ions formed by electrospray ionization. *Anal Chem* 73: 1455–1460
12. Makarov A (2000) Electrostatic axially harmonic orbital trapping: a high-performance technique of mass analysis. *Anal Chem* 72: 1156–1162
13. Hu Q, Noll RJ, Li H, Makarov A, Hardman M, Graham CR (2005) The Orbitrap: a new mass spectrometer. *J Mass Spectrom* 40: 430–443
14. Hardman M, Makarov AA (2003) Interfacing the orbitrap mass analyzer to an electrospray ion source. *Anal Chem* 75: 1699–1705
15. Peng WP, Cai Y, Chang HC (2004) Optical detection methods for mass spectrometry of macroions. *Mass Spectrom Rev* 23: 443–465
16. Mann M, Hojrup P, Roepstorff P (1993) Use of mass spectrometric molecular weight information to identify proteins in sequence databases. *Biol Mass Spectrom* 22: 338–345
17. James P, Quadroni M, Carafoli E, Gonnet G (1993) Protein identification by mass profile fingerprinting. *Biochem Biophys Res Commun* 195: 58–64
18. Pappin DJ, Hojrup P, Bleasby AJ (1993) Rapid identification of proteins by peptide-mass fingerprinting. *Curr Biol* 3: 327–332
19. Lipton MS, Pasa-Tolic L, Anderson GA, Anderson DJ, Auberry DL, Battista JR, Daly MJ, Fredrickson J, Hixson KK, Kostandarithes H et al. (2002) Global analysis of the *Deinococcus radiodurans* proteome by using accurate mass tags. *Proc Natl Acad Sci USA* 99: 11049–11054

20. Strittmatter EF, Ferguson PL, Tang K, Smith RD (2003) Proteome analyses using accurate mass and elution time peptide tags with capillary LC time-of-flight mass spectrometry. *J Am Soc Mass Spectrom* 14: 980–991
21. Eng JK, McCormack AL, Yates JR, III (1994) An approach to correlate tandem mass spectral data of peptides with amino acid sequences in a protein database. *J Am Soc Mass Spectrom* 5: 976–989
22. Perkins DN, Pappin DJ, Creasy DM, Cottrell JS (1999) Probability-based protein identification by searching sequence databases using mass spectrometry data. *Electrophoresis* 20: 3551–3567
23. Zhang W, Chait BT (2000) ProFound: an expert system for protein identification using mass spectrometric peptide mapping information. *Anal Chem* 72: 2482–2489
24. Colinge J, Masselot A, Cusin I, Mahe E, Niknejad A, Argoud-Puy G, Reffas S, Bederr N, Gleizes A, Rey PA et al. (2004) High-performance peptide identification by tandem mass spectrometry allows reliable automatic data processing in proteomics. *Proteomics* 4: 1977–1984
25. Colinge J, Chiappe D, Lagache S, Moniatte M, Bougueleret L (2005) Differential Proteomics via probabilistic peptide identification scores. *Anal Chem* 77: 596–606
26. Biemann K (1990) Appendix 5. Nomenclature for peptide fragment ions (positive ions). *Methods Enzymol* 193: 886–887
27. Roepstorff P, Fohlman J (1984) Proposal for a common nomenclature for sequence ions in mass spectra of peptides. *Biomed Mass Spectrom* 11: 601
28. Spengler B, Kirsch D, Kaufmann R, Jaeger E (1992) Peptide sequencing by matrix-assisted laser-desorption mass spectrometry. *Rapid Commun Mass Spectrom* 6: 105–108
29. Suckau D, Resemann A, Schuerenberg M, Hufnagel P, Franzen J, Holle A (2003) A novel MALDI LIFT-TOF/TOF mass spectrometer for proteomics. *Anal Bioanal Chem* 376: 952–965
30. Collings BA, Stott WR, Londry FA (2003) Resonant excitation in a low-pressure linear ion trap. *J Am Soc Mass Spectrom* 14: 622–634
31. Douglas DJ, Frank AJ, Mao D (2005) Linear ion traps in mass spectrometry. *Mass Spectrom Rev* 24: 1–29
32. Hakansson K, Chalmers MJ, Quinn JP, McFarland MA, Hendrickson CL, Marshall AG (2003) Combined electron capture and infrared multiphoton dissociation for multistage MS/MS in a Fourier transform ion cyclotron resonance mass spectrometer. *Anal Chem* 75: 3256–3262
33. Cooper HJ, Hakansson K, Marshall AG (2005) The role of electron capture dissociation in biomolecular analysis. *Mass Spectrom Rev* 24: 201–222
34. Li W, Hendrickson CL, Emmett MR, Marshall AG (1999) Identification of intact proteins in mixtures by alternated capillary liquid chromatography electrospray ionization and LC ESI infrared multiphoton dissociation Fourier transform ion cyclotron resonance mass spectrometry. *Anal Chem* 71: 4397–4402
35. Mann M, Wilm M (1994) Error-tolerant identification of peptides in sequence databases by peptide sequence tags. *Anal Chem* 66: 4390–4399
36. Keough T, Lacey MP, Youngquist RS (2002) Solid-phase derivatization of tryptic peptides for rapid protein identification by matrix-assisted laser desorption/ionization mass spectrometry. *Rapid Commun Mass Spectrom* 16: 1003–1015
37. Keough T, Lacey MP, Youngquist RS (2000) Derivatization procedures to facilitate *de novo* sequencing of lysine-terminated tryptic peptides using postsource decay matrix-assisted laser desorption/ionization mass spectrometry. *Rapid Commun Mass Spectrom* 14: 2348–2356

38. Munchbach M, Quadroni M, Miotto G, James P (2000) Quantitation and facilitated *de novo* sequencing of proteins by isotopic N-terminal labeling of peptides with a fragmentation-directing moiety. *Anal Chem* 72: 4047–4057
39. Lindh I, Hjelmqvist L, Bergman T, Sjovall J, Griffiths WJ (2000) *De novo* sequencing of proteolytic peptides by a combination of C-terminal derivatization and nano-electrospray/collision-induced dissociation mass spectrometry. *J Am Soc Mass Spectrom* 11: 673–686
40. Hale JE, Butler JP, Knierman MD, Becker GW (2000) Increased sensitivity of tryptic peptide detection by MALDI-TOF mass spectrometry is achieved by conversion of lysine to homoarginine. *Anal Biochem* 287: 110–117
41. Gu S, Pan S, Bradbury EM, Chen X (2002) Use of deuterium-labeled lysine for efficient protein identification and peptide *de novo* sequencing. *Anal Chem* 74: 5774–5785
42. Sonsmann G, Romer A, Schomburg D (2002) Investigation of the influence of charge derivatization on the fragmentation of multiply protonated peptides. *J Am Soc Mass Spectrom* 13: 47–58
43. Schnolzer M, Jedrzejewski P, Lehmann WD (1996) Protease-catalyzed incorporation of 18O into peptide fragments and its application for protein sequencing by electrospray and matrix-assisted laser desorption/ionisation mass spectrometry. *Electrophoresis* 17: 945–953
44. Little DP, Speir JP, Senko MW, O'Connor PB, McLafferty FW (1994) Infrared multiphoton dissociation of large multiply charged ions for biomolecule sequencing. *Anal Chem* 66: 2809–2815
45. Guilhaus M, Selby D, Mlynski V (2000) Orthogonal acceleration time-of-flight mass spectrometry. *Mass Spectrom Rev* 19: 65–107
46. Premstaller A, Oberacher H, Walcher W, Timperio AM, Zolla L, Chervet JP, Cavusoglu N, Van Dorsselaer A, Huber CG (2001) High-performance liquid chromatography-electrospray ionization mass spectrometry using monolithic capillary columns for proteomic studies. *Anal Chem* 73: 2390–2396
47. Peterman SM, Dufresne CP, Horning S (2005) The use of a hybrid linear trap/FT-ICR mass spectrometer for on-line high resolution/high mass accuracy bottom-up sequencing. *J Biomol Tech* 16: 112–124
48. Sjodahl J, Kempka M, Hermansson K, Thorsen A, Roeraade J (2005) Chip with twin anchors for reduced ion suppression and improved mass accuracy in MALDI-TOF mass spectrometry. *Anal Chem* 77: 827–832
49. Mitulovic G, Smoluch M, Chervet JP, Steinmacher I, Kungl A, Mechtler K (2003) An improved method for tracking and reducing the void volume in nano HPLC-MS with micro trapping columns. *Anal Bioanal Chem* 376: 946–951
50. Shen Y, Zhao R, Berger SJ, Anderson GA, Rodriguez N, Smith RD (2002) High-efficiency nanoscale liquid chromatography coupled on-line with mass spectrometry using nanoelectrospray ionization for proteomics. *Anal Chem* 74: 4235–4249
51. Mirgorodskaya E, Braeuer C, Fucini P, Lehrach H, Gobom J (2005) Nanoflow liquid chromatography coupled to matrix-assisted laser desorption/ionization mass spectrometry: sample preparation, data analysis, and application to the analysis of complex peptide mixtures. *Proteomics* 5: 399–408
52. Li X, Gong Y, Wang Y, Wu S, Cai Y, He P, Lu Z, Ying W, Zhang Y, Jiao L et al. (2005) Comparison of alternative analytical techniques for the characterisation of the human serum proteome in HUPO Plasma Proteome Project. *Proteomics* 5: 3423–3441
53. Kalume DE, Molina H, Pandey A (2003) Tackling the phosphoproteome: tools and strategies. *Curr Opin Chem Biol* 7: 64–69
54. Sickmann A, Meyer HE (2001) Phosphoamino acid analysis. *Proteomics* 1: 200–206

55. Medzihradszky KF, Phillipps NJ, Senderowicz L, Wang P, Turck CW (1997) Synthesis and characterization of histidine-phosphorylated peptides. *Protein Sci* 6: 1405–1411
56. Duclos B, Marcandier S, Cozzone AJ (1991) Chemical properties and separation of phosphoamino acids by thin-layer chromatography and/or electrophoresis. *Methods Enzymol* 201: 10–21
57. Meyer HE, Eisermann B, Heber M, Hoffmann-Posorske E, Korte H, Weigt C, Wegner A, Hutton T, Donella-Deana A, Perich JW (1993) Strategies for nonradioactive methods in the localization of phosphorylated amino acids in proteins. *FASEB J* 7: 776–782
58. McLachlin DT, Chait BT (2001) Analysis of phosphorylated proteins and peptides by mass spectrometry. *Curr Opin Chem Biol* 5: 591–602
59. Porath J, Carlsson J, Olsson I, Belfrage G (1975) Metal chelate affinity chromatography, a new approach to protein fractionation. *Nature* 258: 598–599
60. Cao P, Stults JT (1999) Phosphopeptide analysis by on-line immobilized metal-ion affinity chromatography-capillary electrophoresis-electrospray ionization mass spectrometry. *J Chromatogr A* 853: 225–235
61. Heintz D, Wurtz V, High AA, Van Dorsselaer A, Reski R, Sarnighausen E (2004) An efficient protocol for the identification of protein phosphorylation in a seedless plant, sensitive enough to detect members of signalling cascades. *Electrophoresis* 25: 1149–1159
62. Raska CS, Parker CE, Dominski Z, Marzluff WF, Glish GL, Pope RM, Borchers CH (2002) Direct MALDI-MS/MS of phosphopeptides affinity-bound to immobilized metal ion affinity chromatography beads. *Anal Chem* 74: 3429–3433
63. Sano A, Nakamura H (2004) Titania as a chemo-affinity support for the column-switching HPLC analysis of phosphopeptides: application to the characterization of phosphorylation sites in proteins by combination with protease digestion and electrospray ionization mass spectrometry. *Anal Sci* 20: 861–864
64. Larsen MR, Thingholm TE, Jensen ON, Roepstorff P, Jorgensen TJ (2005) Highly selective enrichment of phosphorylated peptides from Peptide mixtures using titanium dioxide microcolumns. *Mol Cell Proteomics* 4: 873–886
65. Meyer HE, Hoffmann-Posorske E, Korte H, Heilmeyer LM Jr (1986) Sequence analysis of phosphoserine-containing peptides. Modification for picomolar sensitivity. *FEBS Lett* 204: 61–66
66. Oda Y, Nagasu T, Chait BT (2001) Enrichment analysis of phosphorylated proteins as a tool for probing the phosphoproteome. *Nat Biotechnol* 19: 379–382
67. McLachlin DT, Chait BT (2003) Improved beta-elimination-based affinity purification strategy for enrichment of phosphopeptides. *Anal Chem* 75: 6826–6836
68. Conrads TP, Issaq HJ, Veenstra TD (2002) New tools for quantitative phosphoproteome analysis. *Biochem Biophys Res Commun* 290: 885–890
69. Thompson AJ, Hart SR, Franz C, Barnouin K, Ridley A, Cramer R (2003) Characterization of protein phosphorylation by mass spectrometry using immobilized metal ion affinity chromatography with on-resin beta-elimination and Michael addition. *Anal Chem* 75: 3232–3243
70. Hunter T (1995) Protein kinases and phosphatases: the yin and yang of protein phosphorylation and signaling. *Cell* 80: 225–236
71. Schlessinger J (1993) Cellular signaling by receptor tyrosine kinases. *Harvey Lect* 89: 105–123
72. Yang X, Wu H, Kobayashi T, Solaro RJ, van Breemen RB (2004) Enhanced ionization of phosphorylated peptides during MALDI TOF mass spectrometry. *Anal Chem* 76: 1532–1536
73. Kjellstrom S, Jensen ON (2004) Phosphoric acid as a matrix additive for MALDI MS analysis of phosphopeptides and phosphoproteins. *Anal Chem* 76: 5109–5117

74. Stensballe A, Jensen ON (2004) Phosphoric acid enhances the performance of Fe(III) affinity chromatography and matrix-assisted laser desorption/ionization tandem mass spectrometry for recovery, detection and sequencing of phosphopeptides. *Rapid Commun Mass Spectrom* 18: 1721–1730
75. Steen H, Kuster B, Mann M (2001) Quadrupole time-of-flight *versus* triple-quadrupole mass spectrometry for the determination of phosphopeptides by precursor ion scanning. *J Mass Spectrom* 36: 782–790
76. Carr SA, Huddleston MJ, Annan RS (1996) Selective detection and sequencing of phosphopeptides at the femtomole level by mass spectrometry. *Anal Biochem* 239: 180–192
77. Steen H, Mann M (2002) A new derivatization strategy for the analysis of phosphopeptides by precursor ion scanning in positive ion mode. *J Am Soc Mass Spectrom* 13: 996–1003
78. Hogan JM, Pitteri SJ, McLuckey SA (2003) Phosphorylation site identification via ion trap tandem mass spectrometry of whole protein and peptide ions: bovine alpha-crystallin A chain. *Anal Chem* 75: 6509–6516
79. Shi SD, Hemling ME, Carr SA, Horn DM, Lindh I, McLafferty FW (2001) Phosphopeptide/phosphoprotein mapping by electron capture dissociation mass spectrometry. *Anal Chem* 73: 19–22
80. Silivra OA, Kjeldsen F, Ivonin IA, Zubarev RA (2005) Electron capture dissociation of polypeptides in a three-dimensional quadrupole ion trap: Implementation and first results. *J Am Soc Mass Spectrom* 16: 22–27
81. Syka JE, Coon JJ, Schroeder MJ, Shabanowitz J, Hunt DF (2004) Peptide and protein sequence analysis by electron transfer dissociation mass spectrometry. *Proc Natl Acad Sci USA* 101: 9528–9533
82. Chamrad DC, Koerting G, Gobom J, Thiele H, Klose J, Meyer HE, Blueggel M (2003) Interpretation of mass spectrometry data for high-throughput proteomics. *Anal Bioanal Chem* 376: 1014–1022
83. Egelhofer V, Gobom J, Seitz H, Giavalisco P, Lehrach H, Nordhoff E (2002) Protein identification by MALDI-TOF-MS peptide mapping: a new strategy. *Anal Chem* 74: 1760–1771
84. Lopez-Ferrer D, Martinez-Bartolome S, Villar M, Campillos M, Martin-Maroto F, Vazquez J (2004) Statistical model for large-scale peptide identification in databases from tandem mass spectra using SEQUEST. *Anal Chem* 76: 6853–6860
85. Qian WJ, Liu T, Monroe ME, Strittmatter EF, Jacobs JM, Kangas LJ, Petritis K, Camp DG, Smith RD (2005) Probability-based evaluation of peptide and protein identifications from tandem mass spectrometry and SEQUEST analysis: the human proteome. *J Proteome Res* 4: 53–62
86. Boehm AM, Grosse-Coosmann F, Sickmann A (2004) Command line tool for calculating theoretical MS spectra for given sequences. *Bioinformatics* 20: 2889–2891
87. Huang L, Jacob RJ, Pegg SC, Baldwin MA, Wang CC, Burlingame AL, Babbitt PC (2001) Functional assignment of the 20 S proteasome from *Trypanosoma brucei* using mass spectrometry and new bioinformatics approaches. *J Biol Chem* 276: 28327–28339
88. Yergey AL, Coorssen JR, Backlund PS Jr, Blank PS, Humphrey GA, Zimmerberg J, Campbell JM, Vestal ML (2002) De novo sequencing of peptides using MALDI/TOF-TOF. *J Am Soc Mass Spectrom* 13: 784–791
89. Fernandez-de-Cossio J, Gonzalez J, Satomi Y, Shima T, Okumura N, Besada V, Betancourt L, Padron G, Shimonishi Y, Takao T (2000) Automated interpretation of low-energy collision-induced dissociation spectra by SeqMS, a software aid for *de novo* sequencing by tandem mass spectrometry. *Electrophoresis* 21: 1694–1699
90. Johnson RS, Taylor JA (2002) Searching sequence databases via *de novo* peptide sequencing by tandem mass spectrometry. *Mol Biotechnol* 22: 301–315

91. Searle BC, Dasari S, Turner M, Reddy AP, Choi D, Wilmarth PA, McCormack AL, David LL, Nagalla SR (2004) High-throughput identification of proteins and unanticipated sequence modifications using a mass-based alignment algorithm for MS/MS *de novo* sequencing results. *Anal Chem* 76: 2220–2230
92. Bruni R, Gianfranceschi G, Koch G (2005) On peptide *de novo* sequencing: a new approach. *J Pept Sci* 11: 225–234
93. Handley J (2002) Software for MS protein identification. *Anal Chem* 74: 159A–162A
94. March RE (1997) An introduction to quadrupole ion trap mass spectrometry. *J Mass Spec* 32: 351–369
95. Johnson RS, Martin SA, Biemann K, Stults JT, Watson JT (1987) Novel fragmentation process of peptides by collision-induced decomposition in a tandem mass spectrometer: differentiation of leucine and isoleucine. *Anal Chem* 59(21): 2621–2625

Plant Systems Biology
Edited by Sacha Baginsky and Alisdair R. Fernie
© 2007 Birkhäuser Verlag/Switzerland

Methods, applications and concepts of metabolite profiling: Primary metabolism

Dirk Steinhauser and Joachim Kopka

Max Planck Institute of Molecular Plant Physiology, Am Muehlenberg 1, 14476 Potsdam-Golm, Germany

Abstract

In the 1990s the concept of a comprehensive analysis of the metabolic complement in biological systems, termed metabolomics or alternately metabonomics, was established as the last of four cornerstones for phenotypic studies in the post-genomic era. With genomic, transcriptomic, and proteomic technologies in place and metabolomic phenotyping under rapid development all necessary tools appear to be available today for a fully functional assessment of biological phenomena at all major system levels of life. This chapter attempts to describe and discuss crucial steps of establishing and maintaining a gas chromatography/electron impact ionization/mass spectrometry (GC-EI-MS)-based metabolite profiling platform. GC-EI-MS can be perceived as the first and exemplary profiling technology aimed at simultaneous and non-biased analysis of primary metabolites from biological samples. The potential and constraints of this profiling technology are among the best understood. Most problems are solved as well as pitfalls identified. Thus GC-EI-MS serves as an ideal example for students and scientists who intend to enter the field of metabolomics. This chapter will be biased towards GC-EI-MS analyses but aims at discussing general topics, such as experimental design, metabolite identification, quantification and data mining.

Introduction

In the 1990s the concept of a comprehensive analysis of the metabolic complement in biological systems, termed metabolomics [1, 2] or alternately metabonomics [3, 4], was established as the last of four corner stones for phenotypic studies in the post-genomic era (e.g., [5–8]). With genomic, transcriptomic, and proteomic technologies in place and metabolomic phenotyping under rapid development all necessary tools appear to be available today for a functional assessment of biological phenomena at all major system levels of life. However, all '-omics' technologies are at different stages of comprehensiveness, sample throughput and accuracy of constituent identification and quantification. While the set of genes in an organism can be exactly defined and described, knowledge of the full inventory of metabolites and a truly comprehensive metabolome analysis remains a vision for the future. The

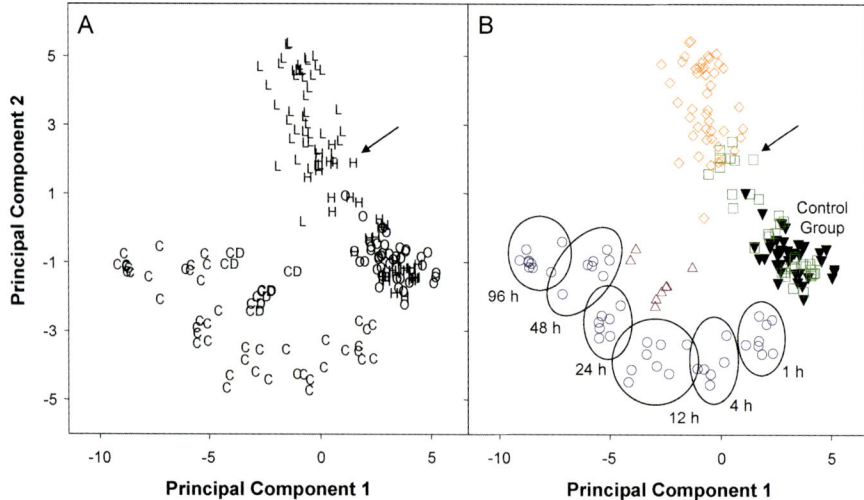

Figure 1. Principal component analysis covering 38.5% and 21.9% of total variance in a dataset of leaf metabolite profiles from *Arabidopsis thaliana* ecotype Columbia. Plants were environmentally challenged by highlight (L, diamonds; long-term adaptation to 560 and 850 µE/m² compared to a control at 120–150 µE/m²), by high temperature (H, squares; up to 4 h at 40°C compared to a control at 20°C) and by low temperature (C, circles; up to 96 h at 4°C compared to a control at 20°C) [17]. Different formatting highlights environmental challenge (A) and time course compared to the control group (B). Note: (1) Highlight and high temperature response exhibits an expected partial overlap (arrows). (2) Cold de-acclimatized plants (CD, triangles; 24 h reversion to 20°C after 96 h at 4°C) show the existence of metabolic memory after reversion to optimum temperature conditions.

highly diverse chemical properties of metabolites which range from gasses, such as O_2 and CO_2, to macromolecules such as starch and complex lipids, is the crucial limiting factor. This high diversity impedes comprehensive metabolomics with single analytical technologies. Thus the current developments in metabolomic technologies focus on establishment and optimization of minimally overlapping, broad-spectrum metabolite profiling methods which have been pioneered decades earlier (e.g., [9–11]).

This chapter attempts to describe and discuss crucial steps of establishing and maintaining a gas chromatography/electron impact ionization/mass spectrometry (GC-EI-MS)-based metabolite profiling platform. GC-EI-MS can be perceived as the first and exemplary profiling technology aimed at simultaneous and non-biased analysis of primary metabolites from biological samples [12, 13]. The potential and constraints of this profiling technology are among the best understood. Most problems are solved as well as pitfalls identified. Thus GC-EI-MS serves as an ideal example for students and scientists who intend to enter the field of metabolomics. This chapter will be biased towards GC-EI-MS analyses but aims at discussing general topics, such as experimental design, metabolite identification, quantification and data mining. For a more detailed review of metabolic inactivation, metabolome sampling, metabolite extraction, chemical derivatization, gas chromatographic separation, mass spectral ionization and detection the reader is referred to previous reviews [14–16].

As detailed bio-analytic aspects are best exemplified with a relevant experiment in mind, most discussions will refer to one data set, which describes the metabolic phenotype of environmentally challenged and genetically modified *Arabidopsis thaliana* plants as summarized by a principal components analysis (Fig. 1). This experiment charts metabolic changes of a model plant in response to common environmental stresses such as variable light and temperature [17].

Experimental design

Pairwise comparison, dose dependency or time-course

Alongside the immediate and full metabolic inactivation at and following time of sampling [14, 15], the crucial issue in a metabolite profiling study is experimental design. It is evident that the result and quality of a profiling experiment depends on a design which is optimally fitted to the question that is about to be addressed. If a genetically modified organism (GMO) or an environmental challenge is first analyzed for metabolic equivalence, metabolite profiling studies can be successfully used to screen for relevant metabolic changes (e.g., [18, 19]). This task is purely descriptive and can be solved by pairwise or multiple comparison. In a comparative experiment only one factor, such as the genotype or one environmental parameter, is changed and all other influences are, ideally, kept constant. Typically each of the compared conditions is replicated within one experiment and in independent consecutive experimental repeats. The aim of repetition is to distinguish true differences from unavoidable experimental errors and basic biological variability (see control samples of Fig 1B; also note that the cold stress experiment was performed in two independent experiments which cannot be distinguished by PCA analysis). By application of statistical significance tests any detected change within the metabolic phenotype can be unequivocally linked to the experimental manipulation, such as mutant *versus* ecotype [12], temperature stress [17], transgene expression or chemical treatment with glucose (e.g., [13, 20]). Functional genomics studies employ multiple comparative analyses for the classification of genes with yet unknown or hypothetical function by similarity of the metabolic phenotypes [8]. However, these comparisons typically result in multiple detected statistically significant changes. Among these the primary mechanistic effect of modified genes or environmental impact can not unambiguously be distinguished from secondary pleiotropic metabolic adaptations to the usually constitutive genetic modification. In other words the permanent presence or absence of transgene expression throughout the life cycle of a GMO may result in unexpected long-term adaptations of primary metabolism, which up to today were overlooked by biased and targeted metabolic analysis.

One strategy to dissect primary metabolic effects from secondary adaptations is the use of dose dependency. In environmental challenges different light intensities, temperatures or concentrations of nutrients and chemicals can be applied. In GMO studies stably modified lines with a range of low, medium to high transgene expression can be selected. Chemically controlled or otherwise inducible promoters can be employed for the same purpose. The use of these promoters may yield different metabolic responses compared to constitutive promoters and generate novel insights into metabolic regulation (e.g., [21]). In all cases sensitive metabolic effects which respond to small doses can be distinguished from effects of high doses that are more prone to cause pleiotropic effects. Moreover, the dose quantity can be linked to a quantitative metabolic effect for example by application of correlation analysis. It can be argued that those metabolic effects which show a strict dose dependency have a strong mechanistic link. Caution needs to be applied in thoroughly controlling dose dependency experiments. For example the effect of a chemical inductor needs to be distinguished from the effect of transgene expression. Also environmental changes may not be independent, for example increased light intensity and heat have similar metabolic effects as is demonstrated by a partial overlap of the heat response and the highlight metabolite phenotypes of *Arabidopsis thaliana* rosette leaves (Fig. 1A).

The best but also most demanding strategy to dissect possible mechanisms of metabolic changes is a time-course design (Fig. 1B). It can be argued that early changes are linked to sensing and represent a direct response mechanism, whereas secondary adaptations will be observed in a long-term transition from the initial to a final metabolic state to, for example, a cold-adapted metabolism (Fig. 1A). Time-course investigations do not only allow comparison of initial and stably adapted metabolic states but also unravel the sequence of metabolic events and transient, i.e., reversible changes, which would otherwise be overlooked, such as early maltose and maltotriose accumulation in *Arabidopsis thaliana* cold adaptation (Fig. 2). The example of cold adaptation in plants also unveils that the history of a biological system may determine the metabolic phenotype. Cold de-acclimatized plants, even after 24 h reversion to optimum temperature, still exhibit a metabolic memory (Fig. 1A). In conclusion, good experimental practice for optimum reproduction of biological experiments not only controls the conditions at the time of sampling but also the history of the biological objects.

Fingerprinting, profiling or exact quantification

The experimental design of GC-EI-MS analyses has a strong impact on the accuracy of metabolome studies. Three major approaches were described and have been extensively discussed, i.e., fingerprinting, metabolite profiling and exact quantification [6–8, 22]. In general, the complexity of information and number of theoretically covered metabolites decreases when moving from fingerprinting to exact quantification [8]. Typically a concomitant increase in experimental complexity is observed, with higher time demand, and requirements for quantitative standardization or compound identification.

Fingerprinting studies appear to be the easiest approach to metabolome analysis. These studies utilize all detector readings for numerical analysis without the at-

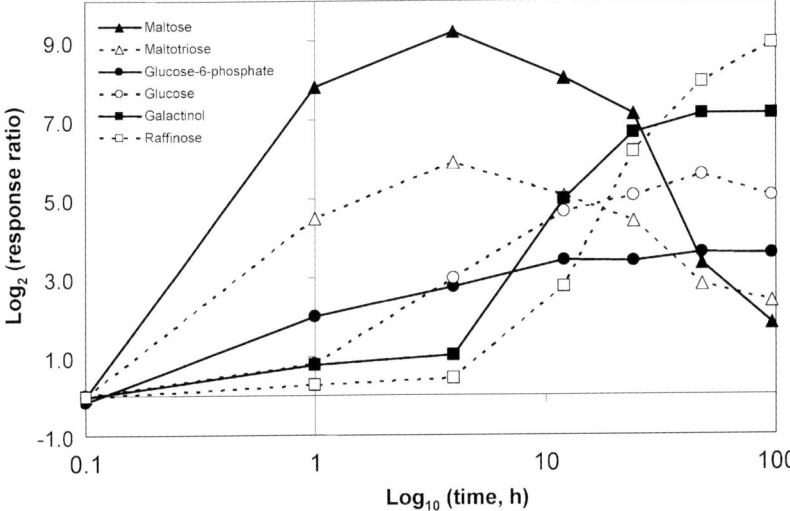

Figure 2. Transition of metabolic states exemplified by the time course of 4°C cold adaptation of *Arabidopsis thaliana* plants, ecotype Columbia. Note that Maltose and Maltotriose exhibit early transitory accumulation followed by sustained increases in Glucose-6-phosphate, glucose, galactinol and ultimately raffinose, a metabolic product of galactinol in plants [17].

tempt, and in some cases even the potential, to unambiguously identify the specific metabolites represented in these experiments. Fingerprints are used for metabolic pattern comparison aimed at the discovery of experimental conditions which result in similar or identical metabolic responses, so-called metabolic phenocopies [20]. This approach is exploited in gene function analysis and has the potential to group genes with known function and orphan genes of unknown or hypothetical function into classes of similar or identical metabolic function [2, 5]. This type of metabolic pattern analysis appears to be especially promising when gene modifications result in 'silent' phenotypes. (For the definition of silent phenotype refer to [18].) This phenomenon is better defined as changes of the metabolic state in organisms, which do not show obvious visual or morphological traits.

Fingerprinting, however, has one fundamental requirement, which results from unavoidable technical drifts in the calibration of mass, retention time and ion current. These decalibration artifacts are inherent to all chromatographic and mass spectrometric analysis technologies. In GC-EI-MS analyses one of the technology breakthroughs was the employment of widely accepted reference substances for the automated mass calibration of the GC-MS systems, such as BFB (4-bromofluorobenzene) and DFTPP (decafluorotriphenylphosphine). These substances are used in so-called tuning procedures which are inbuilt into the maintenance routines of the respective manufacturer. GC-MS tuning of the mass scale is usually performed prior to a series of analyses and allows accurate mass alignment. A rather low resolution of 1 atomic mass unit is sufficient for most of the small molecules which are routinely analyzed by GC-MS. More precise mass calibration can be obtained by reference compounds, which are continuously added to the GC effluent before mass analysis. This so-called 'lock-mass' technology is only useful for the high mass accuracy obtainable with sector field or specialized high-resolution time-of-flight GC-TOF-MS systems. While negligible for the low mass resolution typically achieved by quadrupole, iontrap or fast scanning time-of-flight GC-MS systems, the 'lock-mass' calibration has significantly improved routine LC-MS profiling experiments (e.g., [23]).

Likewise the retention time axis should be calibrated by use of retention time standard substances. One of the most widely accepted procedure utilizes mixtures of n-alkanes [24] and so-called retention time indices (RI) to correct for inevitable retention time shifts within and between series of consecutive chromatograms. The use of retention time indices has been introduced to GC-EI-MS metabolite profiling experiments [12, 13]. In these early studies n-acyl fatty acids were used, which were later substituted for n-alkanes [25] to allow for better comparability with the wealth of previous RI information, which – since 2005 – is commercially provided together with thousands of biologically relevant GC-MS mass spectra [26–28] by the NIST05 mass spectral library (National Institute of Standards and Technology, Gaithersburg, MD, USA; http://www.nist.gov/srd/mslist.htm).

One of the most critical causes for artifacts in fingerprinting studies, in many studies, is the non-calibrated ion current scale. The quantity of metabolic components from GC-MS runs is routinely measured by ion currents detected after chromatography, ionization, and mass separation. The quantity of ions which reaches the final detector system is subject to multiple artifacts. One of the most important effects is exerted through the decrease of detector sensitivity over time. The detector sensitivity is partially corrected by the tuning procedure mentioned above. However, the best approach is the use of quantitative reference substances, so-called internal standards (IS), which are added to the biological sample at constant known quantities prior to metabolite extraction and are carried along throughout the complete analysis. The most versatile IS are stable isotope-labeled substances [12, 22, 29].

Today, software tools which use statistical algorithms for the alignment of mass and time dimensions promise good success by avoiding artifacts through false alignment (for example [30–32] or metAlign, http://www.metalign.nl [33, 34]). However, the limits of both mass and retention time drift successfully corrected by these software tools have still not been thoroughly tested. Therefore, chemical calibration of all three dimensions in hyphenated GC-EI-MS analysis represent the most secure approach towards valid fingerprinting (Fig. 3).

In contrast to fingerprinting, metabolite profiling studies attempt to identify all metabolites which are represented in the dataset. Non-identified components can be discarded or used for fingerprinting. In profiling experiments the analysis is restricted to the selected subset of those analytical detector readings which can be identified. The clear advantage of this approach is the possibility that the metabolic pattern of profiling experiments can be biochemically interpreted. Thus, besides pattern recognition and comparison, metabolite profiling has the potential to provide insight into the mechanism of gene function or the response triggered by envi-

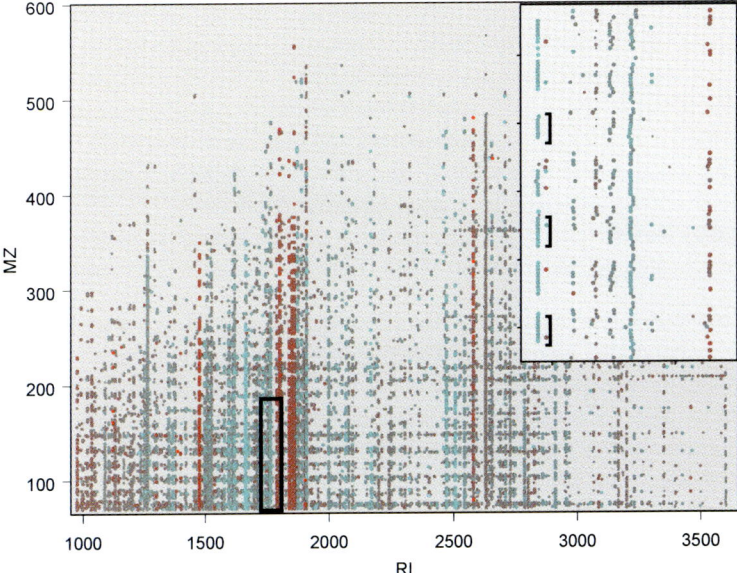

Figure 3. Heat-map display of a comparative GC-EI-MS metabolite fingerprinting study. The heat-map demonstrates the information content of an experiment which compares a treatment to non-treated reference samples. Approximately 13,000 mass fragments are shown. Ion current was corrected by a single quantitative internal standard. The mass fragments are characterized by mass to charge ratio (MZ), retention time index (RI), relative increase (red) or decrease (cyan) in log-transformed response ratios, and significance of the observed change. Large spots indicate p<0.05. The insert demonstrates the high degree of EI-MS fragmentation. Columns of mass fragments, which exhibit the same quantitative change, represent the same substance. Abundant compounds exhibit typical mass isotopomer series resulting mostly from incorporation of the ~1.1% ambient ^{13}C isotope (square brackets). Note the severe co-elution present in complex biological samples.

ronmental changes. For example, part of the early cold stress response in *Arabidopsis* is a massive release of carbohydrates (Fig. 2) in the form of maltose, a process which points towards a fast induction of transitory starch degradation in chloroplasts and the generation of carbon buildings blocks for subsequent metabolic events [17]. In addition sets of metabolites, such as maltose and maltotriose in the above example, can be grouped into modules of substances, which exhibit simultaneous changes. These metabolites can be assumed to be subject to common control mechanisms which may also be beyond pathway connectivity in contrast to this example.

A minor aspect of metabolite profiling but certainly an important asset in avoiding artifact pattern recognition is the opportunity to remove detector readings from subsequent data analysis, which result from laboratory contaminations, intentionally added IS, and electronic or chemical noise.

Because metabolite identification is inherent to profiling experiments, quantitative standardization can be improved compared to fingerprints. If necessary, each metabolite can be provided with an appropriate internal standard, ideally a chemically identical but stable isotope labeled substance. Initially commercially available and expensive, chemically synthesized compounds, such as U-13C or fully deuterated mass isotopomers, have been suggested [12]. Recently this concept has been extended towards fully U-13C-labeled metabolome extracts from organisms which can be grown on exclusive carbon sources and thus are fully labeled *in vivo*. For a short introduction and discussion of the concept of metabolite profiling by mass isotopomer ratios the reader is referred to earlier publications [22, 35–37].

Studies that perform exact quantification of metabolites have only two further, but time consuming, requirements when compared to profiling experiments:

1. The detector reading, such as the observed ion current at a specific mass and chromatographic retention of a metabolite needs to be calibrated to the molar amount or concentration of each quantified compound. This is typically done by dilution series of pure reference substances measured at precise quantities. These calibration series are required because chemical substances exhibit highly different ionization efficiencies and equally variable fragmentation patterns. Quantitative calibration ensures that easily and difficult to ionize compounds as well as abundant and minor mass fragments of the same compound can be used to obtain the same quantitative result.
2. The recovery of each substance needs to be estimated. In comparison to pure reference samples each substance can selectively get lost or may accumulate at all steps from extraction to detection throughout analysis of complex mixtures. Typically the nature and composition of the biological sample influences compound recovery. The effects on specific metabolites are as a rule thumb unpredictable. Therefore, each new type of biological sample needs to be tested for unforeseen changes in metabolite recovery. Typically so-called standard addition experiments are performed [14], which test the apparent quantity of an identical amount of pure reference substance in the presence and the absence of the respective biological sample. When the presence of a biological sample leads to an apparent reduction of the metabolite amount, the term matrix suppression is used. Matrix effects are best estimated by stable isotope labeled mass isotopomers applied as IS. These are absent from typical biological samples and thus recovery experiments do not need to be corrected for the respective endogenous amount of metabolites present in the biological sample.

In conclusion metabolite profiles supplied with stable isotope labeled authentic reference substances already allow correction of variable metabolite recovery and thus are only one step away from fulfilling the prerequisites for exact quantification.

Estimating relative changes in metabolite pool size

While exact quantification of metabolite pools is clearly within the scope of GC-EI-MS profiling experiments (e.g., [20, 38, 39]) accurate quantification is not required for most investigations and screening for relative changes in metabolite pool sizes is performed instead. In the following, all steps in data processing are described which enable detection of quantitative changes such as represented in the heat-map representation of Figure 3.

The first quantitative observation in GC-EI-MS-based profiling is linked to mass fragments or molecular ions, which have the properties, mass (or more precisely mass to charge ratio), chromatographic retention time index (Fig. 3) and an abundance measured as ion current. The so-called response of a mass fragment is obtained by baseline subtraction of ion current caused by electronic and chemical noise and either subsequent integration of chromatographic peaks or determination of peak height (for exact details the interested reader is referred to [40]). These steps are typically performed by chromatography processing software of the respective GC-MS system manufacturers. In a second step responses are normalized to the response of at least one IS and the initial amount of the biological sample, as determined by dry or fresh weight of solid samples or volume of liquid samples. The resulting normalized response takes into account the variation in sample amount, inevitable volume errors, which may occur during extraction, sample preparation and GC-MS injection, and the drift of detector sensitivity discussed above. If the experimental design includes additional substance specific, stable isotope labeled ISs, specific corrections of metabolite recovery can be applied. Additional ISs are especially advised for instable metabolites.

Response ratios are calculated for each metabolite separately using the average normalized response observed in a replicate set of reference or control samples as quotient denominator. If the experiment provides no obvious control condition the response ratio can be calculated utilizing the average normalized response of all samples. Response ratios represent relative changes in metabolite abundance or pools size. However, the fold change may differ from ratios which are calculated after exact quantification, especially when measurements approach upper or lower detection limits. Provided all samples are treated equally the use of reference samples not only allows correction for the inherent technical errors. In addition, randomized or appropriately arrayed reference samples correct for non-controlled factors which might influence the biological experiment, such as unexpected, slight and mostly unnoticed environmental gradients.

Response ratios can furthermore be subjected to numerical transformation (Figs 2 and 3). For example, logarithmic transformation converts factorial into additive numerical changes. Thus a 10-fold increase, a factor of 10, and an equal decrease, a factor of 0.1, gain equally weighted numerical representation, i.e., +1 and -1, respectively. Numerical transformation is advised prior to analyses of statistical significance. Two of the requirements for significance tests, namely normal distribution and homogeneity of variance are typically not met by either normalized responses or response ratios from metabolite profiling analyses. After log-transformation both criteria are usually better approximated or are even fully met.

Metabolite identification

Reference substances, mass spectral tags, and metabolites

The quintessential task of metabolite profiling is the reliable identification of metabolites in complex mixtures. This task has been the limitation of early studies and still is the major bottleneck of today's metabolite profiling studies. The subsequent paragraph will be dedicated to concepts and solutions of this central aspect in metabolome analysis. The presented strategies and concepts apply specifically to ubiquitous primary metabolites and may not be directly transferable to secondary metabolites, which are typically phylum or even species specific. Primary metabolites are best identified by pure references substances (see below). Availability of primary metabolites is satisfying, whereas purified or synthesized reference preparations of secondary metabolites are rare and hard to obtain.

The task of identification is best exemplified by Figure 3. All mass fragments of a profiling experiment need to be linked either to underlying metabolites, ISs or laboratory contaminations. In view of more than 10,000 reliably aligned mass fragments, this task appears to be enormous, if not impossible, to perform. In detail metabolite structures, which are archived in public reference databases such as KEGG [41], BRENDA [42], MetaCyc [43], the PubChem project (http://pubchem.ncbi.nlm.nih.gov/), or the chemical abstracts service (CAS, http://www.cas.org/), need to be linked:

1. To one or multiple alternative analytes. An analyte is the structure of a volatile chemical derivative of a metabolite or the non-modified, volatile metabolite. In short the reagent chemistry applied in routine GC-MS profiling [12, 13, 15] converts carbonyl moieties of metabolites to methoxyamine-moieties, CH_3-$N=C<$, and substitutes exchangeable protons, such as -OH, -COOH, -NRH, and -SH, by trimethylsilyl-moieties, $-Si(CH_3)_3$. Partial derivatization, steric hindrance, and EZ- isomerism of methoxyamines may cause multiple possible analyte structures of the same metabolite [16, 40].
2. The physicochemical properties through which each analyte is represented in GC-EI-MS profiles allow in most cases unambiguous identification. The sum of all relevant properties, in detail, the chromatographic retention time index (RI), the molecular mass to charge ratio (MZ), and the typical, induced EI-MS fragmentation pattern represented by a mass spectrum (MS), was termed mass spectral tag (MST) [40].
3. MSTs comprise multiple mass fragments. Each of these mass fragments needs to be linked unambiguously to one of usually multiple possible co-eluting MSTs and those mass fragments which are selective and specific for single MSTs need to be selected (Fig. 3).
4. Finally, a pure reference substance has to be acquired and identity has to be proven by match of RI and MS. Contaminations of 'pure' reference substances may present a severe source for false identifications. A typical expectation is that the most abundant analyte after chemical derivatization of a pure reference sample indeed represents the metabolite. However, unexpected impurities or laboratory contaminations may compromise this reasoning. For this reason

MSTs need to be interpreted by occurrence of molecular ions, plausible mass fragmentation pattern, or matching to pre-annotated mass spectral compendia, before finally accepting the metabolite identification of a MST.

Identification of mass spectral tags (MSTs)

Single mass fragments without the additional information of MSTs are hard, if not impossible, to unambiguously identify in different laboratories. In contrast, identified MSTs can be exchanged between laboratories [44] and hitherto non-identified MSTs can be identified by standard additional experiments of authenticated reference substances even years after the first MST description, provided the chemometric properties, i.e., molecular mass to charge ratio, chromatographic retention index and an induced mass fragmentation pattern such as an electron impact mass spectrum (EI-MS) are documented together with the respective quantitative profiles.

In the following a MST identification process is described and discussed using the non-trivial identification of hexoaldoses, specifically mannose (D-Man) and galactose (D-Gal) in the presence of abundant glucose (D-Glc) as a test case for isomer identification.

1. Isomers, especially stereoisomers, for example sugar epimers or cis/trans (E/Z-) diastereomers typically exhibit almost identical EI-MS fragmentation pattern and thus cannot be unambiguously distinguished by mass spectrometry alone [25]. The main reason for this limitation of mass spectral matching is the strong impact of analyte concentration on probability-based matching, such as provided by the NIST05 standard software for GC-EI-MS matching [26, 27]. In comparison, diastereomers exert only a small effect on mass fragment abundance.
2. Thus when considering the task of mannose and galactose identification, in addition to the common monosaccharides, all rare hexoaldoses, i.e., talose (D-Tal), gulose (D-Gul), idose (D-Ido), allose (D-All) and altrose (D-Alt) need to be checked.
3. Possible D- and L- enantiomers would further increase the complexity of this test case; however, most GC applications including routine GC-MS profiling are not chiro-selective.
4. For GC-MS analysis anomeric α- and β-structures of reducing sugars are chemically transformed from furanose- or pyranose- rings into open chains. The product is a mixture of E- and Z- >C=N-isomers which is generated at stable ratios and with more than 95% yield (Fig. 4A). As a result major and minor analytes are generated, which exhibit different chromatographic retention (Fig. 4B).
5. Figure 4 shows a typical metabolite profile of an *Arabidopsis* leaf extract in 80% methanol. The characteristic chromatographic region and a selected ion chromatogram at MZ=160, a characteristic mass fragment of aldose derived methoxyamines, is shown. Peaks with mass spectra indicative of aldoses are marked. In addition, the leaf sample was spiked with pure mannose or galactose in standard addition experiments (see above). The resulting chromatograms demonstrate a specific increase of peak size of the major analyte and a shoulder at the respective position of the minor analyte, respectively.

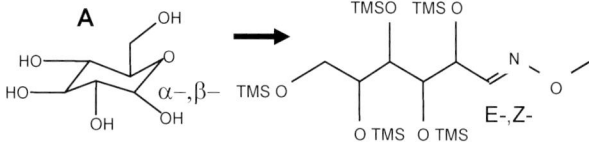
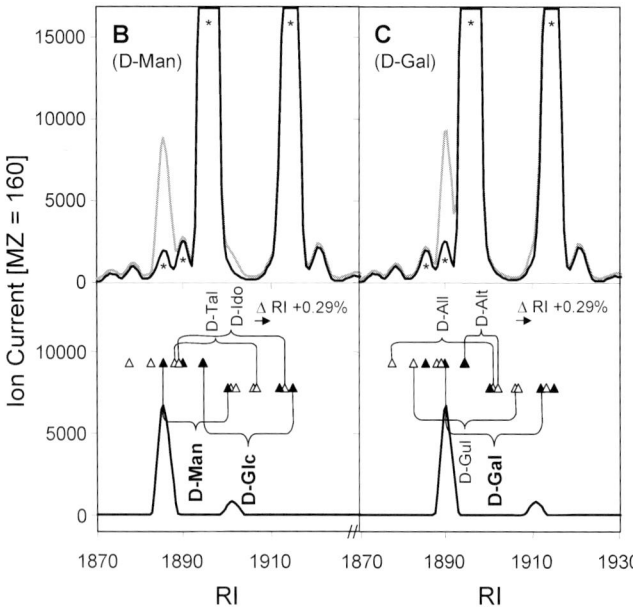

Figure 4. Representation of a MST identification experiment. An 80% methanol extract from *Arabidopsis thaliana* leaf was analyzed. Reducing sugars are routinely converted into methoxyamine structures and per-siliylated (A). RIs of major and minor analytes representing mannose (D-Man), galactose (D-Gal), glucose (D-Glc), closed triangles (B-C), as well as rare talose (D-Tal), gulose (D-Gul), idose (D-Ido), allose (D-All) and altrose (D-Alt), open triangles (B-C), are indicated. A typical standard addition experiment contains a sample of the pure reference substance (bottom), in this case mannose (B) or galactose (C), the reference substance added to a complex biological sample (top, gray), and the biological sample without standard addition (top, black). Mass spectral matching allowed identification of hexoaldoses in general (* indicates Match >800 on a scale of 0–1,000) but no differentiation between sugar epimers. Previously established elution sequences of ubiquitous hexoaldoses and rare isomers are shown. The pure reference substances were used to correct for the RI-offset to the previously established RI sequence (horizontal arrows).

Methods, applications and concepts of metabolite profiling: Primary metabolism 183

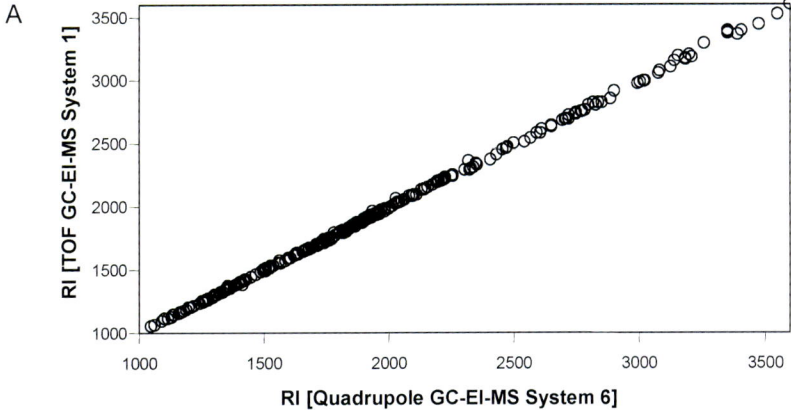

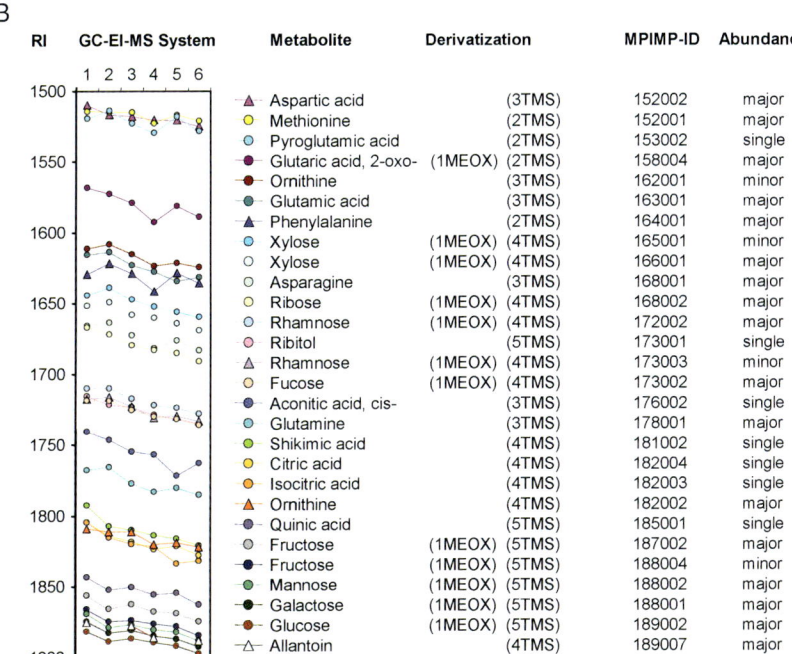

Figure 5. RI-offset between GC-EI-MS systems operated with an identical stationary GC column phase. Authenticated MSTs from pure reference substances exhibit good RI linearity between different GC-EI-MS systems (A) and in general a constant elution sequence (B). Metabolites of identical compound classes exhibit strict repeatability of elution. In contrast, the RI sequence may locally differ between compound classes, for examples refer to allantoin and hexoses, aspartic and pyroglutamic acid, or ornithine and citric acid. GC-EI-MS systems had either TOF (time of flight), 1,2,4, or quadrupole MS technology, 3,5,6. The MPIMP-ID may be used to retrieve further MST information (GMD, http://csbdb.mpimp-golm.mpg.de/gmd.html) [48].

6. Comparison with the elution sequence of all eight possible hexoaldoses, which was previously established on a GC-TOF-MS system [44], shows the best RI fit of mannose. Abundant peaks like glucose in leaf samples can obscure minor isomers. In the absence of clearly visible minor analytes galactose cannot be distinguished from idose and talose (Fig. 4C).
7. Note that previously established RI sequences and RI data determined in other laboratories or on different GC-MS systems (Fig. 5A) exhibit a slight RI-offset, which as a first approximation is best corrected by a factor proportional to the observed RI, such as a percentage (Fig. 4). Late eluting compounds exhibit as a rule a stronger off-set than early eluting analytes. Due to small differences in GC column make and column aging, differences in temperature programming or carrier gas flow and pressure, RIs of different compound classes may exhibit a differential shift. Thus, when alcanes are used for RI standardization hydrocarbons have almost no shift in response to changes in flow or pressure, however different classes of TMS ethers and esters show clear off-sets.
8. The elution sequence within each of the compound classes, however, is fully maintained. RI inversions of co-eluting compounds occur only between different compound classes (Fig. 5B). The correction for RI-offset is best performed by including reference mixtures of pure compounds into every set of routine profiling experiments. These mixtures should ideally contain at least one representative of each of the difficult to identify diastereomer classes. Sugars and respective alcohols or polyhydroxyacids are among the most critical metabolite classes, for example C4-C7 monosaccharides, and respective phosphates, polyols, or acids, such as glucuronic-, glucaric- or gluconic acid.

MS-RI libraries enhance MST identification

The enormous chemical diversity of compounds obtained when analyzing the metabolome of organisms constitutes one of the main challenges in metabolomics [8, 45]. Current estimations vary. However, 4,000–25,000 compounds may represent the metabolome of any given organism [8, 46]. The plant kingdom is believed to comprise in excess of 200,000 metabolites with only a minority of well studied primary metabolites [6, 46].

From what was said above it is evident that the highly diverse chemical characteristics in conjunction with the vast amount of potential compounds have profound implications on any non-biased attempt to apply an analytical technology. Currently only approximately 35% of the MSTs from GC-MS profiling analyses are identified. The majority of known metabolites in GC-EI-MS profiles still are primary metabolites [12, 13, 47]. The huge white parts on the metabolite profiling chart is one of the most puzzling and challenging findings of the metabolite profiling effort.

Did traditional biochemistry overlook a multitude of metabolic products or does metabolite profiling suffer from hard to access or incompletely accessible previous phytochemical research data?

Irrespective of the outcome of the time-consuming peak to peak charting effort in multiple laboratories, it is evident that this task is best performed as a long-term, open access project with contributions of experts on different organisms and pathways. Thus the Golm Metabolome Database (GMD) started to tackle the urgent and necessary need for a public metabolome database that harbors pathway information and the underlying technical details that are prerequisite for metabolome analyses [48]. Because any technology has specific potential and limitations GMD currently focuses on the best understood metabolite profiling technology platform, namely GC-EI-MS profiling of methoxyaminated and trimethyl-silylated extracts of polar metabolites [15, 25, 44]. GMD provides identified and frequently observed yet non-identified MSTs in MS-RI libraries, which are provided in a so-called msp-format, that can be imported either into NIST02/05 or AMDIS mass spectral processing software (National Institute of Standards and Technology, Gaithersburg, MD, USA). AMDIS provides MS deconvolution, a fast automated RI and MS matching algorithm, and allows transfer of mass spectra to NIST02/05, which has a more accurate MS comparison algorithm but no capability for automated RI matching.

Metabolite coverage of GC-MS profiling

Any given protocol for metabolome measurements represents a well-tuned balance between accuracy and metabolite coverage. The coverage of GC-MS based metabolite profiling after methoxyamination and silylation of dried biological extracts is best exemplified by an inventory (Tab. 1) of the environmental stress experiments presented in Figure 1. Table 1 was generated with the GMD custom MSRI library and AMDIS (Version 2.63, 2005). AMDIS settings were peak width 20, adjacent peak substraction 2, resolution and shape requirements low and sensitivity medium. RI windows and penalties were deactivated, multiple identifications allowed and the minimum match factor set to 65. Report files of 15 representative GC-MS profiles from the above experiment were filtered for the best match of each MST present in the GMD library. The RI off-set between library and this GC-MS profiling experiment was corrected by a factor of 0.29 RI% as determined from reference mixture of metabolites. Positive matches were reported within a ± 5.0 RI window. Table 1 reports the quality of identification by signal to noise, RI deviation and reverse match values.

Analytes are characterized by a MPIMP-ID, number of derivatized moieties, possible multiple derivatives, expected RI and five characteristic mass fragments. Additional information on MSTs and identified metabolites can be downloaded from GMD using either name, MPIMP-ID, or mass spectral search options (GMD, http://csbdb.mpimp-golm.mpg.de/gmd.html) [48]. Metabolite identity is established by name, sum formula, and KEGG or CAS identifer and thus linked to pathway and chemometric information. KEGG and CAS metabolite identifiers in this table represent the biologically relevant main enantiomers. GMD pursues the concept of using existing metabolite identification systems rather than creating yet one further redundant metabolite definition. In contrast analytes had to be indexed by GMD, because the majority of analytes are still non-identified and identified products did not always have a CAS index number.

In conclusion, Table 1 clearly shows the high coverage of small primary metabolites which can be classified into organic acids, amino acids, N-containing

Table 1. List of metabolites and analytes from leaf metabolite profiles of *Arabidopsis thaliana* ecotype Columbia. Note that due to changes in metabolite pool size other experiments or plant ecotypes might have a slightly differing inventory. *Maltotriose, was missed by automated AMDIS analysis. These mass spectra were manually deconvoluted at higher sensitivity. Manual matching was performed using NIST05 software; hence the differing range of match values, i.e., 1–1,000. **These compounds may occur as laboratory contaminations and consequently require non-sample background correction.

Metabolite					Analyte							Match		
Name	Sum Formula	KEGG-ID [preferred]	CAS-ID [preferred]		MPIMP-ID	TMS	MEOX	Derivative	Mass Fragments [expected]	RI [expected]	RI [RI-RI(Lib)]	MS Match [reverse]	S/N [total]
Sugars														
Arabinose	C5 H10 O5	C00259	147-81-9	D-	167002	4	1	major 1	307/217/160/103/189	1675.3	1.2	95	75
Ribose	C5 H10 O5	C00353	50-69-1	D-	168002	4	1	major 1	307/217/160/103/189	1690.9	0.6	91	76
Xylose	C5 H10 O5	C00181	58-86-6	D-	166001	4	1	major 1	307/217/160/103/189	1669.2	-0.2	98	123
Xylose	C5 H10 O5	C00181	58-86-6	D-	165001	4	1	minor 2	307/217/160/103/189	1659.7	0.1	88	63
Glucose, 1,6-anhydro, beta-	C6 H10 O5		498-07-7		172001	3		major 1	204/217/333/243/317	1714.9	1.7	100	497
Rhamnose	C6 H12 O5	C00507	3615-41-6	L-	172002	4	1	major 1	117/160/364/277/321	1727.8	1.3	84	63
Fructose	C6 H12 O6	C00095	57-48-7	D-	187002	5	1	major 1	307/217/277/364/335	1874.6	0.7	100	534
Fructose	C6 H12 O6	C00095	57-48-7	D-	188004	5	1	major 2	307/217/277/364/335	1884.5	0.6	100	1418
Galactose	C6 H12 O6	C00124	59-23-4	D-	188006	5	1	minor 1	160/319/229/343/305	1892.3	-0.2	89	155
Galactose	C6 H12 O6	C00124	59-23-4	D-	191002	5	1	minor 2	160/319/229/343/305	1912.4	1.1	76	213
Glucose	C6 H12 O6	C00031	50-99-7	D-	189002	5	1	major 1	160/319/229/343/305	1897.3	0.0	100	735
Glucose	C6 H12 O6	C00031	50-99-7	D-	191001	5	1	minor 2	160/319/229/343/305	1916.4	-0.6	96	243
Mannose	C6 H12 O6	C00936	3458-28-4	D-	188002	5	1	minor 1	160/319/229/343/305	1888.1	1.3	81	95
Mannose	C6 H12 O6	C00936	3458-28-4	D-	189001	5	1	minor 2	160/319/229/343/305	1899.5	3.1	67	2246
Sorbose	C6 H12 O6	C00764	3615-56-3	D-	186008	5	1	major 1	307/217/277/364/335	1873.0	0.0	97	981
Sorbose	C6 H12 O6	C00764	3615-56-3	D-	187007	5	1	minor 2	307/217/277/364/335	1878.2	-0.6	96	2235
Galactopyranoside, 1-O-methyl-, beta-	C7 H14 O6	C03619	97-30-3		185010	4		single	133/204/377/231/290	1856.8	2.8	81	40
Glucopyranoside, 1-O-methyl-, alpha-	C7 H14 O6		97-30-3		186006	4		single	133/204/377/231/290	1866.8	0.9	68	2200
Maltose	C12 H22 O11	C00897	69-79-4		274001	8		major 1	160/480/204/319/300	2744.6	1.9	100	2157
Maltose	C12 H22 O11	C00897	69-79-4		277002	8		minor 2	160/204/361/319/271	2768.4	1.1	100	849
Melibiose	C12 H22 O11	C05402	585-99-9		287003	8		single	160/480/204/319/361	2872.6	2.5	93	68
Melibiose	C12 H22 O11	C05402	585-99-9		290002	8		minor 2	160/480/204/319/361	2903.5	2.3	84	30
Sucrose	C12 H22 O11	C00089	57-50-1		264001	8		major 1	437/451/361/319/157	2653.4	0.3	100	2373
Trehalose, alpha,alpha'-	C12 H22 O11	C01083	99-20-7		274002	8		single	191/169/361/243/331	2749.1	-0.1	100	268
Raffinose	C18 H32 O16	C00492	512-69-6		337002	11		single	437/451/361/217/204	3396.0	0.0	100	318
*Maltotriose	C18 H32 O16	C01835	1109-28-0		355003	11		major 1	204/361/217/480/169	3550.2	-0.1	*924	88
*Maltotriose	C18 H32 O16	C01835	1109-28-0		358001	11		minor 2	204/361/217/271/169	3583.1	-0.1	*891	47
Polyhydroxy Acids														
Dehydroascorbic acid dimer	C6 H6 O6	C05422		L-	185002	4		major 1	316/173/157/245/231	1852.6	1.3	100	533
Ascorbic acid	C6 H8 O6	C00072	50-81-7	L-	195002	4		major 1	332/449/464/117/303	1951.1	1.5	100	448
Glucaric acid-1,4-lactone	C6 H10 O7	C03383	389-36-6		194009	4		single	217/244/480/465/347	1949.6	4.1	72	206
Galactonic acid-1,4-lactone	C6 H10 O6		322328		189003	4		single	217/451/466/332/305	1890.5	1.3	89	103
Gluconic acid-1,4-lactone	C6 H10 O6				189013	4		single	217/244/466/332/305	1897.2	-2.7	75	248
Gluconic acid-1,5-lactone	C6 H10 O6	C00198	90-80-2		189008	4		single	220/229/319/451/129	1857.9	3.3	75	150
Gluconic acid	C6 H10 O7	C00191	1700908		193004	5		major 1	333/160/423/292/364	1937.4	1.7	77	17
Galactaric acid	C6 H10 O8	C01807	526-99-8		204001	5		single	333/292/373/423/305	2045.4	2.5	92	46
Glucaric acid	C6 H10 O8	C00767	87-73-0		201001	6		single	333/292/373/423/305	2013.7	4.4	68	45
Galactonic acid	C6 H12 O7	C00880	576-36-3		196002	6		single	333/292/319/305/157	1997.9	0.7	100	189
Gluconic acid	C6 H12 O7	C00257	526-95-4		200001	6		single	333/292/319/305/157	2002.7	0.6	94	79
Gulonic acid	C6 H12 O7				196001	6		single	333/292/423/433/319	1964.2	1.1	74	44
Sugar Conjugates														
Maltitol	C12 H24 O11			D-	284001	9		single	204/361/345/525/305	2839.2	-0.3	87	41
Galactinol	C12 H22 O11	C01235	565-88-6	D-	299002	9		single	204/191/433/305/169	2993.5	-0.6	89	48
MSTs (exemplary)														
[914; Galactinol (5TMS)]					301005				204/361/433/305/191	3014.2	1.6	100	284
[926; Galactosylglycerol (6TMS)]					231002				337/204/217/3611/129	2328.7	-1.4	97	119
[Benzylglucopyranoside (4TMS)]					241003				91/209/204/217/233	2408.9	1.1	92	50
[Salicylic acid-glucopyranoside (5TMS)]					258003				267/361/217/169/243	2596.4	0.5	94	51

Table 1 (continue)

Metabolite				Analyte							Match			
Name	Sum Formula		KEGG-ID [preferred]	CAS-ID [preferred]	MPIMP-ID	TMS	MEOX	Derivative	Mass Fragments [expected]	RI [expected]	RI [RI-RI(Lib)]	MS Match [reverse]	S/N [total]
Acids														
Oxalic acid	C2 H2 O4		C00209	144-62-7	113002	2		single	219|190|175|147|133	1134.9	4.0	90	422
Glycolic acid**	C2 H4 O3		C00160	79-14-1	106002	2		single	177|205|161|131|103	1064.0	0.5	97	80
Pyruvic acid	C3 H4 O3		C00022	127-17-3	104002	1	1	major 1	174|189|115|89|158	1036.6	0.7	95	115
Malonic acid	C3 H4 O4		C00383	141-82-2	122003	2		major 1	233|248|147|133|109	1211.4	-4.7	86	24
Lactic acid**	C3 H6 O3		C00186	79-33-4	105001	2		single	219|117|191|133|234	1048.9	-2.0	99	106
Glyceric acid	C3 H6 O4		C00258	473-81-4	135003	3		single	292|189|307|205|133	1339.6	-0.3	100	255
Fumaric acid	C4 H4 O4		C00122	110-17-8	137001	2		single	245|115|217|143	1359.8	0.2	99	944
Maleic acid*	C4 H4 O4	D-	C01384	110-16-7	133003	2		single	245|147|170|215	1314.7	0.2	99	167
Erythronic acid-1,4-lactone	C4 H6 O4	Z-		15667-21-7	144008	2		single	247|262|219|233|189	1435.6	3.4	93	78
Succinic acid	C4 H6 O4		C00042	110-15-6	134001	2		single	247|172|147|282|129	1326.0	-0.9	100	258
Threonic acid-1,4-lactone	C4 H6 O4				140005	2		single	247|147|262|217|101	1382.5	0.9	99	143
Malic acid	C4 H6 O5		C00149	97-67-6	149001	3		single	233|245|335|307|217	1492.3	0.2	100	1450
Butyric acid, 4-hydroxy-	C4 H8 O3		C00989	591-81-1	126002	2		single	233|117|204|143|133	1242.6	0.0	81	26
Butyric acid, 2,4-dihydroxy-	C4 H8 O4				143004	3		single	219|321|203|103|147	1419.6	0.9	90	53
Erythritol	C4 H8 O5			15667-21-7	154001	4		single	292|220|117|319|205	1548.7	0.7	99	238
Threonic acid	C4 H8 O5	L-	C01620	7306-96-9	156001	4		single	292|220|205|117|319	1568.2	1.6	99	328
Itaconic acid	C5 H6 O4		C00490	97-65-4	135004	2		single	259|215|133|147|230	1351.7	1.5	84	74
Maleic acid, 2-methyl-	C5 H6 O4	Z-	C02226	498-23-7	137003	2		single	259|184|122|157|231	1358.0	-3.7	83	24
Glutaric acid, 2-oxo-	C5 H6 O5		C00026	328-50-7	158004	2	1	major 1	198|236|304|166|229	1568.7	2.0	80	176
Adipic acid**	C6 H10 O4		C06104	124-04-9	151006	2		single	275|111|147|172|159	1509.0	1.0	80	103
Aconitic acid	C6 H6 O6	Z-	C00417	585-84-2	178002	3		single	229|285|375|211|215	1762.8	0.0	96	182
Citric acid	C6 H8 O7		C00158	77-92-9	182004	4		single	273|375|211|183|257	1827.8	0.7	99	1461
Isocitric acid	C6 H8 O7		C00311	320-77-4	182003	4		single	245|319|390|83|	1831.6	0.0	87	144
Shikimic acid	C7 H10 O5		C00493	138-59-0	181002	4		single	204|462|372|255|357	1820.9	1.7	100	197
Benzoic acid**	C7 H6 O2		C00180	65-85-0	128003	1		single	179|105|135|77|194	1256.7	-0.2	98	88
Benzoic acid, 4-hydroxy-	C7 H6 O3		C00156	99-96-7	164003	2		single	267|223|282|193|126	1639.5	0.2	99	76
Salicylic acid	C7 H6 O3		C00805	69-72-7	152003	2		single	267|209|91|249|135	1511.0	2.2	89	22
Antranilic acid	C7 H7 N O2		C00108	118-92-3	163002	2		major 1	266|281|134|208|232	1630.9	-3.5	92	17
Benzoic acid, 4-amino-	C7 H7 N O2		C00568	150-13-0	184001	2		major 1	266|281|222|192|126	1841.1	2.3	89	16
Ferulic acid	C10 H10 O4	E-			210001	2		major 1	338|249|323|293|308	2058.6	3.8	89	24
Sinapic acid	C11 H12 O5	Z-			207001	2		major 1	338|368|323|353|279	2062.2	4.3	77	23
Sinapic acid	C11 H12 O5	E-	C00482	530-59-6	225001	2		single	368|338|353|323|249	2252.9	1.5	100	278
Amino Acids														
Tryptophan	C11 H12 N2 O2	L-	C00078	73-22-3	223001	3		major 1	202|291|218|303|130	2218.6	1.7	95	44
Glycine	C2 H5 N O2		C00037	56-40-6	133001	3		major 1	174|248|276|100|86	1311.9	-0.2	100	1195
Glycine	C2 H5 N O2		C00037	56-40-6	114001	2		minor 2	102|147|204|176|86	1118.0	-1.2	100	353
Malonic acid, 2-amino-	C3 H5 N O4		C00872	1068-84-4	147001	3		major 1	320|292|218|248|174	1470.8	4.2	95	1409
Alanine	C3 H7 N O2	L-	C00041	56-41-7	110001	2		major 1	116|190|218|100|233	1095.5	-2.7	100	709
Alanine	C3 H7 N O2	L-	C00041	56-41-7	138002	3		minor 2	188|262|290|100|114	1363.9	0.0	100	399
Alanine, beta-	C3 H7 N O2		C00099	107-95-9	144001	3		major 1	248|290|174|160|100	1431.4	1.7	98	155
Cysteine	C3 H7 N O2 S	L-	C00097	52-90-4	156002	3		major 1	294|220|218|100|116	1560.7	1.4	92	47

Table 1 (continue)

Metabolite				Analyte							Match		
Name	Sum Formula	KEGG-ID [preferred]	CAS-ID [preferred]	MPIMP-ID	TMS	MEOX	Derivative	Mass Fragments [expected]	RI [expected]	RI [RI-Ri(Lib)]	MS Match [reverse]	S/N [total]
Amino Acids												
Serine	C3 H7 N O3	L- C00065	56-45-1	138001	3		major 1	204|218|276|306|100	1369.3	-0.1	100	1074
Serine	C3 H7 N O3	L- C00065	56-45-1	128001	2		minor 1	116|132|219|234|159	1265.3	-0.2	100	365
Serine	C3 H7 N O3	L- C00065	56-45-1	158006	4		minor 3	114|290|378|276|223	1578.1	0.0	98	197
Alanine, 3-cyano-	C4 H6 N2 O2	C02512	6232-19-5	138005	2		major 1	141|243|145|202|130	1382.8	0.9	88	147
Aspartic acid	C4 H7 N O4	C00049	56-84-8	152002	2		major 1	232|218|306|202|334	1525.0	0.3	99	600
Aspartic acid	C4 H7 N O4	C00049	56-84-8	174003	3		minor 2	205|304|406|172|100	1744.7	-2.1	92	97
Aspartic acid	C4 H7 N O4	C00049	56-84-8	144003	2		minor 3	160|130|117|245|202	1431.2	2.5	97	400
Asparagine	C4 H8 N2 O3	L- C00152	70-47-3	168001	3		minor 3	116|188|231|258|159	1683.3	1.2	99	204
Asparagine	C4 H8 N2 O3	L- C00152	70-47-3	161004	4		minor 4	188|216|305|420|405	1610.0	1.0	98	130
Asparagine	C4 H8 N2 O3	L- C00152	70-47-3	187001	4		major 1	130|204|216|232|142	1871.1	3.1	92	76
Butyric acid, 2-amino-	C4 H9 N O2	C00334	56-12-2	117002	2		major 1	174|304|216|246|100	1530.7	1.1	85	34
Butyric acid, 4-amino-	C4 H9 N O2	C00263		153003	3		major 1	218|128|292|330|202	1454.0	1.7	89	450
Homoserine	C4 H9 N O3	C00263	672-15-1	146001	3		minor 3	290|200|364|392|158	1682.6	0.8	93	131
Homoserine	C4 H9 N O3	C00188	672-15-1	168006	4		major 1	219|291|218|117|320	1994.0	1.5	82	22
Threonine	C4 H9 N O3	C00188	72-19-5	140001	2		minor 1	117|219|248|130|146	1301.1	0.2	100	460
Threonine	C4 H9 N O3	C00064	72-19-5	132001	2		minor 2	156|245|347|362|203	1785.1	1.2	100	227
Glutamine	C5 H10 N2 O3	C00064	56-85-9	176001	3		major 1	227|317|156|203|128	2000.7	0.9	99	270
Glutamine	C5 H10 N2 O3	C01826	56-85-9	200005	4		minor 2	144|218|156|246|100	1245.5	2.3	100	314
Norvaline	C5 H11 N O2	C00183	6600-40-4	126001	2		major 1	144|218|156|246|100	1220.2	-0.7	71	19
Valine	C5 H11 N O2	C00073	72-18-4	122001	2		major 1	178|128|250|293|202	1521.2	-1.4	99	136
Methionine	C5 H11 N O2 S	C00077	63-68-3	152001	2		major 1	142|174|420|200|258	1821.9	1.5	99	52
Ornithine	C5 H12 N2 O2	C00077	70-26-8	182002	4		major 1	174|186|345|244|142	1756.6	1.1	96	127
Ornithine	C5 H12 N2 O2	C00077	70-26-8	176006	3		minor 1	214|174|375|288|200	2037.7	1.7	93	96
Ornithine	C5 H12 N2 O2	C00077	70-26-8	204003	5		minor 4	142|348|243|216|204	1624.2	0.6	70	17
Pyroglutamic acid	C5 H7 N O3	C02238		162001	2		single	156|258|230|140|273	1528.1	-0.1	86	62
Proline	C5 H9 N O2	C00148	147-85-3	153002	2		major 1	142|130|117|244	1303.4	0.4	99	767
Proline, 4-hydroxy-	C5 H9 N O3	L- E- C01015	51-35-4	132003	2		major 1	230|140|304|158|332	1527.8	1.7	100	749
Glutamic acid	C5 H9 N O4	C00025	56-86-0	163001	3		major 1	246|363|128|348|156	1631.4	0.3	82	1397
Glutamic acid	C5 H9 N O4	C00025	56-86-0	154002	2		minor 2	174|276|158|230|84	1538.9	0.2	100	524
Serine, O-acetyl-	C5 H9 N O4	C00979	5147-00-2	141001	2		major 1	174|132|116|218|100	1402.7	1.1	97	236
Isoleucine	C6 H13 N O2	C00407	73-32-5	132002	2		major 2	158|232|218|102|260	1300.6	0.1	94	66
Isoleucine	C6 H13 N O2	C00407	73-32-5	119002	1		minor 2	86|158|170|146|130	1183.8	0.6	98	101
Leucine	C6 H13 N O2	C00123	61-90-5	129002	2		major 1	158|232|102|260	1278.8	-0.8	85	26
Lysine	C6 H14 N2 O2	C00047	56-87-1	192003	4		major 1	156|174|317|230|434	1920.8	2.7	100	38
Phenylalanine	C9 H11 N O2	C00079	63-91-2	164001	1		minor 2	192|266|218|91|294	1635.4	1.3	91	66
Phenylalanine	C9 H11 N O2	C00079	63-91-2	157001	3		minor 1	120|146|91|204|130			100	215
Tyrosine	C9 H11 N O3	C00082	60-18-4	194002	3		major 1	218|280|354|179|100				
Tyrosine	C9 H11 N O3	C00082	60-18-4	189006	2		minor 2	179|208|219|310|91				

Table 1 (continue)

Metabolite					Analyte								Match		
Name	Sum Formula	KEGG-ID [preferred]	CAS-ID [preferred]	MPIMP-ID	MEOX	TMS	Derivative	Mass Fragments [expexted]	RI [expexted]	RI [RI-RI(Lib)]	MS Match [reverse]	S/N [total]
Fatty Acids[**]												
Hexanoic acid, n-	C6 H12 O2	C01585	142-62-1	106001		1	single	173	188	117	129	145	1561.5	-4.9	82	77
Heptanoic acid, n-	C7 H14 O2		111-14-8	117001		1	single	187	202	117	129	145	1941.4	0.5	95	106
Octanoic acid, n-	C8 H16 O2	C06423	124-07-2	127006		1	single	201	216	117	129	145	1897.6	2.6	67	149
Nonanoic acid, n-	C9 H18 O2	C01601	112-05-0	138003		1	single	215	230	117	129	145				
Decanoic acid, n-	C10 H20 O2	C01571	334-48-5	147004		1	single	229	244	117	201	145	1064.0	-2.3	89	38
Dodecanoic acid, n-	C12 H24 O2	C02679	143-07-7	166003		1	single	257	272	117	201	145	1167.8	0.4	91	31
Tetradecanoic acid, n-	C14 H28 O2	C06424	544-63-8	185004		1	single	285	300	117	201	145	1270.6	1.2	94	70
Hexadecanoic acid, n-	C16 H32 O2	C00249	57-10-3	205001		1	single	313	328	117	201	145	1369.2	0.6	93	78
Octadecanoic acid, n-	C18 H36 O2	C01530	57-11-4	225002		1	single	341	356	117	129	145	1465.9	-3.2	94	49
									1662.5	1.3	92	38
Alcohols									1852.6	0.4	66	55
Benzyl alcohol	C7 H8 O	C00556	100-51-6	115003		1	single	165	135	91	180	105	2050.2	2.4	100	132
Octadecan-1-ol, n-	C18 H38 O		112-92-5	215001		1	single	327	97	111	115	125	2247.0	2.7	98	72
N- Compounds												
Urea	C H4 N2 O	C00086	57-13-6	127002		2	major 1	189	204	171	87	99	1151.6	0.8	85	40
Ethanolamine	C2 H7 N O	C00189	110-60-1	128002		3	major 1	174	86	100	188	262	2154.7	4.1	93	37
Putrescine	C4 H12 N2	C00134	110-60-1	175002		4	major 1	174	361	214	100	200				
Putrescine	C4 H12 N2	C00134		151005		4	major 2	174	142	289	304	115				24
Uracil	C4 H4 N2 O2	C00106	66-22-8	136001		3	single	241	255	99	113	126	1260.1	-3.2	87	219
Pyridine, 3-hydroxy-	C5 H5 N O		109-00-2	114003		1	single	152	167	136	122	92	1269.1	0.7	98	165
Spermidine	C7 H19 N3	C00315	124-20-9	220002		4	minor 2	174	200	418	257	160	1741.6	0.6	100	88
Tyramine	C8 H11 N O	C00483	51-67-2	191004		3	minor 2	174	338	86	100	264	1509.6	0.7	81	85
									1346.6	2.5	95	57
Phosphates									1137.0	3.5	78	44
Glyceric acid-3-phosphate	C3 H7 O7 P	C00197	820-11-1	181003	D-	4	single	387	299	459	357	217	2194.0	0.8	78	40
Glycerol-3-phosphate	C3 H9 O6 P	C00093	29849-82-9	177002	D-	4	single	357	445	299	315	211	1910.4	2.9	85	54
Fructose-6-phosphate	C6 H13 O9 P	C00085	643-13-0	232002	D-	6	major 1	459	315	357	217					
Galactose-6-phosphate	C6 H13 O9 P	C01113		232001	D-	6	major 1	160	387	299	471	357	1817.9	1.5	97	72
Galactose-6-phosphate	C6 H13 O9 P	C01113		235001	D-	6	minor 1	160	387	299	471	357	1775.1	3.2	89	80
Glucose-6-phosphate	C6 H13 O9 P	C00092	56-73-5	233002	D-	6	major 1	160	387	299	471	357	2321.4	0.0	99	225
Glucose-6-phosphate	C6 H13 O9 P	C00092	56-73-5	233002	D-	6	minor 2	160	387	299	471	357	2328.6	4.2	71	39
Mannose-6-phosphate	C6 H13 O9 P	C00275		231001	D-	6	major 1	160	471	387	357	299	2351.0	2.8	80	35
Mannose-6-phosphate	C6 H13 O9 P	C00275		233001	D-	6	minor 2	160	387	299	471	357	2334.5	-0.3	100	246
Phosphoric acid	H3 O4 P	C00009	7664-38-2	129001		3	single	314	299	211	283	225	2352.5	0.6	98	121
									2323.5	-2.2	87	198
Polyols									2334.8	-0.7	97	64
Glycerol	C3 H8 O3	C00116	56-81-5	129003		3	single	293	205	117	103		1281.9	-1.0	100	220
Erythritol	C4 H10 O4	C00503	149-32-6	150002		4	single	217	293	307	205	320				
Threitol	C4 H10 O4		2418-52-2	149002	D-	4	single	217	293	307	205	320				
Arabitol	C5 H12 O5	C01904	488-82-4	171012		5	single	307	319	332	217	205	1282.5	-1.5	99	209
Ribitol	C5 H12 O5	C00474	488-81-3	170001		5	single	319	307	422	217	205	1510.2	0.0	97	141
Xylitol	C5 H12 O5	C00379	87-99-0	171001		5	single	307	319	332	217	205	1501.7	0.4	78	89
Inositol, myo-	C6 H12 O6	C00137	87-89-8	209002		6	single	305	265	318	191	507	1729.7	4.9	94	839
Galactitol	C6 H14 O6	C01697	608-66-2	194001	D-	6	single	319	307	157	217	331	1734.7	0.3	100	1241
Mannitol	C6 H14 O6	C00392	69-65-8	193002	D-	6	single	319	307	157	217	331	1717.6	-0.8	74	445
Sorbitol	C6 H14 O6	C00794	50-70-4	193001	D-	6	single	319	307	157	217	331	2091.9	0.2	100	724
									1941.8	-4.1	75	109
									1929.8	4.2	77	132
									1935.8	0.1	94	130

compounds, sugars, polyols, polyhydroxy acids, and small conjugates. In addition, four hitherto non-identified MST are shown for the purpose of demonstration. These MSTs can be preliminary classified by best mass spectral match to already identified MSTs or by manual mass spectral interpretation. Thus the potential of metabolite profiling to deal with not yet identified MST and the option to link future precise metabolite identifications to past measurements is demonstrated. While automated analysis is already fairly powerful, it is not perfect and manual identification still allows extension of automated inventories, for example maltotriose (Tab. 1). Validation of usually rare or usually absent metabolites such as sorbose in this example, or *Arabidopsis* leaf, is still required. In ambiguous cases repeated standard addition experiments are advised. A completed inventory finally allows choice of selective metabolite derivatives and mass fragments for the quantitative analysis [48].

Limitations of metabolite coverage in GC-MS profiling

GC-MS profiling technology is perhaps the best understood platform for metabolome analyses. Our understanding not only comprises metabolome coverage but also detailed information about limitations. The most obvious limitation of GC-MS profiling is analyte volatility. Small compounds close to the volatility of the reagent and solvent are lost as are high molecular weight compounds which have boiling points exceeding the temperature range of gas chromatography. A good overview of the current size limitations is provided by RI and sum formula information of Table 1. Besides these obvious limitations a small number of specific pitfalls exist in GC-MS profiling which are well understood and arise mainly from metabolite instability, conversion of different metabolites into the same analyte through action of the chemical reagent, or co-elution of chemically distinct diastereomers and enantiomers without option for selective choice of mass fragments. In the following exemplary cases will be discussed.

Metabolite instability is a general problem for metabolite analysis. A typical example is ascorbic acid. Ascorbic acid can be analyzed by GC-MS or traditional HPLC based technologies provided oxygen is eliminated by degassing and argon or nitrogen enriched atmosphere. Without these precautions ascorbic acids yields more than 10 distinctive products in routine GC-MS metabolite profiling, the most abundant among these is – not unexpected – dehydroascorbic acid. Recovery experiments using chemically synthesized isoascorbic acid demonstrate a sample dependent loss of this instable stereoisomer of vitamin C which unexpectedly can be chromatographically separated from ascorbic acid in routine GC-MS profiling experiments. Applying GC-MS profiling without protective gasses results in 20–30% recovery of isoascorbic acid from potato leaves; in comparison potato tubers have only 5–10% recovery and the compound is completely lost from potato root samples.

Analyte conversion is specific for the reagent chemistry applied. A typical example is the loss of N-aminoiminomethyl- (guanidino-; $-NH-CNH-NH_2$) and N-carbamoyl- (ureido-; $-NH-CO-NH_2$) moieties, which result in conversion of arginine, and citrulline to ornithine and of agmatine to putrescine.

A general restriction brought about by methoxyamination is the conversion of alpha- and beta- conformations of cyclic hemiacetals – present in reducing sugars – into the respective methoxyamine, and the loss of phosphate moieties linked to hemiacetals, such as glucose-1-phosphate. In contrast, glycosidic bonds maintain conformation and structural integrity. A borderline case between analyte conversion and metabolite instability is pyroglutamate, which is formed from glutamine through loss of NH_3 and by far smaller proportion from glutamate by loss of H_2O. These cycle formation processes occur in aqueous solution and are enhanced by prolonged TMS derivatization protocols.

Co-elution is a specific chromatographic problem. As long as co-eluting analytes can be distinguished by specific and selective mass fragments, co-elution presents no problem for compound specific quantification. In general routine capillary GC columns such as employed for metabolite profiling are not enantio-selective. Thus L-amino acids and D-sugars cannot be distinguished from the rare D- and L- enantiomers. Identifications such as the preferred metabolite IDs given in Table 1 represent an approximation based on expected enantiomer abundance. Library updates of GMD are in preparation, which will list all frequent and rare metabolites which are currently known to be represented by each of the included analytes.

Diastereomers such as the different hexoaldoses can usually be chromatographically separated. However the high number of possible structures inevitably leads to co-elution of analytes (Fig. 4). Co-elution problems are today addressed by GC-MS technology extensions. One strategy utilizes two capillary columns with alternate separation properties. This ultimately highly powerful approach is called GCxGC-TOF-MS technology and can be employed for two-dimensional chromatographic separation in metabolite profiling experiments (e.g., [49–51]). The future will show if repeatability of 2D-separation and the higher apparent sensitivity of GCxGC-TOF-MS can indeed be utilized for a high-throughput routine profiling technology of approximately 2,000 MSTs as reported by a recent publication [52].

Acknowledgements

The highlight experiments were provided by P Doermann, Max Planck Institute of Molecular Plant Physiology, Potsdam-Golm, Germany. The underlying data set is available upon request to the contact author. The authors acknowledge N Schauer and AR Fernie, Max Planck Institute of Molecular Plant Physiology, Am Muehlenberg 1, D-14476 Golm, Germany, S Strelkov and D Schomburg, University of Cologne, CUBIC – Institute of Biochemistry, Zuelpicher Str. 47, D-50674 Cologne, Germany, T Moritz and K Lundgren, Umea Plant Science Centre, Department of Forest Genetics and Plant Physiology, Swedish University of Agricultural Sciences, SE-901 83 Umea, Sweden, U Roessner and MG Forbes, University of Melbourne, School of Botany, 3010 Victoria, Australia, A Barsch, M Puehse and M Persicke, Bielefeld University, Department of Molecular Phytopathology (Prof. Niehaus), D-33501 Bielefeld, Germany, for making MST information publicly available.

This work was supported by the Max-Planck Society, and the Bundesministerium für Bildung und Forschung (BMBF), grant PTJ-BIO/0312854.

References

1. Tweeddale H, Notley-McRobb L, Ferenci T (1998) Effect of slow growth on metabolism of *Escherichia coli*, as revealed by global metabolite pool ('metabolome') analysis. *J Bacteriol* 180: 5109–5116
2. Oliver SG, Winson MK, Kell DB, Baganz F (1998) Systematic functional analysis of the yeast genome. *Trends Biotechnol* 16: 373–378
3. Nicholson JK, Lindon JC, Holmes E (1999) 'Metabonomics': understanding the metabolic responses of living systems to pathophysiological stimuli via multivariate statistical analysis of biological NMR spectroscopic data. *Xenobiotica* 29: 1181–1189
4. Stoughton RB, Friend SH (2005) Innovation – How molecular profiling could revolutionize drug discovery. *Nat Rev Drug Dis* 4: 345–350
5. Trethewey RN, Krotzky AJ, Willmitzer L (1999) Metabolic profiling: a Rosetta stone for genomics? *Curr Opin Plant Biol* 2: 83–85
6. Fiehn O (2002) Metabolomics – the link between genotypes and phenotypes. *Plant Mol Biol* 48: 155–171
7. Sumner LW, Mendes P, Dixon RA (2003) Plant metabolomics: large-scale phytochemistry in the functional genomics era. *Phytochem* 62: 817–836
8. Fernie AR, Trethewey RN, Krotzky AJ, Willmitzer L (2004) Metabolite profiling: from diagnostics to systems biology. *Nat Rev Mol Cell Biol* 5: 763–769
9. Jellum E, Helland P, Eldjarn L, Markwardt U, Marhofer J (1975) Development of a computer-assisted search for anomalous compounds (CASAC). *J Chromatogr* 112: 573–580
10. Jellum E (1977) Profiling of human-body fluids in healthy and diseased states using gas-chromatography and mass-spectrometry, with special reference to organic-acids. *J Chromatrogr B* 143: 427–462
11. Jellum E (1979) *Application of mass-spectrometry and metabolite profiling to the study of human-diseases.* Philosophical Transactions of the Royal Society of London Series A-Mathematical Physical and Engineering Sciences, 293: 13–19
12. Fiehn O, Kopka J, Dörmann P, Altmann T, Trethewey RN, Willmitzer L (2000) Metabolite profiling for plant functional genomics. *Nat Biotechnol* 18: 1157–1161
13. Roessner U, Wagner C, Kopka J, Trethewey RN, Willmitzer L (2000) Simultaneous analysis of metabolites in potato tuber by gas chromatography-mass spectrometry. *Plant J* 23: 131–142
14. Kopka J, Fernie AF, Weckwerth W, Gibon Y, Stitt M (2004) Metabolite profiling in plant biology: platforms and destinations. *Genome Biol* 5(6): 109–117
15. Erban A, Schauer N, Fernie AR, Kopka J (2006) Non-supervised construction and application of mass spectral and retention time index libraries from time-of-flight GC-MS metabolite profiles. In: W Weckwerth (ed.): *Methods in Molecular Biology* Vol. 358. Humana Press Inc., Totowa, USA, pp 19–38
16. Kopka J (2006) Gas chromatography mass spectrometry, Chapter 1.1. In: K Saito, R Dixon, L Willmitzer (eds): Plant Metabolomics (Biotechnology in Agriculture and Forestry Vol. 57), Springer-Verlag, Heidelberg, pp 3–20
17. Kaplan F, Kopka J, Haskell DW, Zhao W, Schiller KC, Gatzke N, Sung DY, Guy CL (2004) Exploring the temperature-stress metabolome of *Arabidopsis*. *Plant Physiol* 136: 4159–4168
18. Weckwerth W, Loureiro ME, Wenzel K, Fiehn O (2004) Differential metabolic networks unravel the effects of silent plant phenotypes. *Proc Natl Acad Sci USA* 18: 7809–7814

19. Catchpole GS, Beckmann M, Enot DP, Mondhe M, Zywicki B, Taylor J, Hardy N, Smith A, King RD, Kell DB et al. (2005) Hierarchical metabolomics demonstrates substantial compositional similarity between genetically modified and conventional potato crops. *Proc Natl Acad Sci USA* 102: 14458–14462
20. Roessner U, Luedemann A, Brust D, Fiehn O, Linke T, Willmitzer L, Fernie AR (2001) Metabolic profiling allows comprehensive phenotyping of genetically or environmentally modified plant systems. *Plant Cell* 13: 11–29
21. Junker BH, Wuttke R, Tiessen A, Geigenberger P, Sonnewald U, Willmitzer L, Fernie AR (2004) Temporally regulated expression of a yeast invertase in potato tubers allows dissection of the complex metabolic phenotype obtained following its constitutive expression. *Plant Mol Biol* 56: 91–110
22. Birkemeyer C, Luedemann A, Wagner C, Erban A, Kopka J (2005) Metabolome analysis: the potential of *in vivo* labeling with stable isotopes for metabolite profiling. *Trends Biotechnol* 23: 28–33
23. Bino RJ, de Vos CHR, Lieberman M, Hall RD, Bovy A, Jonker HH, Tikunov Y, Lommen A, Moco S, Levin I (2005) The light-hyperresponsive high pigment-2(dg) mutation of tomato: alterations in the fruit metabolome. *New Phytologist* 166: 427–438
24. Kovàts ES (1958) Gas-chromatographische charakterisierung organischer verbindungen: teil 1. retentionsindices aliphatischer halogenide, alkohole, aldehyde und ketone. *Helv Chim Acta* 41: 1915–1932
25. Wagner C, Sefkow M, Kopka J (2003) Construction and application of a mass spectral and retention time index database generated from plant GC/EI-TOF-MS metabolite profiles. *Phytochem* 62: 887–900
26. Ausloos P, Clifton CL, Lias SG, Mikaya AI, Stein SE, Tchekhovskoi DV, Spark-man OD, Zaikin V, Zhu D (1999) The critical evaluation of a comprehensive mass spectral library. *J Am Soc Mass Spectrom* 10: 287–299
27. Stein SE (1999) An integrated method for spectrum extraction and compound identification from gas chromatography/mass spectrometry data. *J Am Soc Mass Spectrom* 10: 770–781
28. Halket JM, Waterman D, Przyborowska AM, Patel RKP, Fraser PD, Bramley PM (2005) Chemical derivatization and mass spectral libraries in metabolic profiling by GC/MS and LC/MS/MS. *J Exp Bot* 56: 219–243
29. Gullberg J, Jonsson P, Nordström A, Sjöström M, Moritz T (2004) Design of experiments: an efficient strategy to identify factors influencing extraction and derivatization of *Arabidopsis thaliana* samples in metabolomic studies with gas chromatography/mass spectrometry. *Anal Biochem* 331: 283–295
30. Duran AL, Yang J, Wang L, Sumner LW (2003) Metabolomics spectral formatting, alignment and conversion tools (MSFACTs). *Bioinformatics* 19: 2283–2293
31. Jonsson P, Gullberg J, Nordström A, Kusano M, Kowalczyk M, Sjöström M, Moritz T (2004) A strategy for identifying differences in large series of metabolomic samples analyzed by GC/MS. *Anal Chem* 76: 1738–1745
32. Jonsson P, Johansson AI, Gullberg J, Trygg J, Grung B, Marklund S, Sjöström M, Antti H, Moritz T (2005) High-throughput data analysis for detecting and identifying differences between samples in GC/MS-based metabolomic analyses. *Anal Chem* 77: 5635–5642
33. Bino RJ, de Vos CHR, Lieberman M, Hall RD, Bovy A, Jonker HH, Tikunov Y, Lommen A, Moco S, Levin I (2005) The light-hyperresponsive high pigment-2dg mutation of tomato: alterations in the fruit metabolome. *New Phytol* 166: 427–438
34. Vorst O, de Vos CHR, Lommen A, Staps RV, Visser RGF, Bino RJ, Hall RD (2005) A non-directed approach to the differential analysis of multiple LC-MS derived metabolic profiles. *Metabolomics* 1: 169–180

35. Luedemann A, Erban A, Wagner C, Kopka J (2004) Method for analyzing metabolites. International patent application (PCT/EP2004/014450) published under the patent cooperation treaty (WO 2005/059556 A1)
36. Mashego MR, Wu L, Van Dam JC, Ras C, Vinke JL, Van Winden WA, Van Gulik WM, Heijnen JJ (2004) MIRACLE: mass isotopomer ratio analysis of U-13C-labeled extracts. A new method for accurate quantification of changes in concentrations of intracellular metabolites. *Biotech Bioeng* 85: 620–628
37. Wu L, Mashego MR, van Dam JC, Proell AM, Vinke JL, Ras C, van Winden WA, van Gulik WM, Heijnen JJ (2005) Quantitative analysis of the microbial metabolome by isotope dilution mass spectrometry using uniformly 13C-labeled cell extracts as internal standards. *Anal Biochem* 336: 164–171
38. Roessner-Tunali U, Hegemann B, Lytovchenko A, Carrari F, Bruedigam C, Granot D, Fernie AR (2003) Metabolic profiling of transgenic tomato plants overexpressing hexokinase reveals that the influence of hexose phosphoryla-tion diminishes during fruit development. *Plant Physiol* 133: 84–99
39. Schauer N, Zamir D, Fernie AR (2005) Metabolic profiling of leaves and fruit of wild species tomato: a survey of the *Solanum lycopersicum* complex. *J Exp Bot* 56: 297–307
40. Kopka J (2005) Current challenges and developments in GC-MS based metabolite profiling technology. *J Biotechnol* 124: 312–322
41. Kanehisa M, Goto S, Kawashima S, Okuno Y, Hattori M (2004) The KEGG resource for deciphering the genome. *Nucleic Acid Res* 32: D277–280
42. Schomburg I, Chang A, Ebeling C, Gremse M, Heldt C, Huhn G, Schomburg D (2004) BRENDA, the enzyme database: updates and major new developments. *Nucleic Acid Res* 32: D431–433
43. Krieger CJ, Zhang P, Mueller LA, Wang A, Paley S, Arnaud M, Pick J, Rhee SY, Karp PD (2004) MetaCyc: a multiorganism database of metabolic pathways and enzymes. *Nucleic Acid Res* 32: D438–442
44. Schauer N, Steinhauser D, Strelkov S, Schomburg D, Allison G, Moritz T, Lundgren K, Roessner-Tunali U, Forbes MG, Willmitzer L et al. (2005) GC-MS libraries for the rapid identification of metabolites in complex biological samples. *FEBS Letters* 579: 1332–1337
45. Oksman-Caldentey K-M, Inzé D, Orešič M (2004) Connecting genes to metabolites by a systems biology approach. *Proc Natl Acad Sci USA* 101: 9949–9950
46. Trethewey RN (2004) Metabolite profiling as an aid to metabolic engineering in plants. *Curr Opin Plant Biol* 7: 196–201
47. Fiehn O, Kopka J, Trethewey RN, Willmitzer L (2000a) Identification of uncommon plant metabolites based on calculation of elemental compositions using gas chromatography and quadrupole mass spectrometry. *Anal Chem* 72: 3573–3580
48. Kopka J, Schauer N, Krueger S, Birkemeyer C, Usadel B, Bergmüller E, Dörmann P, Gibon Y, Stitt M, Willmitzer L et al. (2005) GMD@CSBDB: The Golm Metabolome Database. *Bioinformatics* 21: 1635–1638
49. Sinha AE, Fraga CG, Prazen BJ, Synovec RE (2004a) Trilinear chemometric analysis of two dimensional comprehensive gas chromatography-time-of-flight mass spectrometry data. *J Chromatogr A* 1027: 269–277
50. Sinha AE, Hope JL, Prazen BJ, Nilsson EJ, Jack RM, Synovec RE (2004b) Algorithm for locating analytes of interest based on mass spectral similarity in GC × GC–TOF-MS data: analysis of metabolites in human infant urine. *J Chromatogr A* 1058: 209–215
51. Sinha AE, Prazen BJ, Synovec RE (2004c) Trends in chemometric analysis of comprehensive two-dimensional separations. *Anal Bioanal Chem* 378: 1948–1951
52. Kell DB, Brown M, Davey HM, Dunn WB, Spasic I, Oliver SG (2005) Metabolic footprinting and systems biology: The medium is the message. *Nat Rev Microbiol* 3: 557–565

Plant Systems Biology
Edited by Sacha Baginsky and Alisdair R. Fernie
© 2007 Birkhäuser Verlag/Switzerland

Methods, applications and concepts of metabolite profiling: Secondary metabolism

Lloyd W. Sumner, David V. Huhman, Ewa Urbanczyk-Wochniak and Zhentian Lei

Plant Biology Division, The Samuel Roberts Noble Foundation, 2510 Sam Noble Parkway, Ardmore, OK 73401, USA

Abstract

Plants manufacture a vast array of secondary metabolites/natural products for protection against biotic or abiotic environmental challenges. These compounds provide increased fitness due to their antimicrobial, anti-herbivory, and/or alleopathic activities. Secondary metabolites also serve fundamental roles as key signaling compounds in mutualistic interactions and plant development. Metabolic profiling and integrated functional genomics are advancing the understanding of these intriguing biosynthetic pathways and the response of these pathways to environmental challenges. This chapter provides an overview of the basic methods, select applications, and future directions of metabolic profiling of secondary metabolism. The emphasis of the application section includes the combination of primary and secondary metabolic profiling. The future directions section describes the need for increased chromatographic and mass resolution, as well as the inevitable need and benefit of spatially and temporally resolved metabolic profiling.

Introduction

Secondary metabolites represent a diverse and vast array of compounds that have evolved over time and are found throughout a wide range of terrestrial and marine species [1–8]. Plants contain an especially rich source of natural products and approximately 100,000 unique plant natural products have been identified to date [9]. However, there are still a large number that have not been identified and overall estimates exceeding 200,000 throughout the plant kingdom are common [5, 6]. A representative list of secondary metabolite classes is provided in Table 1. The large number and diversity of plant secondary metabolites can be attributed to the broad substrate specificity and the generation of multiple reactions products that are typical of natural product enzymes. These enzymatic traits enhance the probability of generating chemical diversity and hence beneficial compounds. The selection and retention of chemical diversity is a critical factor in an organism's adaptation and fitness [10–12] and a primary reason for the large number of natural products.

Table 1. Representative secondary metabolite classes

Artemisinins	Hydroxycinnamic acids
Acetophenones	Isoflavonoids
Alkaloids *(imidazole, isoquinoline, piperidine/pyridine, purine, pyrrolizide, quinoline, quinolizidine, terepene, tropane, and tropolone alkaloids)*	Isothiocyanates
	Lignins/Lignans
	Non protein amino acids
	Phenanthrenes
Amines	Phenolics
Anthranoids/Anthraquinones	Phenols *(phloroglucinols, acylphloro glucinols, etc.)*
Anthocyanidins	
Aristolochic acids	Phenylpropanoids
Aurones	Polyacetylenes
Azoxyglycosides	Polyines
Benzenoids	Polyketides
Coumarins	Steroidal and Triterepenoid Saponins
Cyanogenic glycosides	Stilbenes
Condensed tannins	Taxols
Dibenzofurans	Terepenoids *(hemi, mono, sesqui, di, tri, and tetra)*
Flavonoids *(flavanols, flavones, flavanones, etc.)*	
	Thiosulfinates
Glucosinolates	Xanthones
Hyrdroxybenzoic acid	

Plants manufacture a vast array of secondary metabolites/natural products for protection against biotic or abiotic environmental challenges [5]. Thus, these compounds provide increased fitness due to their antimicrobial, anti-herbivory, and/or alleopathic activities. These toxic chemical weapons thwart potential damage by pathogenic viruses/bacteria/fungi/herbivores and/or minimize competition with other plants. For example, select secondary metabolites produce unfavorable responses in targeted plant predators such as bloat (saponins) in cattle and infertility in sheep (isoflavones). Many natural products also have other beneficial biological functions such as flavor/fragrance/color attractants [13–15], UV-protectants, antioxidants, signaling compounds associated with ecological interactions and symbiotic nodulation [16–18], and nutraceutical/pharmacological properties related to human and animal health [16–25]. In fact, natural products account for approximately 30% of all the sales of human therapeutics [26]. The anticancer utility of taxol [27, 28] and the antimalarial properties of artemisinin [29–31] are good examples.

In addition to the large diversity in basic chemical structures, many natural products are further conjugated with a variety of sugars and/or organic acids. The conjugation process is believed to be an import part of the cellular detoxification and storage mechanisms. However, they can also dramatically impact the biological activity of these compounds. Additional derivatives of natural products are achieved through the attachment of chemical moieties, such as acylation or prenylation, which continue to add to the chemical diversity of the metabolome and impact biological activity [32–34].

Methods

The vast numbers of plant secondary metabolites represent an extreme challenge for large-scale metabolite profiling, i.e., metabolomics, and a singular tool for profiling all primary or secondary plant metabolites currently does not exist. Most present strategies involve 'divide and conquer' strategies. This is achieved by employing a series of parallel targeted profiling methods focused on singular or multiple metabolite classes. Natural product classes are selectively extracted through the use of optimized solvents and often analyzed separately or in parallel. If specific natural products are of particular low abundance, enrichment methods such a solid phase extraction may also be employed.

There exist a growing number of successful technical methods that are employed in metabolic profiling of secondary metabolites [35, 36] and the selection of any specific method is usually a compromise between sensitivity, selectivity and speed [37]. GC/MS is capable of profiling many of the smaller and volatile secondary metabolites including the isoprenoids [38], triterepenoids such as β-amyrin [39], and phenylpropanoid aglycones such as ferulic acid [39]. However, a large number of secondary metabolites are conjugated with sugars as described above and are not amenable to GC/MS even following derivatization. Therefore, high performance liquid chromatography (HPLC) coupled to ultraviolet (UV) and mass spectrometry (MS) detection [40, 41], capillary electrophoresis-MS [42–44], NMR [45], and/or HPLC-NMR [46–49] are heavily relied upon in most approaches for metabolic profiling of secondary metabolism. The use of various established metabolomics technologies have been reviewed previously [35] and will not be replicated here. However, a detailed discussion of emerging technologies that offer significant enhancements in metabolic profiling of secondary metabolites will be discussed in the 'Future directions' section below.

Applications

Functional genomics and systems biology approaches based upon high density microarray analyses have traditionally been pursued in a limited number of model plant species such as *Arabidopsis*, rice, and *Medicago* as these species offer the major genomic and transcript sequence resources. Fortunately, the quantity of sequence information in the form of genomic or expressed sequence tags (ESTs) is growing exponentially for a vast number of plant species (http://www.tigr.org/tdb/tgi/plant.shtml) which is making cDNA or oligonucleotide arrays for these species possible. However, these resources are coming at additional costs. Metabolomics and/or metabolic profiling on the other hand are less species dependent as most primary and some secondary metabolites such as flavonoids are observed across major portions of the plant kingdom. Thus, metabolomics offers greater diversity in its application to various plant species relative to transcriptomics and proteomics platforms without the additional costs. Accordingly, metabolic profiling has been significantly utilized in the study of primary metabolism of model species [13, 50–54] and also in many other crop plants such as potato [55–58], tomato [59], and cucurbits [60]. However, the study of secondary metabolism in model species has been less actively pursued [61, 62].

Metabolic profiling as a tool to study secondary metabolism has traditionally been focused on two major areas. First, it was traditionally a phytochemical tool for the rigorous separation, isolation, and identification of individual and unknown secondary metabolites [63]. For example, LC/MS might be used to obtain a nominal or accurate mass of a highly purified unknown metabolite to aid in structural determination. Secondly, metabolic profiling has been used as a tool to study the molecular aspects of secondary metabolism [15, 64, 65]. These efforts often focus upon a limited number of secondary metabolites related to the specific pathway being studied and less attention is directed toward the cumulative differential profiles. More recently, the scale and scope of metabolic profiling related to secondary metabolism have dramatically broadened towards a larger-scale and more comprehensive nature [39, 41, 44, 66, 67]. However, these larger-scale functional genomics applications are still somewhat limited.

The most exciting applications of metabolomics are not focused solely on specific natural product classes, but are bridging the gap by profiling both primary and secondary metabolites to better understand the interrelationship between these two important areas. For example, von Roepenack-Lahaye and colleagues have developed a capillary HPLC coupled to quadrupole time-of-flight mass spectrometry (LC-QtofMS) method for profiling both primary and secondary metabolites and used it to evaluate chalcone synthase deficient *tt4* mutants in *Arabidopsis* [68]. Hirai and colleagues have also used an integrated approach composed of multiple technologies to show that sulfur and nitrogen metabolism were coordinately modulated with the secondary metabolism of glucosinolates and anthocyanins [42, 69, 70]. Further, these pioneers also integrated metabolomic and mRNA expression data to render gene-to-metabolite networks used in the identification of gene function and subsequent improvement in the production of useful compounds in plants. Similarly, Nikiforova and colleagues determined the impact of sulfur deprivation on primary metabolism and flavonoid levels and used this information to reconstruct the coordinating network of their mutual influences [71].

Colleagues at The Noble Foundation are currently applying metabolic profiling in both genomic and functional genomic approaches for discovery of new genes and for new insight into the biosynthetic mechanisms related to secondary metabolism. A major area of focus includes triterpene saponins. Although the biosynthetic pathway is poorly understood, these compounds have a large diversity of important biological activities including anti-herbivory (i.e., hemolytic and cause bloat), antifungal, antimicrobial, alleopathic, lowering of cholesterol, anticancer, and utility as adjuvants. Recently, Achnine and coworkers utilized EST mining, *in vitro* assays, and metabolic profiling to identify putative glycosyltransferases (GTs) involved in triterpenoid

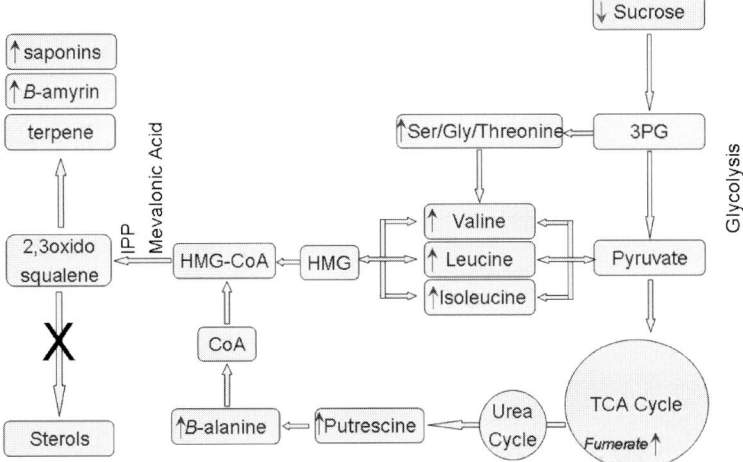

Figure 1. A proposed mechanistic model of the metabolic response of *Medicago truncatula* cell suspension cultures to methyl jasmonate elicitation [39]. The data suggest a major reprogramming of metabolism in which as carbon normally destined for sucrose is redirected towards secondary metabolism (triterpene saponin).

saponin biosynthesis [41]. In this report, two new uridine diphosphate GTs were identified and characterized that possessed saponin specificity. This project continues with a large number of additional putative GTs under investigation.

In a separate study on biotic stress, Broeckling and colleagues reported a major reprogramming of carbon flow from primary towards secondary saponin metabolism in response to methyl jasmonate elicitation in *Medicago truncatula* [39, 72]. Based on metabolic profiling of both primary and secondary metabolism, a mechanistic response model was proposed and is presented in Figure 1, which involves a major reprogramming of carbon from primary metabolism towards secondary metabolism (i.e., triterpene saponins). The response includes increased levels of serine/glycine/threonie metabolism which is believed to result in increased levels of branched chain amino acids suggesting increased hydroxylmethylgluturate (HMG) levels. The increased levels of the polyamine beta-alanine and putrescine imply increased levels of the HMG-CoA ester which serves as the source of carbon for triterpene saponin and sterol production. However, no increase in sterol accumulation was observed supporting carbon flow directed toward saponin production which was confirmed by LC/MS metabolic profiling. Although the HMG-CoA ester was not observed in the metabolic profiles, microarray data (Naoumkina et al., unpublished) reveal increased levels of HMG-CoA synthase and HMG-CoA reductase that further support this model and will be presented in detail elsewhere. Continued efforts are underway that will further integrate transcript, protein, and metabolite data consistent with a systems biology approach.

Future directions

The separation of complex secondary metabolome mixtures is still quite challenging, and there exists a need for greater differentiation and resolution in metabolomics approaches at both the technical and biological levels. We are actively pursuing these needs by increasing chromatographic resolution and by increasing spatially/temporally resolved biological sampling. These efforts are amplifying the biological context of our metabolic profiling efforts.

Increased chromatographic resolution

Currently, analytical HPLC commonly used in many secondary metabolic profiling approaches has an upper peak capacity (i.e., theoretical number representing the maximum peaks resolvable by the system based on optimum performance) of approximately 300. Based on this estimate, a maximum of 300 components could be resolved in a best case scenario; however in practice, this value is seldom achieved and more realistic peak capacities are between 100 and 200. Thus, current HPLC technologies are limiting the comprehensive scope of metabolomics. Separation efficiencies can be improved by altering selectivity, increasing column lengths, decreasing column diameters, reducing particle sizes, increasing temperature, and/or utilization of alternative column materials. These approaches have been recently reviewed [73] and we are currently evaluating alternative techniques, including capillary/nano-HPLC-QtofMS and ultra-performance liquid chromatography mass spectrometry (UPLC-MS) in an effort to increase the comprehensive coverage of metabolic profiling. Both methods have yielded increased separation efficiencies. For example, average separation efficiencies exceeding 225,000 plates per meter were obtained by capillary column (300 μm in diameter) HPLC-QtofMS analysis of a saponin extract from *Medicago truncatula* (see Fig. 2). This represents an approximate three-fold increase in efficiency as compared to an average efficiency of 87,000 plates per meter for analytical HPLC (4.6 x 250 mm, Agilent 1100) system coupled to a quadrupole ion trap mass spectrometer (LC-QITMS) [40]. All separation gradients and sample loadings were identical. Unfortunately, the standard deviation was higher for the capillary system (16.6%) relative to the analytical system (8.8%). The higher variability was attributed to the passive flow splitting associated with the LC Packings Ultimate HPLC pump; however, active splitting modules are now available that should significantly lower this variability.

We have also completed preliminary evaluations of ultra-performance liquid chromatography mass spectrometry (UPLC-MS) for the analysis of phenolics and saponins. These efforts yielded impressive results as illustrated in Figure 3. The average peak widths were approximately 6 seconds at half height and represent an average separation efficiency of approximately 500,000 plates per meter. These results illustrate that high resolution and separation efficiencies are possible for high pressure liquid chromatography and compare favorably to those obtained by capillary GC/MS. Further, these high efficiencies were reached using faster separa-

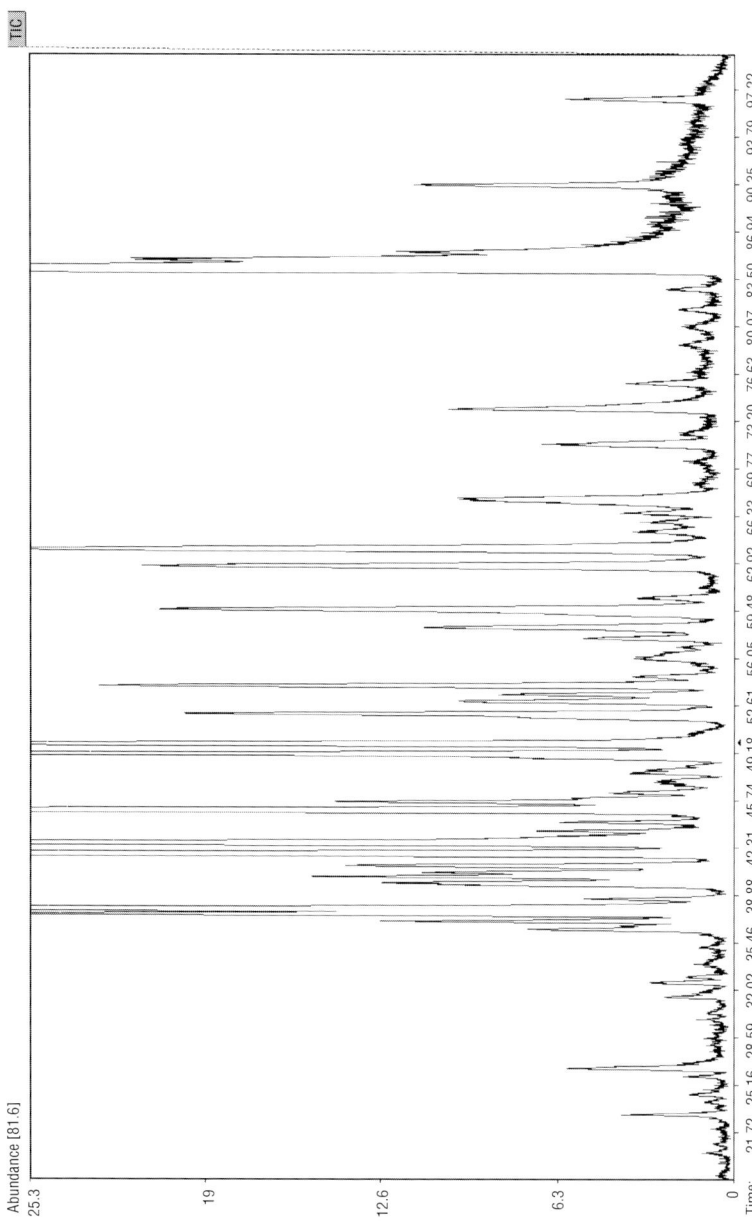

Figure 2. Representative base-peak ion chromatogram obtained by capillary HPLC-QtofMS analysis of 8 μg total saponin extract from *Medicago truncatula* (cv Jemalong A17). Separation gradients were similar to those reported previously [40, 74], and utilized a 300 μm x 250 mm id, 5 μm, 100 Å, C18, PepMap (LC Packings) column operating at a flow rate of 4 μl/min. Mass spectra were recorded on an ABI QSTAR Pulsar *i* (Applied BioSystems).

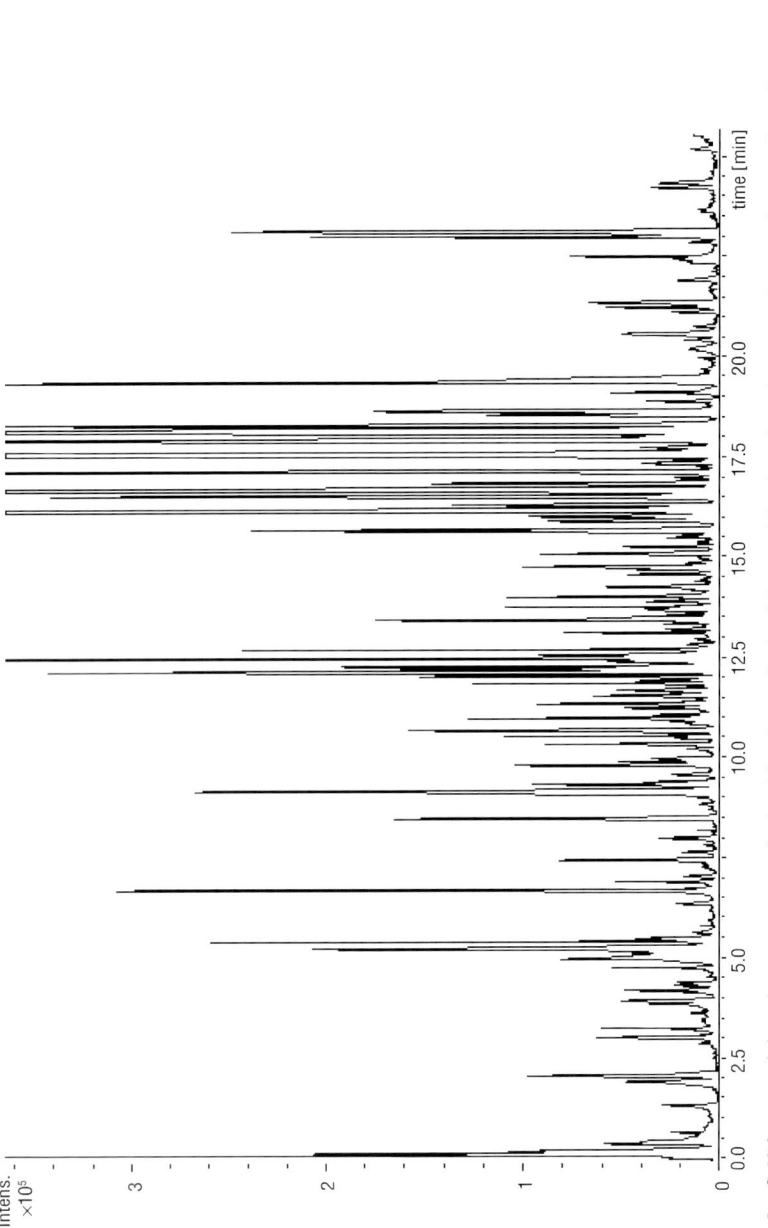

Figure 3. UPLC-QtofMS base-peak ion chromatograms obtained for the analysis of the combined methanol extracts from soybean and *Medicago truncatula* (cv Jemalong A17). Separation gradients were similar to those reported previously [40, 74]; however the analysis time was cut in half to 30 min by increasing the slope of the gradient by approximately two-fold. Separations were achieved using a Waters Acquity UPLC 2.1 × 100 mm, BEH C18 column with 1.7 μm particles and a flow of 600 μl/min. Mass spectra were collected on a Waters QTOFMS Premier.

tions than previously reported [40, 74] thereby increasing throughput at the same time.

Although the above techniques can be used to achieve enhanced chromatographic resolution, the resolution enhancements are still far from that which is needed for complex metabolomics mixtures. It is expected that the maximum peak capacities obtainable by capillary HPLC or UPLC methods will reach a maximum in the range of 600 to 1,000. However, peak capacities of thousands to tens of thousands are necessary to separate complex metabolome mixtures. Currently, only multidimensional chromatographic methods offer peak capacities of this magnitude [75, 76]. Multidimensional chromatography utilizes combinations of two or more orthogonal separation mechanisms based on different selectivity, e.g., ion-exchange and reverse-phase or capillary electrophoresis and reverse-phase LC. These systems offer enhanced resolution due to the utilization of multiple columns with independent chemistries and selectivity which can dramatically improve resolution. The maximum peak capacity of a multidimensional system is the product of the two or more individual separation dimensions. For example, a realistic system that has a peak capacity in the first dimension (n_x) of 150 and the peak capacity in the second dimension (n_y) of 50, then the total maximum peak capacity of the multidimensional system is $n_x \times n_y = 150 \times 50 = 7,500$. If one considers that an individual metabolome consists of 15,000 metabolites, then this is a considerable increase in comprehensive coverage relative to existing methods.

Multidimensional LCxLC separations have been utilized in proteomics research and are commonly referred to as multidimensional protein identification technology (i.e., MUDPIT; [77, 78]. Multidimensional LC separations have not been applied to secondary metabolism, but GC×GC/time-of-flight-MS has been used with a focus on primary metabolism [79]. Unfortunately, these complex separations often come with increased analysis times, but we believe that the additional depth of coverage provided by these experiments will be worth the additional temporal costs.

If higher resolution chromatography is obtained, mass analyzers must also be employed with compatible scan speeds to record data for compounds eluting in very short temporal periods. It is expected that LC peak widths of 1–5 s will be routine in the very near future. For accurate quantification, it is commonly accepted that the sampling rate should be sufficient to capture 10 data points across the eluting peak to provide a statistically valid representation of the peak profile and higher sampling rates are beneficial. Thus, sampling rates should be less than 0.1 s or greater than 10 Hz. This is achievable with current time-of-flight mass analyzers (TOF-MS). It is worth mentioning that quadrupole-based mass analyzers, including traps, can approach these speeds; however, TOF mass spectrometers equipped with delayed extraction and ion-reflectrons also offer improved mass accuracy over quadrupoles.

Improvements in the accuracy of the mass analyzer can further enhance metabolite differentiation, provide elemental compositions useful in identification, and allow for the profiling of greater numbers of metabolites. Mass accuracy is directly related to the mass resolution or the ability of the mass analyzer to resolve compounds of different m/z values. Mass resolution is defined in Equation 1 and is a function of mass (M) divided by the peak width (ΔM) which is most commonly defined at half-height:

$$R_m = \frac{M}{\Delta M} \qquad \text{(Eq. 1)}$$

Often, LC/MS is performed with quadrupole ion-traps or linear quadrupole mass analyzers that yield mass accuracies in the range of 1.0–0.1 Da. Unfortunately, many metabolites have similar nominal masses which can not be differentiated at this level of mass accuracy. For example, the important natural products genistein and medicarpin have similar nominal masses of 270, but have different accurate masses of 270.2390 ($C_{15}H_{10}O_5$) and 270.2830 ($C_{16}H_{14}O_4$) respectively, due to different chemical compositions. If the mass can be measured with sufficient accuracy, then these compounds can be differentiated in the mass domain even if they cannot be physically separated in the chromatographic domain. This mass differentiation can be achieved at a mass resolution (M/ΔM) greater than 6136. Compounds with closer accurate masses such as rutin ($C_{27}H_{30}O_{16}$ = 610.5180) and hesperidin ($C_{28}H_{34}O_{15}$ = 610.5620) would require a higher mass resolution of 13,864 for their differentiation. Mass resolutions on the order of 10,000 can be achieved with modern TOF-MS analyzers, and resolutions in excess of 100,000 with sub-part-per-million mass accuracies (i.e., less than 0.001 at m/z of 1,000 Da) are achievable with Fourier transform ion cyclotron mass spectrometry (FTMS). Newer technologies, such as Thermo Electron Corporation's Orbitrap mass analyzer are currently surfacing that also offer high-resolution (100,000) solutions. Although high resolution accurate mass measurements have great advantages, this technology is still rather costly.

Interestingly, a significant argument can be made that accurate mass measurements significantly reduce the need for ultra-high resolution separations due to the enhanced separation in the mass domain. However, if the chromatography step is omitted or compressed significantly, then ion suppression, competitive ionization, and other matrix affects become increasingly more influential. We personally believe that both improved chromatographic resolution and accurate mass measurements offer the best solution and that the combination of these techniques will provide greater comprehension and confidence in our ability to profile the metabolome. Further, we also believe that the needed magnitude of enhancements in chromatographic resolution can only be achieved with multidimensional approaches at this point in time.

Spatially and temporally resolved metabolomics

Higher organisms localize both primary and secondary biochemistry into cellular compartments, tissues, and organs; however traditional sampling strategies for the majority of metabolomics or functional genomic applications have involved the pooling of tissues, organs, and/or organisms. This sampling approach dramatically reduces the resolving power of the experiment and related conclusions due to dilution of specific biochemical responses that are often spatially segregated within the organism. For example, the differential accumulation of specific conjugated

Methods, applications and concepts of metabolite profiling: Secondary metabolism 205

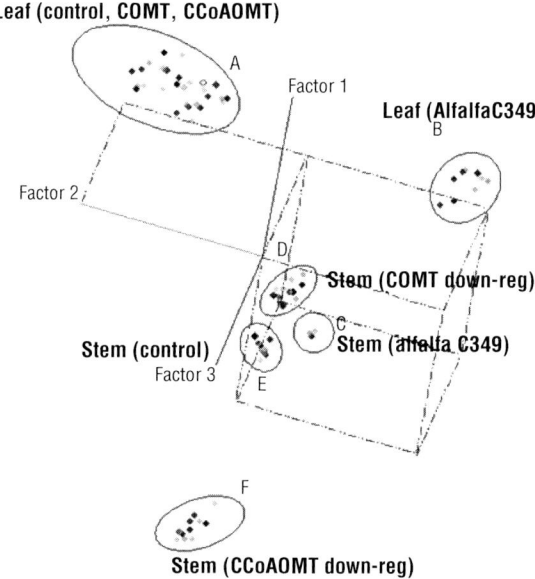

Figure 4. Principal component analyses of HPLC/UV data collected for soluble phenolic compounds extracted from stem and leaf tissues of wild-type (Regen SY control) and lines of alfalfa downregulated in expression of caffeic acid 3-O-methyltransferase (COMT) and caffeoyl CoA 3-O-methyl-transferase (CCoAOMT) [67].

forms of triterpene saponins in various tissues of *Medicago truncatula* has been observed [74] suggesting specialized roles of these individual components that were not previously observable using a pooled sampling strategy [40]. Spatially resolved phenolic metabolite profiles were also used to differentiate tissues in transgenic alfalfa modified in lignin biosynthesis [67] as shown in Figure 4. GC/MS and HPLC have also been used to evaluate metabolism in other specialized organs such as glandular and non glandular trichomes. Using this approach, gross differences in the metabolic profiles were observed as illustrated in Figure 5 which dramatically enhance opportunities for increased understanding of localized biochemical processes [80]. Recent technologies including laser microdissection [81, 82] and fluorescent cell sorting [83] will continue to advance the utility and information content of spatially resolved metabolomics.

Spatially resolved sampling is more time consuming and requires considerable, additional effort to yield sufficient quantities of tissue for metabolic profiling. Thus, if spatially resolved metabolomics is to be successful, then scalable or more sensitive methods will be required. For example, previously reported methods that utilized milligram quantities of starting material for GC/MS metabolic profiling have been scaled down to the microgram level (see Fig. 6).

The biosynthesis and accumulation of primary and secondary metabolites are also temporally regulated. The temporal accumulation of secondary metabolites can be correlated with normal development and/or programmed responses to biotic and

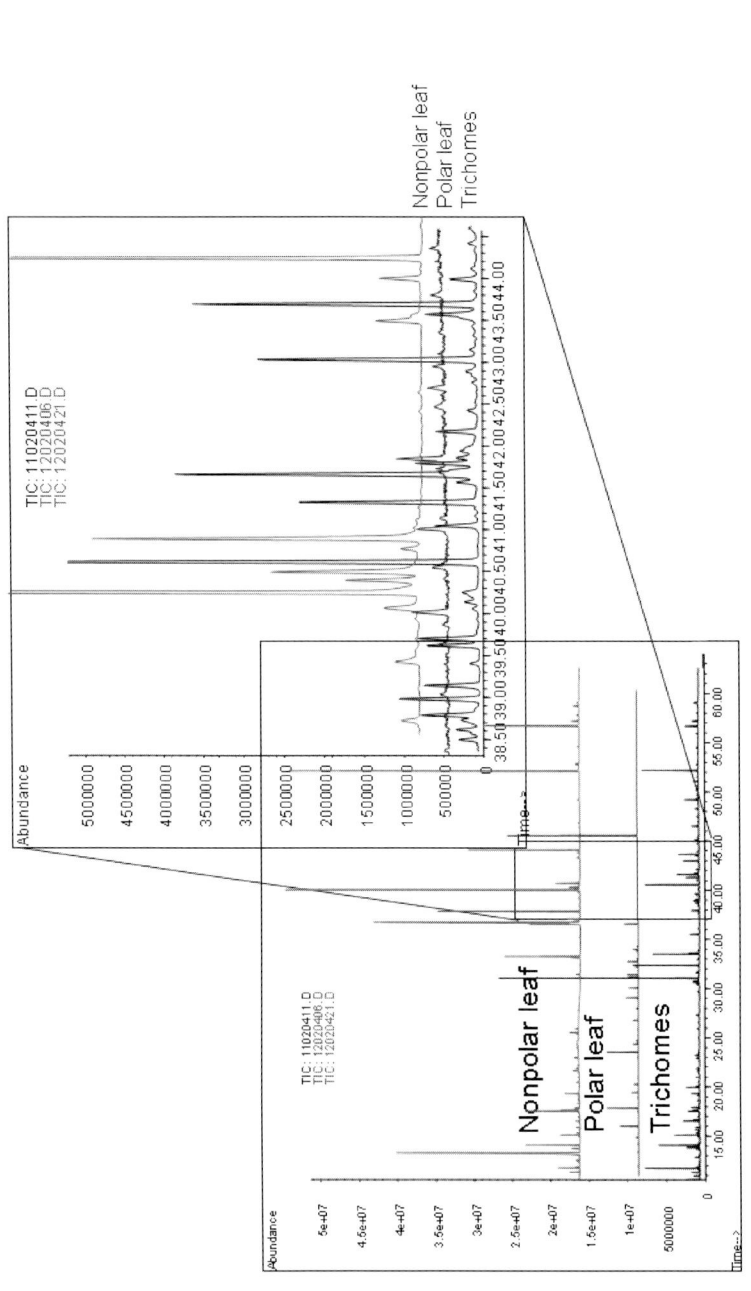

Figure 5. Superimposed GC-MS profiles of alfalfa trichome and leaf metabolites illustrating major quantitative and qualitative differences between the different tissues. Separations were achieved using methods describe previously [39].

Methods, applications and concepts of metabolite profiling: Secondary metabolism

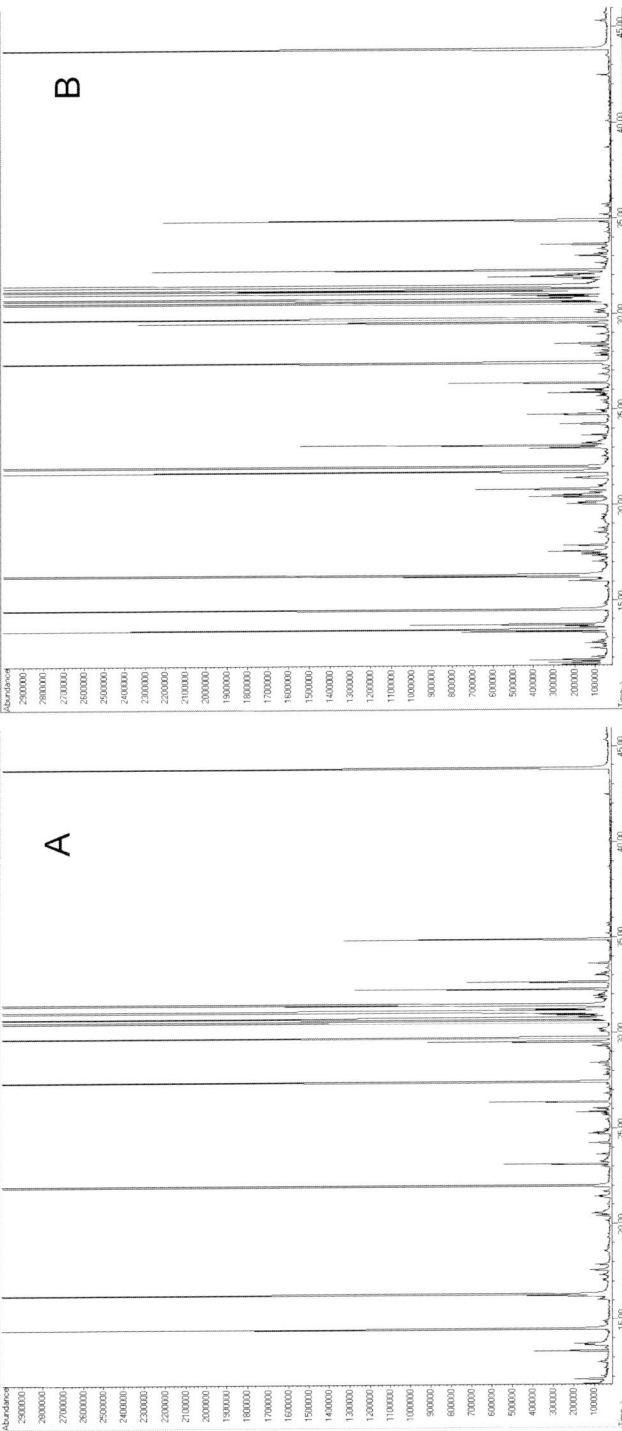

Figure 6. Representative base-peak GC/MS ion chromatograms of polar extracts obtained from 6.04 mg (panel-A) and 580 µg (panel-B) dry weight of 5 weeks old internodes of *Medicago truncatula* (cv Parabinga). These data illustrate the comparability and scalability of current methods toward lower material quantities. The GC/MS method was similar to that reported previously [39] except that the volume of the polar extraction solvent was reduced proportionally to the quantity of material extracted (1 ml for 6 mg and 100 µl for 580 µg respectively), and 1 µl samples were injected and analyzed for both.

abiotic stress [39, 72]. Several examples were also provided above in relationship to glucosinolate [42, 70] and triterpenoid metabolism [39].

Summary

We believe that there still exists tremendous opportunities in the use of metabolomics in the pursuit of advanced understanding of the biochemical and molecular aspects of secondary metabolism. Our current integrated functional genomics approach is yielding a significant number of new gene discoveries and mechanistic insight. We will continue to push forward this important area of research for the advancement of plant productivity and for the improvement of human and animal nutrition and health.

References

1. Field B, Cardon G, Traka M, Botterman J, Vancanneyt G, Mithen R (2004) Glucosinolate and amino acid biosynthesis in *Arabidopsis*. *Plant Physiol* 135: 828–839
2. Keller N, Turner G, Bennett J (2005) Fungal secondary metabolism – from biochemistry to genomics. *Nat Rev Microbiol* 3: 937–947
3. Muller WEG, Schroder HC, Wiens M, Perovic-Ottstadt S, Batel R, Muller IM (2004) Traditional and modern biomedical prospecting: Part II – The benefits: approaches for a sustainable exploitation of biodiversity (secondary metabolites and biomaterials from sponges). *Evid Based Complement Altern Med* 1: 133–144
4. Wink ME (1999) *Biochemistry of plant secondary metabolism*, vol. 2, CRC Press, Boca Raton
5. Dixon RA (2001) Natural products and disease resistance. *Nature* 411: 843–847
6. Dixon RA, Sumner LW (2003) Legume natural products: understanding and manipulating complex pathways for human and animal health. *Plant Physiol* 131: 878–885
7. Dixon RA (2004) Phytoestrogens. *Ann Rev Plant Biol* 55: 225–261
8. Goossens A, Hakkinen ST, Laakso I, Seppanen-Laakso T, Biondi S, De Sutter V, Lammertyn F, Nuutila AM, Soderlund H, Zabeau M et al. (2003) A functional genomics approach toward the understanding of secondary metabolism in plant cells. *PNAS* 100: 8595–8600
9. Wink ME (1999) *Functions of plant secondary metabolites and their exploitation in biotechnology*, vol. 3, CRC Press, Boca Raton
10. Firn R, Jones C (2003) Natural products–a simple model to explain chemical diversity. *Nat Prod Rep* 20: 382–391
11. Firn R, Jones C (1999) Secondary metabolism and the risks of GMOs. *Nature* 400: 13–14
12. Firn R, Jones C (2000) The evolution of secondary metabolism – a unifying model. *Mol Microbiol* 37: 989–994
13. Rohloff J, Bones A (2005) Volatile profiling of *Arabidopsis thaliana* – putative olfactory compounds in plant communication. *Phytochemistry* 66: 1941–1955
14. Verdonk J, Ric de Vos C, Verhoeven H, Haring M, van Tunen A, Schuurink R (2003) Regulation of floral scent production in petunia revealed by targeted metabolomics. *Phytochemistry* 62: 997–1008
15. Frydman A, Weisshaus O, Bar-Peled M, Huhman DV, Sumner LW, Marin FR, Lewinsohn E, Fluhr R, Gressel J, Eyal Y (2004) Citrus fruit bitter flavors: Isolation and functional characterization of the gene encoding a 1,2 rhamnosyltransferase, a key enzyme in the biosynthesis of the bitter flavonoids of citrus. *Plant J* 40: 88–100

16. D'Haeze W, Holsters M (2002) Nod factor structures, responses, and perception during initiation of nodule development. *Glycobiology* 12: 79R–105
17. Relic B, Perret X, Estrada-Garcia M, Kopcinska J, Golinowski W, Krishnan H, Pueppke S, Broughton W (1994) Nod factors of Rhizobium are a key to the legume door. *Mol Microbiol* 13: 171–178
18. Oldroyd GED (2001) Dissecting symbiosis: developments in Nod factor signal transduction. *Ann Bot* 87: 709–718
19. Deavours BE, Dixon RA (2005) Metabolic engineering of isoflavonoid biosynthesis in Alfalfa. *Plant Physiol* 138: 2245–2259
20. Aerts RJ, Barry TN, McNabb WC (1999) Polyphenols and agriculture: beneficial effects of proanthocyanidins in forages. *Agriculture Ecosystems & Environment* 75: 1–12
21. Bagchi D, Bagchi M, Stohs SJ, Das DK, Ray SD, Kuszynski CA, Joshi SS, Pruess HG (2000) Free radicals and grape seed proanthocyanidn extract: importance in human health and disease prevention. *Toxicology* 148: 187–197
22. Setchell KDR, Cassidy A (1999) Dietary isoflavones: Biological effects and relevance to human health. *J Nutrition* 129: 758S–767S
23. MerzDemlow BE, Duncan AM, Wangen KE, Xu X, Carr TP, Phipps WR, Kurzer MS (2000) Soy isoflavones improve plasma lipids in normocholesterolemic, premenopausal women. *Am J Clin Nutr* 71: 1462–1469
24. Manach C, Scalbert A, Morand C, Remesy C, Jimenez L (2004) Polyphenols: food sources and bioavailability. *Am J Clin Nutr* 79: 727–747
25. Gidley M (2004) Naturally functional foods – challenges and opportunities. *Asia Pac J Clin Nutr* 13: S31
26. Grabley S, Thiericke R (2000) *Drug Discovery from Nature*, 366, Springer-Verlag, New York
27. Rowinsky EK, Donehower RC (1995) Paclitaxel (Taxol). *N Engl J Med* 332: 1004–1014
28. Khayat D, Antoine E, Coeffic D (2000) Taxol in the management of cancers of the breast and the ovary. *Cancer Invest* 18: 242–260
29. Sriram D, Rao V, Chandrasekhara K, Yogeeswari P (2004) Progress in the research of artemisinin and its analogues as antimalarials: an update. *Nat Prod Res* 18: 503–527
30. Price R (2000) Artemisinin drugs: novel antimalarial agents. *Expert Opin Investig Drugs* 9: 1815–1827
31. Jung M, Lee K, Kim H, Park M (2004) Recent advances in artemisinin and its derivatives as antimalarial and antitumor agents. *Curr Med Chem* 11: 1265–1284
32. Botta B, Vitali A, Menendez P, Misiti D, Delle Monache G (2005) Prenylated flavonoids: pharmacology and biotechnology. *Curr Med Chem* 12: 717–739
33. Stevens J, Page J (2004) Xanthohumol and related prenylflavonoids from hops and beer: to your good health! *Phytochemistry* 65: 1317–1330
34. Cos P, De Bruyne T, Apers S, Vanden Berghe D, Pieters L, Vlietinck A (2003) Phytoestrogens: recent developments. *Planta Med* 69: 589–599
35. Sumner L, Mendes P, Dixon R (2003) Plant metabolomics: large-scale phytochemistry in the functional genomics era. *Phytochemistry* 62: 817–836
36. Kopka J, Fernie A, Weckwerth W, Gibon Y, Stitt M (2004) Metabolite profiling in plant biology: platforms and destinations. *Genome Biol* 5: 109
37. Trethewey R (2004) Metabolite profiling as an aid to metabolic engineering in plants. *Curr Opin Plant Biol* 7: 196–201
38. Lange BM, Ketchum REB, Croteau RB (2001) Isoprenoid biosynthesis. Metabolite profiling of peppermint oil gland secretory cells and application to herbicide target analysis. *Plant Physiol* 127: 305–314

39. Broeckling CD, Huhman DV, Farag MA, Smith JT, May GD, Mendes P, Dixon RA, Sumner LW (2005) Metabolic profiling of *Medicago truncatula* cell cultures reveals the effects of biotic and abiotic elicitors on metabolism. *J Exp Bot* 56: 323–336
40. Huhman D, Sumner L (2002) Metabolic profiling of saponins in *Medicago sativa* and *Medicago truncatula* using HPLC coupled to an electrospray ion-trap mass spectrometer. *Phytochemistry* 59: 347–360
41. Achnine L, Huhman D, Farag M, Sumner L, Blount J, Dixon R (2005) Genomics-based selection and functional characterization of triterpene glycosyltransferases from the model legume *Medicago truncatula*. *Plant J* 41: 875–887
42. Hirai MY, Klein M, Fujikawa Y, Yano M, Goodenowe DB, Yamazaki Y, Kanaya S, Nakamura Y, Kitayama M, Suzuki H et al. (2005) Elucidation of gene-to-gene and metabolite-to-gene networks in *Arabidopsis* by integration of metabolomics and transcriptomics. *J Biol Chem* 280: 25590–25595
43. Soga T, Ohashi Y, Ueno Y, Naraoka H, Tomita M, Nishioka T (2003) Quantitative metabolome analysis using capillary electrophoresis mass spectrometry. *J Proteome Res* 2: 488–494
44. Sato S, Soga T, Nishioka T, Tomita M (2004) Simultaneous determination of the main metabolites in rice leaves using capillary electrophoresis mass spectrometry and capillary electrophoresis diode array detection. *Plant J* 40: 151–163
45. Mesnard F, Ratcliffe R (2005) NMR analysis of plant nitrogen metabolism. *Photosynth Res* 83: 163–180
46. Wolfender J, Queiroz E, Hostettmann K (2005) Phytochemistry in the microgram domain – a LC-NMR perspective. *Magn Reson Chem* 43: 697–709
47. Zanolari B, Wolfender J, Guilet D, Marston A, Queiroz E, Paulo M, Hostettmann K (2003) On-line identification of tropane alkaloids from *Erythroxylum vacciniifolium* by liquid chromatography-UV detection-multiple mass spectrometry and liquid chromatography-nuclear magnetic resonance spectrometry. *J Chromatogr A* 1020: 75–89
48. Wolfender J, Ndjoko K, Hostettmann K (2003) Liquid chromatography with ultraviolet absorbance-mass spectrometric detection and with nuclear magnetic resonance spectroscopy: a powerful combination for the on-line structural investigation of plant metabolites. *J Chromatogr A* 1000: 437–455
49. Exarchou V, Krucker M, van Beek T, Vervoort J, Gerothanassis I, Albert K (2005) LC-NMR coupling technology: recent advancements and applications in natural products analysis. *Magn Reson Chem* 43: 681–687
50. Kaplan F, Kopka J, Haskell DW, Zhao W, Schiller KC, Gatzke N, Sung DY, Guy CL (2004) Exploring the temperature-stress metabolome of *Arabidopsis*. *Plant Physiol* 136: 4159–4168
51. Fiehn O, Kopka J, Dormann P, Altmann T, Trethewey RN, Willmitzer L (2000) Metabolite profiling for plant fuctional genomics. *Nat Biotechnol* 18: 1142–1161
52. Steinhauser D, Usadel B, Luedemann A, Thimm O, Kopka J (2004) CSB.DB: a comprehensive systems-biology database. *Bioinformatics* 20: 3647–3651
53. Taylor J, King RD, Altmann T, Fiehn O (2002) Application of metabolomics to plant genotype discrimination using statistics and machine learning. *Bioinformatics* 18: 241S–248
54. Cook D, Fowler S, Fiehn O, Thomashow MF (2004) A prominent role for the CBF cold response pathway in configuring the low-temperature metabolome of *Arabidopsis*. *PNAS* 101: 15243–15248
55. Roessner U, Wagner C, Kopka J, Trethewey RN, Willmitzer L (2000) Simultaneous analysis of metabolites in potato tuber by gas chromatography-mass spectrometry. *Plant J* 23: 131–142

56. Roessner U, Luedemann A, Brust D, Fiehn O, Linke T, Willmitzer L, Fernie AR (2001) Metabolic profiling allows comprehensive phenotyping of genetically or environmentally modified plant systems. *Plant Cell* 13: 11–29
57. Roessner-Tunali U, Urbanczyk-Wochniak E, Czechowski T, Kolbe A, Willmitzer L, Fernie AR (2003) *De novo* amino acid biosynthesis in potato tubers is regulated by sucrose levels. *Plant Physiol* 133: 683–692
58. Urbanczyk-Wochniak E, Baxter C, Kolbe A, Kopka J, Sweetlove L, Fernie A (2005) Profiling of diurnal patterns of metabolite and transcript abundance in potato (*Solanum tuberosum*) leaves. *Planta* 221: 891–903
59. Urbanczyk-Wochniak E, Fernie AR (2005) Metabolic profiling reveals altered nitrogen nutrient regimes have diverse effects on the metabolism of hydroponically-grown tomato (*Solanum lycopersicum*) plants. *J Exp Bot* 56: 309–321
60. Fiehn O (2003) Metabolic networks of *Cucurbita maxima* phloem. *Phytochem* 62: 875–886
61. D'Auria J, Gershenzon J (2005) The secondary metabolism of *Arabidopsis thaliana*: growing like a weed. *Curr Opin Plant Biol* 8: 308–316
62. Romeo JT (2004) *Secondary metabolism in model systems, volume 38: recent advances in phytochemistry*, vol. 38, Elsevier Science, San Diego, CA
63. Blount J, Masoud S, Sumner L, Huhman D, Dixon R (2002) Over-expression of cinnamate 4-hydroxylase leads to increased accumulation of acetosyringone in elicited tobacco cell-suspension cultures. *Planta* 214: 902–910
64. Liu C, Huhman D, Sumner L, Dixon R (2003) Regiospecific hydroxylation of isoflavones by cytochrome p450 81E enzymes from *Medicago truncatula*. *Plant J* 36: 471–484
65. Frydman A, Weisshaus O, Huhman D, Sumner L, Bar-Peled M, Lewinsohn E, Fluhr R, Gressel J, Eyal Y (2005) Metabolic engineering of plant cells for biotransformation of hesperedin into neohesperidin, a substrate for production of the low-calorie sweetener and flavor enhancer NHDC. *J Agric Food Chem* 53: 9708–9712
66. Hirai MY, Klein M, Fujikawa Y, Yano M, Goodenowe DB, Yamazaki Y, Kanaya S, Nakamura Y, Kitayama M, Suzuki H et al. (2005) Elucidation of gene-to-gene and metabolite-to-gene networks in *Arabidopsis* by integration of metabolomics and transcriptomics. *J Biol Chem* 280: 25590–25595
67. Chen F, Duran AL, Blount JW, Sumner LW, Dixon RA (2003) Profiling phenolic metabolites in transgenic alfalfa modified in lignin biosynthesis. *Phytochem* 64: 1013–1021
68. von Roepenack-Lahaye E, Degenkolb T, Zerjeski M, Franz M, Roth U, Wessjohann L, Schmidt J, Scheel D, Clemens S (2004) Profiling of *Arabidopsis* secondary metabolites by capillary liquid chromatography coupled to electrospray ionization quadrupole time-of-flight mass spectrometry. *Plant Physiol* 134: 548–559
69. Hirai MY, Saito K (2004) Post-genomics approaches for the elucidation of plant adaptive mechanisms to sulphur deficiency. *J Exp Bot* 55: 1871–1879
70. Hirai MY, Yano M, Goodenowe DB, Kanaya S, Kimura T, Awazuhara M, Arita M, Fujiwara T, Saito K (2004) Integration of transcriptomics and metabolomics for understanding of global responses to nutritional stresses in *Arabidopsis thaliana*. *PNAS* 101: 10205–10210
71. Nikiforova VJ, Kopka J, Tolstikov V, Fiehn O, Hopkins L, Hawkesford MJ, Hesse H, Hoefgen R (2005) Systems rebalancing of metabolism in response to sulfur deprivation, as revealed by metabolome analysis of *Arabidopsis* plants. *Plant Physiol* 138: 1887–1896
72. Suzuki H, Reddy MS, Naoumkina M, Aziz N, May GD, Huhman DV, Sumner LW, Blount JW, Mendes P, Dixon RA (2005) Methyl jasmonate and yeast elicitor induce differential transcriptional and metabolic re-programming in cell suspension cultures of the model legume *Medicago truncatula*. *Planta* 220: 696–707

73. Sumner LW (2006) Current status and forward looking thoughts on LC/MS metabolomics, *In* Saito K, Dixon RA, Willmitzer L (ed.) Biotechnology in Agriculture and Forestry, vol. 57. Springer-Verlag, Berlin, 21–32
74. Huhman DV, Berhow M, Sumner LW (2005) Quantification of saponins in aerial and subterranean tissues of *Medicago truncatula*. *J Ag Food Chem* 53: 1914–1920
75. Mondello L, Lewis AC, Bartle KD (2002) *Multidimensional chromatography*, John Wiley & Sons Ltd, Chichester, UK
76. Evans C, Jorgenson J (2004) Multidimensional LC-LC and LC-CE for high-resolution separations of biological molecules. *Anal Bioanal Chem* 378: 1952–1961
77. Washburn M, Wolters D, Yates J (2001) Large-scale analysis of the yeast proteome by multidimensional protein identification technology. *Nat Biotechnol* 19: 242–247
78. Wolters D, Washburn M, Yates J (2001) An automated multidimensional protein identification technology for shotgun proteomics. *Anal Chem* 73: 5683–5690
79. Welthagen W, Shellie RA, Spranger J, Ristow M, Zimmermannn R, Fiehn O (2005) Comprehensive two-dimensional gas chromatography-time-of-flight mass spectrometry (GC x GC-TOF) for high resolution metabolomics: biomarker discovery on spleen tissue extracts of obese NZO compared to lean C57BL/6 mice. *Metabolomics* 1: 65–73
80. Aziz N, Paiva NL, May GD, Dixon RA (2005) Transcriptome analysis of alfalfa glandular trichomes. *Planta* 221: 28–38
81. Asano T, Masumura T, Kusano H, Kurita S, Shimada H, Kadowaki KI (2002) Construction of a specialized cDNA library from plant cells isolated by laser capture microdissection: toward comprehensive analysis of the genes expressed in the rice phloem. *Plant J* 32: 401–408
82. Nakazono M, Qiu F, Borsuk LA, Schnable PS (2003) Laser-capture microdissection, a tool for the global analysis of gene expression in specific plant cell types: identification of genes expressed differentially in epidermal cells of vascular tissues of maize. *Plant Cell* 15: 583–596
83. Birnbaum K, Shasha DE, Wang JY, Jung JW, Lambert GM, Galbraith DW, Benfey PN (2003) A gene expression map of the *Arabidopsis* root. *Science* 302: 1956–1960

Plant Systems Biology
Edited by Sacha Baginsky and Alisdair R. Fernie
© 2007 Birkhäuser Verlag/Switzerland

Metabolic flux analysis: Recent advances in carbon metabolism in plants

Martine Dieuaide-Noubhani[1], Ana-Paula Alonso[3], Dominique Rolin[1], Wolfgang Eisenreich[2] and Philippe Raymond[1]

[1] UMR 619 'Biologie du Fruit', INRA Université Bordeaux 2, IBVM, BP 81, 33883 Villenave d'Ornon Cedex, France
[2] Lehrstuhl für Organische Chemie und Biochemie, Technische Universität München, Lichtenbergstraße 4, 85747 Garching, Germany
[3] Department of Plant Biology, Michigan State University, 166 Plant Biology Building, East Lansing, MI 48824, USA

Abstract

Isotopic tracers are used to both trace metabolic pathways and quantify fluxes through these pathways. The use of different labeling methods recently led to profound changes in our views of plant metabolism. Examples are taken from primary metabolism, with sugar interconversions, carbon partitioning between glycolysis and the pentose phosphate pathway, or metabolite inputs into the tricarboxylic acid (TCA) cycle, as well as from secondary metabolism with the relative contribution of the plastidial and cytosolic pathways to the biosynthesis of terpenoids. While labeling methods are often distinguished according to the instruments used for label detection, emphasis is put here on labeling duration. Short time labeling is adequate to study limited areas of the metabolic network. Long-term labeling, when designed to obtain metabolic and isotopic steady-state, allows to calculate various fluxes in large areas of central metabolism. After longer labeling periods, large amounts of label accumulate in structural or storage compounds: their detailed study through the retrobiosynthetic method gives access to the biosynthetic pathways of otherwise undetectable precursors. This chapter presents the power and limits of the different methods, and illustrates how they can be associated with each other and with other methods of cell biology, to provide the information needed for a rational approach of metabolic engineering.

Introduction

Curiosity about metabolic pathways arises from the need to understand the biological mechanisms of plant life or from intents to improve the yield or quality of a plant product like wood, fruits or flowers, or the production of particular compounds. The first answers can be obtained from the analysis of metabolites, either by specific assays or by comprehensive methods of metabolite profiling. More specific questions that may require the use of tracers arise after the observations of changes in the levels of a metabolite of interest in relation to the genotype, development stages or the environment, or from unexpected results of carbon balance calculations. In recent years, labeling experiments have been used to unravel the function of regulatory or structural proteins in genetic engineering experiments.

Isotopic tracers are used to study metabolic pathways both qualitatively, to identify fluxes, and quantitatively, to quantify the fluxes in the pathways. The tracers may be either radioactive (^{14}C) or stable (^{13}C) isotopes. A wide range of enrichments is used for [^{13}C] labeled precursors, from about 100%, as in most of the works reviewed here, to around 1% with natural substrates when small variations around the natural abundance of ^{13}C are studied [1, 2]. Analyses are performed either by nuclear magnetic resonance (NMR) [3], or by mass spectrometry [4]. The combination of tracers, tracer concentrations and detection methods constitute a large number of methods. In addition, it must be noted that time is an essential parameter in labeling experiments because the duration of labeling determines how the labeling results can be handled and, more specifically, which type of model is adequate for the quantitative interpretation of enrichments in terms of flux values.

The experimental setup for a labeling experiment may be 'hypothesis free', but the interpretation of labeling data benefits from computational modeling of the metabolic pathways, which is necessarily based on hypotheses on the occurrence of certain metabolic pathways. The basic principles of modeling were established many years ago [5–8]. Establishing the set of metabolic pathways is the first step of setting up a model: the preliminary metabolic scheme is derived from published data on enzyme activities and compartmentation obtained from the literature. It should be noted that as long as the model fits the experimental data, the proposed pathways are validated, but the model itself does not lead to pathway discovery. The systematic search for pathways by methods such as elementary flux mode analysis [9] will provide more certainty in including all the pathways that may account for the observed label distribution. In addition, as underlined in [10], various sets of reactions may lead to similar label distribution from one given substrate. Therefore, fitting the model with experimental data is no proof that the metabolic scheme is valid. Redundancy is required in tracer experiments, i.e., a conclusion must be obtained through various means: by complementary labeling experiments with precursors labeled on different positions or with different labeling times, or by different methods like enzyme assays, enzyme inhibition, gene disruption or overexpression, etc.

Properties of labeling methods according to the length of labeling

Short-term labeling

In a typical short-term experiment (Fig. 1), the flow of tracer can be followed along the pathway: the amount of label in the pools, expressed as a percentage of the total incorporated label, decreases along the sequence. Similarly, the enrichment, or spe-

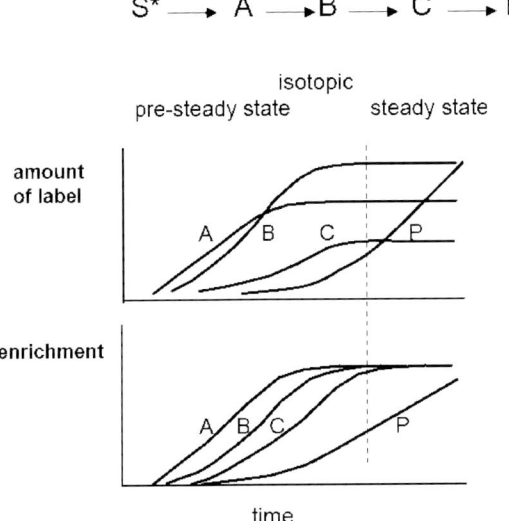

Figure 1. Labeling of pools in a pathway as a function of time. In labeling experiments, a pool may be a group of metabolites (proteins), a metabolite from a given cell compartment, or a particular moiety, or atom, of a metabolite. A purified metabolite may be a mixture of different pools of this compound from different cellular compartments, or from different cells of a tissue, each with different metabolic fates.

The results of tracer experiments are expressed as the amount of tracer in a given pool of metabolite (A) or as enrichment of the pools (B). ^{13}C enrichment is expressed as % and varies between 1.1%, the natural enrichment of carbon, and 100%, the enrichment of commercial tracers. For ^{14}C and other radioactive isotopes, enrichment is expressed as specific radioactivity which is an amount of radioactivity per mol (dpm (or Bq)/mol). In early pre-steady-state, both the amount of label per pool or the enrichment decrease along the pathway: both can be used as indicators of the position of the metabolites in the pathways (pool compartmentation, or branching pathways are possible complications). Unidirectional fluxes are calculated as the ratio (amount of label accumulated)/(enrichment of the precursor); underestimation may happen where labeling time is so long that label is lost from the product of interest. At isotopic and metabolic steady-state, the labeling and concentration of the intermediates remain constant: in a linear pathway, as illustrated here, the amount of label per pool is proportional to pool size, which brings no information on the pathway itself.

cific radioactivity, of the different pools decreases along the pathway. Short-term experiments are useful to solve three types of problems:

1. to establish the sequence of metabolites in a pathway; for example, the C3 and C4 photosynthesis types were named from the first metabolite found to be labeled after a few seconds of labeling with $^{14}CO_2$.
2. to quantify the absolute flux in the pathway: the number of moles of a metabolite, or group of metabolites, produced is calculated by dividing the amount of label accumulated by the enrichment of the precursor in the pathway (Fig. 1).

3. to deduce kinetic parameters of enzymes in the pathway from the kinetics of label distribution, by using models that include kinetic parameters of the enzymes. However, many kinetic parameters that are typically calculated from *in vitro* experiments with isolated enzymes may fail to meet the actual values under *in vivo* conditions of a compartmentalized plant cell or whole plant. Therefore, on the basis of the current technologies, modeling short-term labeling data in plant cells is intended with only limited areas of the metabolic network. As an example, this method was used for the identification of constraints in the accumulation of glycine betaine in plants [11, 12].

Steady-state labeling

As labeling time increases, isotopic steady-state is established in the pathway. In plants labeled with glucose, this was found to take a few hours. At this stage, the enrichments of different pools in the pathway are found to be constant, but the whole cells are not yet uniformly labeled. This was called 'relative steady-state' [13]. When a uniformly labeled substrate is provided, the steady-state enrichment in a linear pathway is uniform. This provides no information on fluxes in the pathway. However, where entering fluxes of unlabeled endogenous substrates lead to a dilution of label, the relative values of the labeled and unlabeled fluxes can be quantified from the decreased enrichment induced at the entry step (see Fig. 2). With non-uniformly labeled substrates, such as [1-^{13}C]glucose, the redistribution of the labeled atom(s) provides additional qualitative and quantitative information on substrate cycles in the pathway. This steady-state labeling method has been applied to the relatively large network formed by central metabolism (see below).

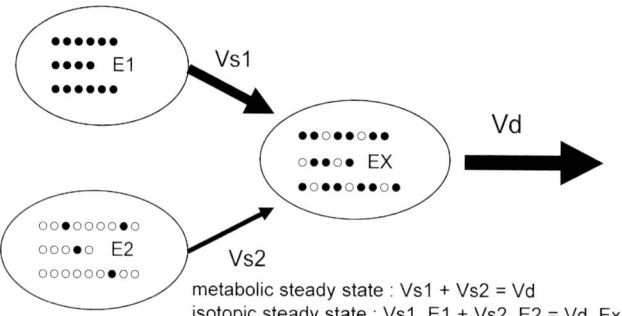

Figure 2. Modeling label distribution at metabolic and isotopic steady-state. Labeling to metabolic and isotopic steady-state enrichments provides information on joining pathways. For each pool (metabolite, or part of a metabolite) formed from two or more precursors, enrichment depends on both the enrichment of and the relative flux from each of the precursors. Two sets of equations can be written for each pool of metabolite or metabolite moiety: the metabolic steady-state equations state that C input = C output; the isotopic steady-state equations state that label input = label output. These equations link fluxes to enrichments. Relative values of Vs fluxes are calculated from measured enrichments of the precursors (E1 and E2) and product (EX).

Modeling of the isotopic and metabolic steady-state uses relatively simple linear equations, which link enrichment ratios with relative rates (Fig. 2). The amount of experimental data required to feed the model is lower after steady-state than after short-term labeling because after the long labeling times used, rapidly exchanging pools of a metabolite that are present in two or more compartments can be considered to have the same labeling. The review by Roscher et al. [14] discusses the effects of compartmentation and of transient conditions in long-term labeling experiments. In most experimental conditions, near steady-state rather than true steady-state conditions is obtained: applying steady-state models creates a problem when transient situations are studied, because the metabolic steady-state condition is not verified [3]. When the changes occur slowly, the turnover of the metabolites may be sufficient to ensure that changes in labeling in one step will be transmitted to the whole system. When changes in the level of a metabolite cannot be neglected, the metabolic steady-state equation must be modified to take this particular flux into account.

Long-term labeling for retrobiosynthetic analysis

After longer labeling time, final metabolites like protein amino acids become strongly labeled. Information is obtained from the relative abundances of different isotopologs in the sink metabolites (e.g., amino acids from proteins, starch, lipids) in these experiments. The isotopolog profiles of their respective precursors can be reconstructed by retrobiosynthetic analysis. The wealth of the method is that, on this basis, otherwise inaccessible metabolic intermediates can be analyzed that also constitute the central nodes of a metabolic network.

This chapter shows how labeling methods of metabolic flux analysis have recently led to a renewal of our views of the pathways of central metabolism, from sugars and hexose-P to the TCA cycle, and of isoprenoid biosynthesis. Clearly, many fields where sound approaches were developed are not treated here. The aims of this limited presentation are to illustrate the basic principles as well as the power and limits of the different methods, and to show how the qualitative and quantitative information provided by labeling experiments may contribute to the global approaches of systems biology.

Sucrose, glucose and hexose-P interconversions in heterotrophic cells

Heterotrophic cells import sugars, usually sucrose, from photosynthetic tissues. Sucrose enters the cell as sucrose or as glucose and fructose after hydrolysis by cell wall invertase. In the cell, sucrose can be hydrolyzed to glucose and fructose by invertase or cleaved to UDP-Glc and fructose by sucrose synthase. Intracellular glucose is also formed by substrate cycles similar to the turnover of sucrose, starch or cell wall polysaccharides. The operation of sucrose cycling was deduced after pulse/chase labeling experiments with labeled Glc where the decrease of the radioactivity measured in sucrose was more rapid than the decrease in the amount of sucrose [15]. It was deduced that sucrose was simultaneously synthesized (incorpo-

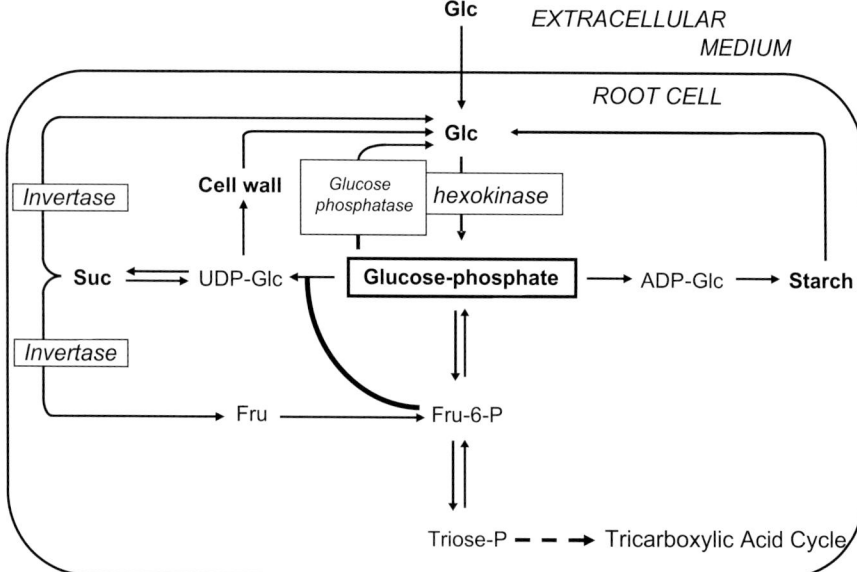

Figure 3. The sources of intracellular Glc in non photosynthetic plant cells. Glc is imported from the apoplast (extracellular medium). It is also a product of the turnover of intracellular oligo- and polysaccharides. This global flux was calculated after steady-state labeling experiments. The flux of Glc import and the fluxes of Glc formation from cell walls, starch and sucrose were measured by short-time labeling experiments. The occurrence of a Glc-phosphatase reaction results from the comparison of the global and individual fluxes towards intracellular Glc [22].

ration of label during the pulse) and degraded (decrease of labeling during the chase). In contrast, starch was found to be stable. The turnover of sucrose and starch was then quantified in other tissues: *Chenopodium* cells [16], ripening banana [17], potato tubers [18], and tomato fruit [19]. Using an approach of steady-state labeling in maize root tips [20], and in tomato cells [21], a high rate of cycling between hexose-P and glucose was observed and, based on enzyme activity data, it was suggested that this cycle was the result of sucrose turnover. More recently [22], a combination of short time and steady-state labeling approaches led to an evaluation of the respective role of the different pathways that may be involved in the Glc-P to Glc conversion (see Fig. 3). This work is presented in more detail here as an illustration of the properties of these two methods of labeling.

Short-term labeling estimations of free Glc formation in plant cells

Short-term labeling experiments were used, together with metabolite measurements, to evaluate the flux of external Glc uptake and the fluxes of Glc formation from the turnover of sucrose, starch and cell wall polysaccharides (Fig. 3). The approach was similar to that used in [16], and consists of:

1. Measuring the unidirectional flux of synthesis (Vs) using short-term labeling experiments.
2. Calculating the net flux of sugar (i.e., sucrose or starch) accumulation (Va), as the variation in sugar content, measured by a method of quantitative analysis of metabolites, over a time period: Va = Δ sugar content/Δt.
3. Deducing the unidirectional flux of degradation (Vd) as: Vd = Va-Vs.

The unidirectional flux of synthesis of a compound is calculated as the rate of incorporation of radioactivity (V_{RA}) divided by the specific radioactivity of its precursor. The precursors of sucrose and starch are UDP-Glc and ADP-Glc, respectively. Because measuring their specific radioactivity is difficult, glucose [16], or hexose-P [15, 23] were used as indicators because they give more certainty and they were expected to be in rapid exchange with UDPGlc and ADPGlc. In maize root tips, it was verified that Glc-6P and UDP-Glc were identically labeled, even after a very short time of labeling [22]. On the other hand, intracellular Glc was not identically labeled to UDPGlc, which may be explained by the slow labeling of the Glc vacuolar pool [20].

In growing maize root tips, short-term labeling experiments showed that the turnover of cell walls and starch were low compared to sucrose turnover and could therefore be neglected as sources of intracellular glucose. Steady-state labeling was used to examine whether sucrose turnover accounts for Glc-6P turnover.

Steady-state labeling measurements of Glc-P cycling

At isotopic steady-state the labeling of intracellular Glc results from the relative values of the flux of external Glc uptake (external Glc is labeled on C1 only) and the sum of the intracellular fluxes of Glc production from cellular oligo- and polysaccharides. The Glc molecules formed from these reactions derive from the hexose-P pool: they are less labeled on C1 than external Glc, and more labeled on C6. The enrichment of C1 and C6 of intracellular Glc and of the sucrose glucosyl was measured by ^{1}H and ^{13}C NMR. Resolution of the equations for either C1 or C6 leads to estimations of the flux ratio of total intracellular flux of Glc production (called Vrem) to the flux of Glc uptake. The absolute value of Vrem was then calculated using this ratio and the absolute value of Glc uptake measured in the short-term experiment. Vrem was found to be very much higher than the flux of Glc production from sucrose turnover determined by short time labeling. This result pointed to the operation of another substrate cycle in maize root tips, possibly the direct hydrolysis of Glc-P to Glc by a Glc-phosphatase [22].

This work illustrates how short- and steady-state labeling are complementary approaches to a better insight into central metabolism.

Partitioning of Glc-P through the pentose phosphate pathway and glycolysis

Glucose 6P can be catabolized through glycolysis or the oxidative pentose phosphate pathway (OPPP) which plays an important role in cell biosyntheses and defence through the production of NADPH. Measuring the partition of hexose-P between OPPP and glycolysis is important to establish the function of the pathways. This is difficult in all organisms because the two pathways are interconnected through the exchange of fructose-6-P and triose-P. In addition, in plants, both pathways are present in two compartments, the cytosol and the plastids.

Classic assays with [1-^{14}C]- and [6-^{14}C]glucose

The approaches used to compare the fluxes in glycolysis and the OPPP have been elaborated by Katz and collaborators [6]. A model was set up to calculate the contribution of each pathway by using [^{14}C]glucose labeled on C1 or C6, through the specific yields of evolved $^{14}CO_2$ (the C1/C6 ratio) or the enrichments ratios of the triose-P and their derivatives (alanine, malate, etc.). Glucose labeled on C2 or C3 was also used to obtain complementary information through the redistribution of label in the Glc molecule. The specific yield of CO_2 is higher, and the enrichment of triose-P is usually found to be lower with [1-^{14}C]glucose than with [6-^{14}C]glucose. This is explained by the different fates of the Glc-C1 and -C6 through the OPPP. For Glc-6-P that enters the OPPP, C1 is lost as CO_2 at the second step of this pathway, whereas the C6 is incorporated into fructose-P or glyceraldehyde-3-P via the non-oxidative part of the pentose phosphate pathway. It may either be lost as CO_2 much further along the metabolic pathway, after two turns in the TCA cycle, or be retained in biosynthetic products, the most important, quantitatively, being the proteinogenic amino acids. Conversely, the fate of Glc-6-P C1 and C6 through glycolysis, is the same. Therefore, the differences observed in the labeling of CO_2 or triose-P derivatives are attributed to the OPPP. In fact, two distinct mechanisms affect the production of $^{14}CO_2$ from [1-^{14}C]glucose or [6-^{14}C]glucose: with [1-^{14}C]glucose, $^{14}CO_2$ evolves earlier as can be seen in short-term experiments, and in higher amounts when in an isotopic steady-state. Very often the two effects are confused. For example, the fact that the C1 of Glc-1-P is lost earlier in the OPPP does not explain that the specific yield of CO_2 is higher with [1-^{14}C]glucose than with [6-^{14}C]glucose, because the specific yields, (which give the C1/C6 ratio) are measured in near steady-state conditions. Indeed, if glucose was fully oxidized to CO_2, the C1/C6 ratio (at steady-state) would be 1, whatever the flux through the OPPP. The difference in specific CO_2 yields essentially depends on the incomplete oxidation of the triose-P derivatives [6, 8].

The problem is then to derive flux quantification from the observed differences in specific yields or enrichments. The method most often used because of its apparent simplicity was to incubate the tissues with either [1-^{14}C]glucose or [6-^{14}C]glucose and measure the specific yields of CO_2 and calculate the C1/C6 ratio: the C1/C6 ratio higher than 1 was used as an indicator of the operation of the OPPP [6]. The application to plants has been critically analyzed by ap Rees [8]. It was noted that, in plants, the pathway of pentosan synthesis which releases the Glc carbon 6 as CO_2 would be a cause of error. The results obtained on maize root tips show that this method is effectively unreliable with plant tissues: the same production of $^{14}CO_2$ was measured from [1-^{14}C]glucose and [6-^{14}C]glucose, which confirmed previous data that had been interpreted as an indication that the OPPP was not active in this material [20]. However, the decreased enrichment of triose-P derivatives compared to that of hexose-P after steady-state labeling experiment [20] (see below) strongly suggested that the OPPP was highly active. In addition, this is consistent with the high biosynthetic activity of the growing root tips, which requires a source of NADPH. It was suggested that the C1/C6 ratio was disturbed by the pathway of pentosan synthesis. This example demonstrates that the method based on $^{14}CO_2$ yields is not reliable with plant tissues, as previously indicated [8]. It may be noted that, in the same labeling conditions, the observation of triose derivatives, instead of CO_2, would be less prone to errors.

As an improvement to this method, Garlick et al. [24] replaced [1-^{14}C]glucose with [1-^{14}C]gluconate. They showed that plant cells can take up [1-^{14}C]gluconate and metabolize it essentially by direct phosphorylation into [1-^{14}C]6-phosphogluconate which is then decarboxylated. Therefore, the release of $^{14}CO_2$ from [1-^{14}C]gluconate is a reliable indicator of the occurrence of a flux through the OPPP. The C1*/C6 ratio, with [1-^{14}C]gluconate and [6-^{14}C]glucose, respectively, was used. The method was found to be broadly applicable to plants, and showed that the OPPP was active in a number of plant materials, including maize root tips. However, it would be difficult to make this method quantitative. The C1*/C6 ratio depends on both the flux through the OPPP relative to that of glycolysis, and on the fraction of triose-P oxidized to CO_2. Therefore, a variation in the C1*/C6 ratio would not be reliably interpreted as a change in the flux through the OPPP relative to glycolysis, since it may also reflect a change in the fraction of triose-P retained in stored products. A quantification of the absolute flux through the OPPP could be made in short-term labeling experiments from the rate of $^{14}CO_2$ evolution if the specific radioactivity of the pool of 6-phosphogluconate could be measured; however, as discussed in [24], the cellular location of the reaction, cytosolic or plastidial, is not known.

Assays through NMR measurements of carbon enrichments

Steady-state labeling of plant tissues with stable isotopes ([1-^{13}C]-, [2-^{13}C]-, [1,2-$^{13}C_2$]-, or [U-$^{13}C_6$]-Glc) associated with NMR or MS label measurements of metabolites provides a great deal of information about the reactions of intermediary metabolism. Estimations of the partitioning of hexose-P between glycolysis and the OPPP can be obtained after steady-state labeling with [1-^{13}C]glucose, through the analysis of sucrose, starch and alanine. The labeling of sucrose and starch reflects that of the cytosolic and plastidial hexose-phosphates, respectively, and the labeling of alanine reflects that of pyruvate, which derives from the triose-P. The information that was obtained by the comparison of specific CO_2 yields with [1-^{14}C]- or [6-^{14}C]Glc can be obtained with [1-^{13}C]glucose alone because, in the latter case, the carbon enrichments of hexose-P and triose-P can be compared. However, redundancy through the use of other tracers is still useful.

This approach was used to study the intermediary metabolism of maize root tips [20] and in tomato cells [21]. After incubation with [1-^{13}C]Glc up to isotopic steady-state, the enrichments of carbon atoms in glucose, sucrose, starch and alanine were determined. Initially, the qualitative analysis of data were used to determine which metabolic pathways had to be included in the model, an important step before writing the equations that relate fluxes (the unknowns) to enrichments (experimental data). As an example, the OPPP was included in the model after the observation that alanine C3 was less labeled than the average of Glc C1 and C6. In a second step, fluxes were calculated to fit experimental enrichments. The carbon flux entering the OPPP was found to be higher than the flux of glycolysis measured at the PEP formation step [20, 21].

It is characteristic of steady-state labeling studies that fluxes can be quantified but the pathway involved cannot be identified with certainty. Since, in maize root tips, the ratio of enrichments of C6 to C1 was higher in starch than in sucrose, the plastidial OPPP was considered as a possibility to explain the loss of label from the Glc-P C1 position. In a complementary experiment with [2-^{14}C]Glc, the transfer of label to Glc C1, which characterizes the operation of the OPPP, was sought in the glucosyl units of sucrose and starch: it was found essentially in starch, thus confirming the plastidial location of the OPPP. In maize root tips, it was possible to fit the model with a null flux through the cytosolic OPPP [20]. In tomato cells the situation was found to be different: sucrose and starch were identically labeled, which was interpreted as a rapid exchange between the cytosolic and plastidial hexose-P; consequently, it was not possible to estimate the flux of the OPPP in each of these subcellular compartments [21]. It must be observed that in these two studies [20, 21] not all the possible reactions in the non-oxidative branch of the PPP were considered: the ribose-5P isomerase and ribulose-5P isomerase reaction were assumed to function close to equilibrium.

A more complete description of the pentose phosphate pathway was obtained by the complete analysis of the intramolecular labeling of sucrose and starch in *Brassica napus* embryos incubated to isotopic steady-state with [U-^{13}C$_6$]glucose, [1-^{13}C]glucose, [6-^{13}C]glucose, [U-^{13}C$_{12}$]sucrose, and [1,2-^{13}C$_2$]glucose [25]. Labeling with [2-^{13}C]Glc was used to evaluate the reversibility of the transketalose and transaldolase reactions. The labeling in amino acids, lipids, sucrose and starch was measured by GC-MS and NMR. The similar labeling of cytosolic and plastidial metabolites was interpreted as a rapid exchange of metabolites between these compartments. The measured fluxes were used to evaluate the split of hexose-P towards glycolysis and the OPPP: the latter was found to have a contribution to the supply of reductant for fatty acid biosynthesis lower than usually estimated. In a further study [26], the balance of carbohydrate to oil conversion was found to be much higher than would be expected from established pathways. Metabolic and isotopic steady-state experiments and modeling, using [1-^{13}C]alanine and [U-^{13}C]alanine as substrates, showed that a significant fraction of the CO_2 lost in the pyruvate dehydrogenase reaction, which forms the acetyl-CoA used for fatty acid biosynthesis, is recycled by Rubisco in a light dependent manner, but without Calvin cycle.

Using steady-state labeling, metabolic pathways and fluxes were also analyzed in developing maize kernels [27–29]. The *in vitro* culture of maize kernels represents a system to study the metabolism in intact kernels at different developmental stages under defined conditions. Typically, the kernels were supplied with culture media containing a mixture of [U-$^{13}C_6$]glucose and unlabeled glucose. After growth on the labeled medium for several days, glucose was isolated from the starch hydrolysate and analyzed by NMR spectroscopy.

Due to the use of totally ^{13}C-labeled glucose as a tracer, highly complex signal patterns were detected in the ^{13}C-NMR spectra that reflect couplings between ^{13}C-atoms in a given molecule. Due to the inherently restricted coupling information in complex molecules (typically, ^{13}C-^{13}C couplings can only be observed via 1–3 bonds) and due to limited spectral resolution, isotopolog groups (so-called X-groups) [30] give sets of individual glucose isotopologs. Numerical deconvolution can then be used to determine the abundances of individual carbon isotopologs from the abundances of the X-groups.

As a major finding, the relative abundances of the [U-$^{13}C_6$]-isotopolog were low showing that the carbon skeleton of the vast majority of the applied labeled glucose had been broken and reassembled at least once. The observed [1,2,3-$^{13}C_3$]- and [4,5,6-$^{13}C_3$]-isotopologs reflected glycolytic cycling via triose phosphates. The [1,2-$^{13}C_2$]-isotopologs showed cycling via the transketolase reaction of the pentose phosphate pathway, and the [2,3-$^{13}C_2$]- and [4,5-$^{13}C_2$]-isotopologs have been explained by cycling involving the tricarboxylic acid cycle.

As outlined in more detail below, the isotopolog compositions can then be balanced by numerical or computational methods affording relative metabolic fluxes in the biosynthesis of the metabolites under study. In the kernel experiments, a computational approach [29, 31] was used that assessed the contributions and interconnections of glycolysis, glucogenesis, the pentose phosphate pathway, and the citrate pathway in considerable detail. Interestingly, minor modulations of the flux pattern were found during different phases of kernel development probably as an answer to the specific demands for metabolic precursors during kernel development [29].

Carbon inputs into the TCA cycle

The tricarboxylic acid cycle (TCA cycle) is the major pathway of respiration in all eukaryotic cells. It is well known for its energetic and biosynthetic roles. Acetyl-CoA, usually produced in the mitochondrion by the PDH reaction, is condensed with OAA to form citrate. In one 'turn' of the cycle, two carbons are lost as CO_2 and a new OAA molecule is formed: this is equivalent to the complete oxidation of the acetyl unit, but the entering acetyl carbons remain present in the OAA molecule. The intermediates of the TCA cycle are also used as building blocks for biosyntheses, particularly, in quantitative terms, the biosynthesis of amino acids of the glutamate and aspartate families. For each molecule taken out of the TCA cycle, so-called 'anaplerotic' reactions provide the OAA required as acetyl-unit acceptor. In plants, the PEP carboxylase reaction, which produces OAA in the cytosol, plays this role (Fig. 4). Equivalent anaplerotic substrates are four carbon compounds derived from the catabolism of amino acids of the aspartate family, or succinate produced by the glyoxylic acid cycle; the five C compound alpha-ketoglurate, which is de-

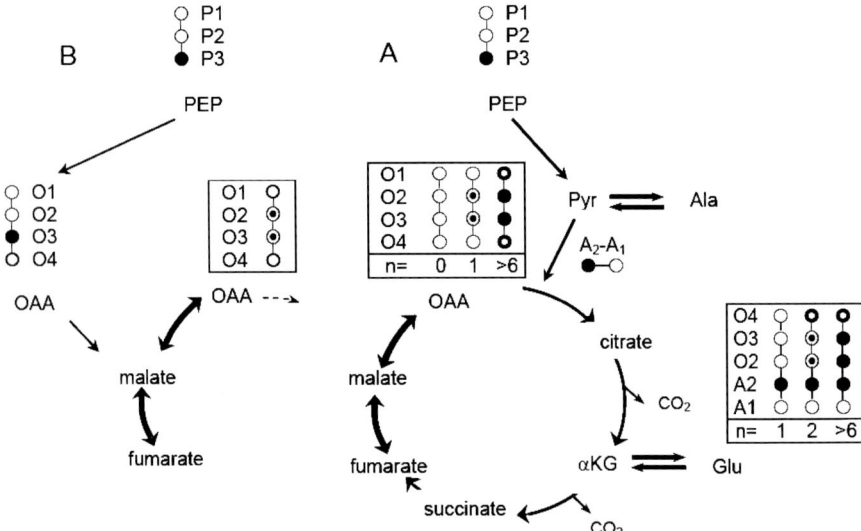

Figure 4. Glycolytic carbon input into the TCA cycle. Glc labeled on C1 or C6 produces PEP, pyruvate and alanine labeled on their C3 (●), with the other two carbons unlabeled (○). A: pyruvate dehydrogenase produces acetyl units labeled on their C2 (A2). A2 then forms the C4 of glutamate carbons. During the first turn of the TCA cycle (n=1), A2 and O3 are incorporated into the methylene carbons of succinate; because succinate is symmetrical, A2 goes to either of the central carbons of OAA. As the number of 'turns' increases, the enrichments of the OAA carbons O2 and O3 increases up that of A2 (shown here for n>6). B: The PEP carboxylase reaction forms OAA labeled on its C3 (O3), and the near equilibrium reactions between malate, fumarate and OAA randomize this label between O2 and O3 of OAA; O4 is also labeled, according to the enrichment of cytosolic CO_2. The OAA metabolized in the TCA cycle, as observed in the Glu molecule, is a mixture of the OAA formed in the TCA cycle (A) and that formed by the PEPC reaction (B).

rived from the catabolism of amino acids of the glutamate family also plays this role. The full oxidation of OAA is possible after its conversion to pyruvate through the malic enzyme reaction.

Major questions about the TCA cycle are the following:

- Among sugars, proteins and lipids, what is the substrate of respiration?
- In sugar-fed cells, where glycolysis provides both pyruvate and OAA to the TCA cycle:
  – how is the glycolytic flux partitioned between these two branches?
  – is OAA used as anaplerotic substrate only, or is it converted to pyruvate, via the malic enzyme (ME) reaction, to feed respiration?

Short-term labeling has been used for pathway identification, and steady-state labeling experiments have provided quantitative information about fluxes. The origin and fate of some carbon atoms in intermediates of the TCA cycle will be described first, because this knowledge helps to deduce qualitative information from labeling patterns and to design experiments that can produce the information needed, even if the final, quantitative, interpretation of the data needs comprehensive modeling of the pathways.

Glutamate as the indicator molecule in studies of the TCA cycle

In steady-state labeling studies of the tricarboxylic acid cycle, the essential molecule to examine is glutamate, the indicator molecule for alpha-ketoglurate. Glutamate is a stable compound, it is usually abundant and its enrichments can be easily measured by ^{1}H and ^{13}C NMR spectroscopy (for example, see [20]). The glutamate carbons 4 and 5 are made of the acetyl units incorporated into citrate by the citrate synthase reaction, whereas the other three carbons are derived from oxaloacetate (OAA, Fig. 4A). During the first turn of the TCA cycle, the C4 and C5 glutamate carbons are incorporated into the methylene and carboxylic carbons, respectively, of succinate. Because succinate is symmetrical, the labeled methylene carbon goes to either of the central carbons of OAA; the carboxylic carbons go to either of the corresponding positions in OAA. A simple model of this sequence of reactions (input of one acetyl unit and loss of two CO_2 per turn) shows that, at steady-state, the acetyl-C2 forms the C2-C3-C4 moiety of glutamate. Therefore, after labeling with [1-^{13}C]glucose or [2-^{13}C]acetate, each of these central glutamate carbons would have the same enrichment as the acetyl-C2.

In plants, however, the NMR analysis of glutamate most often shows that the glutamate C2 and C3 are less enriched than C4. This accounts for the anaplerotic input of OAA, which is usually attributed to the PEPC reaction (see discussion below). In labeling experiments with [2-^{13}C]acetate, the OAA produced by the PEPC reaction is not labeled. In labeling experiments with [1-^{13}C]glucose, the PEPC reaction labels the OAA C3, but this label is randomized between C3 and C2 in the OAA-fumarate-succinate exchange that occurs in the TCA cycle (Fig. 4B). The average enrichment of C2 and C3 in the OAA molecules from the PEPC reaction is about half of that found in glutamate C4. Small differences observed between the C2 and C3 of glutamate have been attributed to incomplete randomization of the OAA produced by the PEC flux [20, 32]. The alternative mechanism is partial channeling of the TCA cycle flux, but there is no evidence for channeling at this step in plants [20].

In labeling experiments where labeled Glc or acetate are used as substrate, the dilution of the glutamate C2-C3 relative to C4 at isotopic steady-state can be used to calculate the anaplerotic flux, but the dilution itself does not indicate which of the different potential anaplerotic pathways is responsible for this flux. The choice of the PEPC reaction as that being responsible for the anaplerotic flux in sugar-fed tissues does not result from the observed labeling but from indications that PEPC activity is related to N assimilation [33] and protein synthesis, or to malate overproduction (see references below). On the other hand, the alternative anaplerotic pathways, proteolysis or the glyoxylic acid cycle, are found in special cases such as decaying or sugar-starved tissues [34].

Partitioning of the glycolytic flux at the PEP branch point

In plants, cytosolic glycolysis produces pyruvate or OAA, through the pyruvate kinase (PK) or the phosphoenolpyruvate carboxylase (PEPC) reactions, respectively. The partitioning of glycolysis at this branch point was studied by both short time or steady-state labeling experiments.

Changes in the PEPC/PK flux with development measured by short-term labeling

In the developing seeds of barley, at the stage of maximum fresh weight, the endosperm acidifies rapidly as it receives malic acid formed in the aleurone layer. This was found to be accompanied by a five-fold rise in the PEPC activity in the aleurone, which suggested that the increase in malic acid production was linked to an increased flux through the PEPC reaction. Alternative hypotheses included either a change of the fate of OAA produced by PEPC from amino acid synthesis to malic acid formation, or an increase in the glyoxylic acid cycle. The hypothesis of an increased PEPC/PK flux was tested by a short-term labeling experiment where uniformly labeled glucose was used as substrate, and the incorporation of radioactivity was monitored for up to 10 min in the major products of the two branches of glycolysis: alanine for the PK branch and malate + aspartate for the PEPC branch, as well as in the common products of the pathways, the TCA cycle intermediates citrate and glutamate [35]. Among the carboxylic acids and amino acids, the greater amounts of label were found in the compounds analyzed, with comparatively little label in citrate and glutamate. This showed that malate was not significantly labeled through the TCA cycle. Since, in the time period studied, most of the label was still present in the products of interest, the quantitative comparison of the PEPC and PK fluxes could be made by comparing the amounts of label incorporated in malate, aspartate and alanine. The PEPC/PK flux ratio was found to increase from 1.6 in aleurone of young seeds, to 7.5 in older, acidifying seeds. The kinetics of labeling also showed that the pattern of labeling changes in old compared to young aleurone. Alanine, aspartate and malate are labeled to similar extents in young seeds, whereas malate is the major product of glycolysis in old seeds. It should be noted that only ratios of tracer amounts were compared between materials. Amounts of incorporated label were not compared as they also depend on a number of factors that may differ according to development stages, such as the rate of tracer (Glc) input into the tissues, the size of the intracellular Glc pool, etc.

Changes of the PEPC/PK ratio according to growth conditions studied by steady-state labeling

The PEPC flux was also measured after steady-state labeling, based on its effect on the differential enrichments of the glutamate carbons. In maize root tips [20, 36] and in tomato cells [21] labeled at isotopic steady-state, the enrichments of Ala-C3 was the same as that of Glu-C4. This indicates that Pyr-C3 is the only source of Glu-C4, in agreement with the generally accepted view that sugars are the major respiratory substrate in plant cells. The lower labeling of glutamate carbons C2 and C3 compared to C4 was related to the PEPC flux. As illustrated in details [36], the effect of the PEPC flux on the labeling of TCA cycle intermediates depends on where the carbon drain for biosyntheses occurs in the TCA cycle. In [20] the fluxes towards amino acids of the glutamate and aspartate families were assumed to be equal; this was confirmed in tomato cells in culture by analyzing the amino acid composition of the proteins [21]. From the steady-state models, the PEPC/PK flux ratio was calculated to be 0.5 in maize roots and 0.4 in tomato cells during the exponential growth phase. This means that of three PEP molecules formed by glycolysis, one goes through PEPC and two through the PK branch of glycolysis.

Changes induced in the metabolism of maize root tips submitted to sugar starvation were studied [34] by providing [1-^{13}C]glucose for 4 h, then incubating them in the absence of glucose (i.e., sugar starvation was induced in pre-labeled tissue). Modeling of these data was not intended because the system was clearly far from both isotopic and metabolic steady-state. However, the labeling data could be interpreted in qualitative terms. At the end of the 4 h labeling period, the carbons of alanine and glutamate were less enriched than at steady-state (16 h labeling) but, as expected in glucose-fed tissue, the alanine C3 and glutamate C4 enrichments were similar, and the glutamate C2-C3 were clearly less enriched than the C4, reflecting the PEPC activity. After 5 h of glucose starvation, the C2, C3 and C4 had become equal and remained so, although at a lower value at 16 h. This was interpreted as an indication that the PEPC flux had stopped as a consequence of glucose starvation.

Similarly, during the culture cycle of tomato cells, the C2-C3 *versus* C4 difference was found to decrease at the same time as protein accumulation rate decreased towards the end of the exponential growth phase [21]. At this stage, the PEPC/PK ratio had decreased to 0.25, indicating that only one PEP molecules out of five formed from hexose-P was used in the PEPC reaction. This is in keeping with the decreased rate of protein accumulation, compared to earlier stages of the culture. Together, these results support the view that the PEPC flux is linked with the biosynthetic activities of the cell. Moreover, as described below, the detailed study of the fate of OAA showed that the PEPC flux is essentially anaplerotic.

Quantification of the malic enzyme flux: The fate of oxaloacetate

How much of the PEPC flux is used for biosyntheses or is converted to pyruvate to feed respiration? OAA can be converted to pyruvate (Pyr) in the malic enzyme (ME) reaction. During [1-^{14}C]glucose labeling, the ME reaction produces Pyr and alanine molecules that are equally labeled on their C2 and C3, whereas glycolysis produces Pyr labeled on carbon 3 only. In most experiments with aerobic plant cells [21, 34, 36], the enrichment of alanine C2 was 2–3%, whereas that of Ala C3 was around 30%. The low labeling of Ala-C2 compared to Ala-C3 shows that little conversion of OAA to Pyr occurs *in vivo*. Using a comprehensive [20] or a simplified [36] model, the malic enzyme flux was found to provide only 3% or 8% of the Pyr flux to the TCA cycle. This result was contrasted to previous studies of malate respiration by isolated mitochondria and of ME activity that suggested that the PEPC-ME couple might supply Pyr to the mitochondrial pyruvate dehydrogenase [36]. The labeling experiments *in vivo* established unambiguously that ME catalyses a minor flux in normal conditions; therefore, the PEPC flux is essentially anaplerotic.

The ME/PK ratio was found to increase six-fold under severe hypoxia, as calculated from the increase in the enrichment of Ala-C2 from 1.6 to 4.2 above natural abundance [30]. This increased ME activity is consistent with the decrease in malic acid content that occurs in most plant tissues transferred to anoxic or deeply hypoxic conditions and was explained by the rapid decrease in pH that occurs as oxygen is depleted [36].

The beta-oxidation of fatty acids as an alternative source of acetyl-CoA for respiration

A different configuration of the TCA cycle was observed in the particular case of germinating fatty seeds. In fatty seeds, the massive consumption of oil reserves starts about one day after radicle emergence. At this stage, the fatty acids are converted to sugars that are transported to the growing seedling through the concerted action of the beta-oxidation of fatty acids, the glyoxylic acid cycle and gluconeogenesis. What happens earlier, in the pre-emergence phase was less clear. The respiratory metabolism was thought to depend on sugars, with glycolysis and the pentose phosphate pathway playing a major role. However, fatty seeds such as lettuce or sunflower were found to have a very low fermentation rate under anoxia [37], which was not consistent with the known activation of glycolysis under anoxic conditions. This led to an examination of the pathways of respiration in germinating fatty seeds, using radioactive glucose, acetate and fatty acids. It was found that, similar to glucose and acetate, short chain or long chain fatty acids label the TCA cycle intermediates.

Three possible pathways were considered. The alpha oxidation of labeled fatty acids would produce CO_2 which would be incorporated by the PEPC reaction into OAA, and then be transferred to other TCA cycle intermediates. The other two pathways involved the beta-oxidation of fatty acids which produces acetyl units. The beta-oxidation of fatty acids associated with the glyoxylic acid cycle is active in growing seedlings might also present some activity in early germination. The third possibility was the beta-oxidation of fatty acids feeding the TCA cycle directly, as occurs in animal tissues.

The operation of the TCA cycle and of the glyoxylic acid cycle can be distinguished from each other by short time labeling with acetate or fatty acids because there is only one entry point for acetyl unit in the TCA cycle, the citrate synthase reaction, whereas there are two entry points in the glyoxylic acid cycle, the citrate synthase and the malate synthase reactions. In the classic experiments of Canvin and Beevers [38] which established the occurrence of the glyoxylic acid cycle in the endosperm of castor bean seedlings, more label had accumulated in malate than in citrate, and more in aspartate than in glutamate, after 2 min of labeling with [^{14}C]acetate.

Evidence for a direct entry of acetyl-CoA into the TCA cycle by short-term labeling

When lettuce embryos were labeled with [^{14}C]palmitic acid or [^{14}C]hexanoic acid for 1–10 min, the amount of radioactivity measured in organic acids and amino acids was found to be the highest in citrate, followed by glutamate, succinate and malate [32]. This sequence clearly reflects the operation of the TCA cycle (Fig. 4), and is not consistent with either the glyoxylic acid cycle or alpha-oxidation. It shows that the acetyl units produced from fatty acids by beta-oxidation are incorporated into citrate through a citrate synthase reaction. This tells nothing of the quantitative importance of this pathway in the respiratory metabolism. Because of the multiplicity of acetyl-CoA pools in plant cells, the measurement of this flux through short time labeling experiments would be very difficult, as previously underlined after studies with animal systems [13].

Quantification of non-glycolytic carbon input by steady-state labeling

A quantitative estimation of the glycolytic and non-glycolytic origins of acetyl units into the TCA cycle was obtained from a steady-state labeling experiment with uniformly labeled glucose, i.e., only the glycolytic acetyl-units were labeled. Glutamate labeling was examined in two ways: its specific radioactivity was compared with that of aspartate, and the labeling of glutamate C1 was compared with glutamate C5 after selective decarboxylations of the molecule. It was found that the C4-C5 moiety of glutamate, which originates from the acetyl unit incorporated at the citrate synthase step, was only slightly labeled compared to the C1-C3 moiety derived from the OAA molecule. Modeling of the pathway, and assuming that the non-glycolytic pathway is essentially beta-oxidation, indicated that the beta-oxidation of fatty acids provides more than 90% of the acetyl-CoA entering the TCA cycle. The enrichments of the glutamate carbons, particularly the non-carboxylic carbons, are now easily measured by ^{13}C- and ^{1}H-NMR analysis. However, whereas [^{14}C]glucose can be used at tracer (micromolar) concentrations, [^{13}C]glucose must be provided at a high concentration which may lead to an artifactual increase in the activity of glycolysis.

Similar experiments showed that the beta-oxidation of fatty acids plays a similar role in sugar-starved tissues [34]. Experiments aimed at providing a confirmation of these labeling experiments showed that an isolated peroxisomal fraction from germinating sunflower seeds converts labeled palmitic acid to acetyl-CoA and, when OAA is added, to citrate [39]. It was proposed that the acetyl units produced by the peroxisomal beta-oxidation of fatty acids are exported to the mitochondria as citrate.

Given the quantitative importance of fatty acid beta-oxidation during germination, mutations that affect beta-oxidation could be expected to strongly affect the germination process. Clear phenotypes were observed on seedling growth but only in two cases on germination itself [40]. The mutation of a transporter that imports acyl-CoAs into the peroxisome and a double mutation that suppresses the citrate synthase activity in peroxisomes produce seeds that do not germinate normally but can be made to germinate by removing the seed coat and supplying sucrose. The normal development of mutants affected on other genes in this pathway is explained by the multiplicity and overlapping functions of these genes. Of the different methods used to establish the function of beta-oxidation, the labeling experiments were the most important in establishing its quantitative importance in respiration in early germination. They could not, however, resolve its cellular localization, either peroxisomal or mitochondrial. The data obtained by the molecular genetic methods indicated that the peroxisome is the major, if not unique, site of beta-oxidation in germinating seeds [40].

Steady-state model solving

The resolution of isotopic and metabolic steady-state models, which relate fluxes and enrichments through linear equations, is relatively simple. Model solving was obtained using a matricial approach with the software Excel [20], or using the resolution of simultaneous algebraic equations using the software Mathematica [21].

As the amount of experimental data increases, specific softwares such as ^{13}C-Flux [25] or 4F [29, 31] are needed. The use of ^{13}C-Flux requires writing the forward and backward reactions of glycolysis and the OPPP, specifying the transition of carbon atoms from one metabolite to another for each reaction. ^{13}C-Flux makes it possible to simulate the steady-state distribution and to calculate the isotopomers for each intermediate of these pathways. Using an optimization algorithm, flux calculations are then fitted with the labeling measurements. In addition to the simulation and optimization tools, ^{13}C-Flux provides statistical output, including a sensitivity matrix that shows which fluxes have influence in which measurements, a covariance matrix that can be derived into confidence intervals for each flux value, and a parameter sensitivity matrix that shows the impact of the change of single measurements on the estimated fluxes [41, 42]. With the large quantity of experimental data from the different ^{13}C-substrates and the GC-MS and NMR measurements used in the study of *Brassica napus* embryos [25], an overdetermination of the flux parameters was obtained, which provides an improved reliability in flux calculations. Indeed, it was possible to accurately quantify the fluxes through glycolysis and the OPPP, including the reverse fluxes of TA and TK. The development of software packages that can automatically generate and handle the equations of complex metabolic networks and manage a large quantity of experimental data offers huge advances in flux quantification.

Retrobiosynthetic analysis: The origin of plant terpenoids

Steady-state labeling experiments have a long history in the discovery and analysis of metabolic pathways. Experiments using general ^{13}C-labeled precursors (e.g., glucose, acetate) in conjunction with the retrobiosynthetic concept provided a solid basis to reconstruct the metabolic pathways in microorganisms [43]. As already mentioned above, the use of general tracers is also a powerful method to assign and to quantify metabolic routes in plant cell cultures, organs of plants or even whole plants grown on medium supplemented with the ^{13}C-labeled tracer. As a consequence of the general nature of the precursor used, the label is typically diverted to every metabolite through the metabolic network of the plant cell. Whereas the obtained isotopolog profiles are highly complex and typically show mixtures of several isotopologs, they nevertheless reflect the metabolic history of every metabolite under study, and provide a concise data matrix for the quantitative analysis of the pathways and fluxes between the metabolites under study. The concept will be illustrated in the following chapter in light of the discovery of a novel pathway for the biosynthesis of terpenes.

Well above 20,000 plant terpenoids have been reported [44]. A subgroup comprising sterols, carotenoids, chlorophylls, geraniol and dolichol serve essential functions in all plants. On the other hand, the vast majority of plant terpenes can be classified as secondary metabolites, serving specialized functions such as pollinator attraction or defense against predators. All plant terpenoids studied up to about 1990 had been assigned a mevalonate origin (for review, see [45]). Many of these assignments were incorrect in light of more recent evidence. It is important to understand the reasons for the earlier mis-assignments of many compounds. As described in more detail below, a major reason lies in the incomplete compartmental separation of a recently discovered mevalonate-independent pathway, a phenomenon which has been addressed as a crosstalk between the two pathways and compartments, respectively.

It is now common knowledge that plants invariably use the cytosolic mevalonate pathway as well as the plastidic mevalonate-independent pathway (non-mevalonate pathway, deoxyxylulose phosphate pathway or MEP pathway) for the biosynthesis of isopentenyl diphosphate (IPP) and dimethylallyl diphosphate (DMAPP). These precursors serve as the basic building blocks for all terpenoids. The genes, proteins and intermediates of the novel non-mevalonate pathway (cf. Fig. 5) have been determined over the last 10 years by a combination of bioinformatic studies, *in vitro* approaches including cloning of the genes and expression of the enzymes, as well as isotope labeling techniques (for reviews, see [46, 47]. In line with the intracellular topology of the two pathways, the open reading frames of all non-mevalonate pathway genes from plants encode N-terminal sequences which fulfill the criteria for chloroplast targeting sequences. On the other hand, the mevalonate pathway genes of plants do not specify targeting sequences, in line with their cytoplasmic location [46, 47]. Since both biosynthetic machineries for the formation of IPP/DMAPP are present in plants, it is crucial to evaluate the biogenesis of plant terpenoids on a quantitative basis.

The origin of the biosynthetic precursors (i.e., IPP and DMAPP) of different plant terpenoids is best approached by *in vivo* studies with whole plants, plant tissue or cultured cells. A powerful strategy for elucidation of the biosynthetic origin of specific plant terpenoids uses stable isotope labeled glucose as precursor. Since glucose is a general intermediary metabolite, the isotope from the proffered carbohydrate can

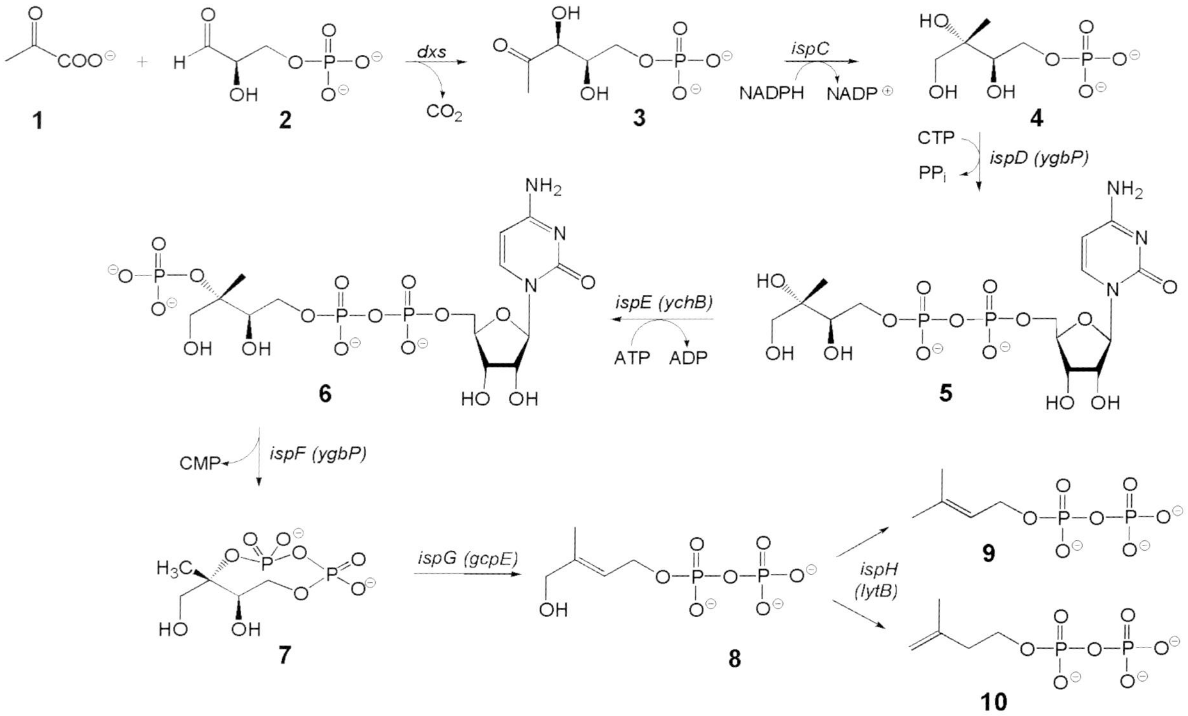

Figure 5. Biosynthetic pathways of DMAPP (9) and IPP (10), the universal precursors of terpenoids. The pathway starts with the formation of 1-deoxyxylulose 5-phosphate (3) from pyruvate (1) (via hydroxyethyl-TPP) and glyceraldehyde 3-phosphate (2). Rearrangement and reduction yields 2-C-methylerythritol 4-phosphate (4), which is then converted into 4-diphosphocytidyl-2-C-methylerythritol (5). Phosphorylation leads to the formation of the respective 2-phosphate (6), which is then converted, into the cyclic diphosphate (7). Ring opening and reduction provides the hydroxymethylbutenyl diphosphate (8), which is finally reduced to IPP (10) and DMAPP (9).

be diverted to virtually all metabolic compartments of plant cells. Biosynthetic information derives from the positional aspects of the label distribution in the target molecule rather than from the net transfer of isotope. This procedure is in sharp contrast with many earlier studies where the transfer of isotope from mevalonate into a given target compound was taken as *bona fide* evidence for mevalonate origin.

Two different techniques for data interpretation will be briefly discussed below. Even on a superficial level of interpretation, it is obvious that carbon atoms 2, 4 and 5 of IPP or DMAPP, respectively, are all derived from acetate methyl groups in case of a mevalonate origin (indicated by b in Fig. 6A), and carbon atoms 1 and 3 of IPP and DMAPP are derived from the carboxylic group of acetate units (indicated by a in Fig. 6A). Irrespective of the nature of the biosynthetic precursor, carbon atoms derived from C-2, 4 and 5 of IPP/DMAPP should have the same isotope abundances in case of a mevalonate origin. Likewise, all atoms derived from C-1 and 3 of DMAPP/IPP should show identical isotope abundance. Moreover, the mevalonate pathway can at best transfer blocks of two labeled carbon atoms to the target molecule, whereas a block of three labeled carbon atoms can be transferred via the deoxyxylulose pathway, albeit under bond breakage and fragment religation brought about by 1-deoxyxylulose phosphate reductoisomerase (IspC protein) (cf. Fig. 5). Using ^{13}C NMR spectroscopy, the ^{13}C enrichment for all non-isochronous carbon atoms can be determined with high precision. Moreover, NMR can diagnose the joint transfer of ^{13}C atom groups, even in the case of an intermolecular rearrangement, by a detailed analysis of the ^{13}C coupling pattern via one- and two-dimensional experiments.

In a more rigorous approach, the entirety of all metabolic precursors in a given experimental system is treated as a network with hundreds to thousands of nodes where an isotope label can spread in every direction. If the isotope distribution in such a system is experimentally determined at a sufficient number of nodes (e.g., biosynthetic amino acids and nucleotides), then the label distribution can be assessed with high precision at a quantitative basis. As examples, the labeling patterns of the central metabolites acetyl-CoA, hydroxyethyl-TPP and glyceraldehyde phosphate can be reconstructed from the labeling patterns of leucine, valine and tyrosine on the basis of well-known pathways of amino acid biosynthesis in plants (Fig. 6). These data can then be used to construct labeling patterns of IPP/DMAPP via different hypothetical pathways, e.g., the mevalonate and non-mevalonate pathway, respectively, and the predicted patterns can be compared with the experimentally determined labeling patterns in the downstream products.

The biosynthetic origin of a considerable number of primary and secondary plant terpenoids has been reinvestigated recently using the technology described above. The experimental systems included members of the gymnosperm and angiosperm families of higher plants as well as liverworts as examples for lower plants. The data show that sterols are invariably synthesized in the cytoplasm via the mevalonate pathway [27]. Ubiquinone is biosynthesized in plant mitochondria using mevalonate-derived precursors from the cytoplasm [48].

Representative examples shown to be derived by the non-mevalonate pathway are given in Figure 7. A wide variety of monoterpenes and diterpenes is now known to be biosynthesized via the non-mevalonate pathway [49, 50]. They include com-

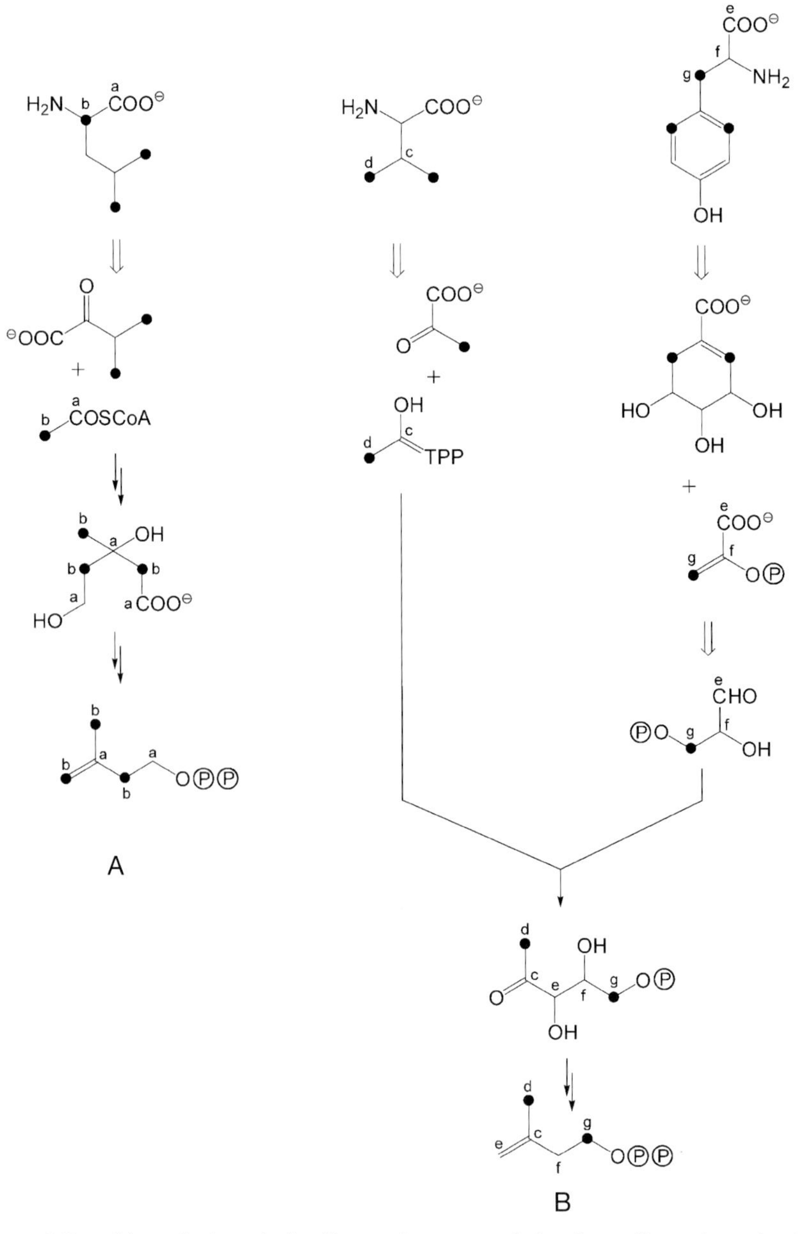

Figure 6. Retrobiosynthetic analysis of isotopolog patterns in leucine, valine and tyrosine. The isotopolog profiles of acetyl-CoA, glyceraldehyde phosphate and hydroxyethyl-TPP are reconstructed on the basis of known pathways for amino acid biosynthesis. Small characters indicate biosynthetically equivalent positions. The isotopolog compositions in the terpene building block IPP is then predicted **A**, via mevalonate or **B**, via 1-deoxyxylulose 5-phosphate, respectively. Filled dots indicate labeled positions from [1-^{13}C]glucose. It is immediately obvious that the labeling patterns differ via the two respective pathways.

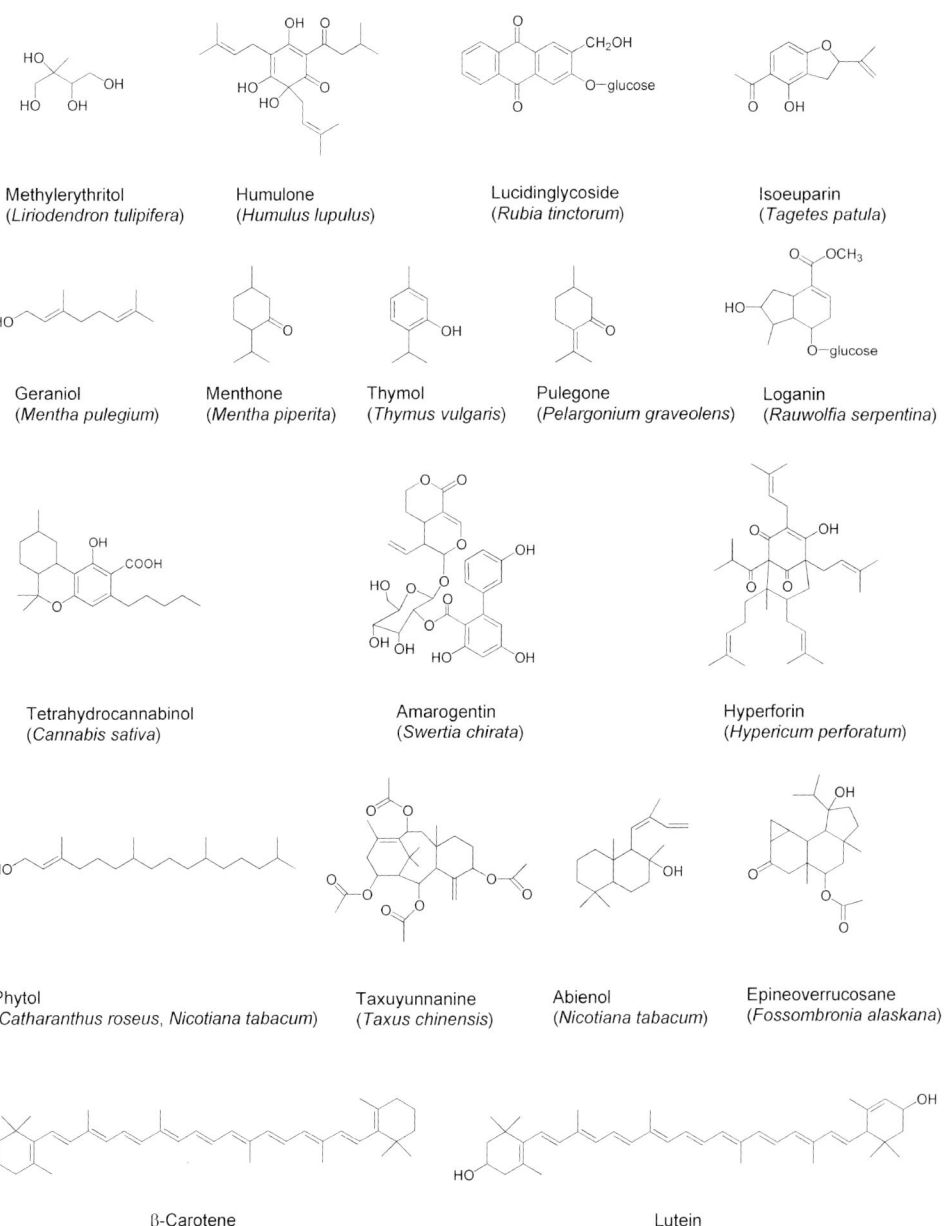

Figure 7. Examples for plant terpenoids that are predominantly or entirely derived via the non-mevalonate pathway. The biosynthetic routes of the displayed terpenoids were assigned by the retrobiosynthetic approach with the species indicated in parentheses.

pounds with central physiological significance for all plants as well as a much larger number of compounds that occur in specific taxonomic groups. Most notably, the phytol side chain which recruits chlorophyll, the most abundant organic pigments on earth, to the thylakoid membrane, is a deoxyxylulose derivative [51]. Carotenoids which play a central role in all green plants as light-protecting and light-assembling agents as well as specific roles as pigments in flowers are derived from the deoxyxylulose pathway [51].

Other examples of plant metabolites derived entirely or predominantly from the deoxyxylulose pathway include loganin, which is a basic precursor for many indole alkaloids [52], verrucosane-type compounds from liverworts [53], and taxoids from yew which play a dominant role as cytostatic agents [49]. They also comprise the isoprenoid moieties in various meroterpenoids including anthraquinone [54], benzofuran [55], tetrahydrocannabinol [56], or humulone from hops [57], the antidepressant hyperforin from St. John's wort [58], as well as the bitter-tasting amarogentin [59] (Fig. 7).

The ^{13}C incorporation studies performed with these compounds are not limited to delineating the origin of the building blocks but are also conducive to an unequivocal identification of the precursor modules. Since the biosynthesis of many terpenes involves one or more skeletal rearrangement, dissecting the isoprenoid building blocks affords important clues with regard to the downstream biosynthetic mechanism; for example, the regiochemistry in the formation of cyclic terpenes. This approach has its maximum impact for deoxyxylulose-derived compounds since universally ^{13}C-labeled 3-carbon blocks can be contributed from appropriate precursors such as [U-$^{13}C_6$]glucose and can be diagnosed in the complex metabolic products by ^{13}C homocorrelation NMR experiments. In favorable cases, very complex mechanisms of terpene formation can be extracted reliably from a small number of experiments (for a representative example, see [53]).

As mentioned above, many plant terpenoids had been incorrectly attributed in the past to the mevalonate pathway on the basis of isotope incorporation experiments with mevalonate or acetate. Whereas these experiments proceeded with minimal incorporation rates attributed to permeability barriers, the label distribution, when analyzed carefully, was in line with the mevalonate paradigm. In light of the more recent evidence described above, it is now clear that these earlier results were experimentally correct yet inappropriately interpreted. The recent studies have established that the compartmental separation between the two isoprenoid pathways is not an absolute one. Minor amounts of unidentified metabolite(s) common to both pathways can be exchanged in both directions via the chloroplast/chromoplast membranes. Thus, minor fractions of deoxyxylulose-derived isoprenoid moieties can be diverted to the cytoplasm where they can become part of sterol molecules. Likewise, a small fraction of isoprenoid moieties derived from the mevalonate pathway find their way into the chloroplast compartment where they become part of mono- and diterpenes which are predominantly obtained via the chloroplast-based deoxyxylulose pathway [60–63].

The retrobiosynthetic concept described above is a powerful tool in order to avoid pitfalls such as pathway crosstalk since it provides a quantitative dissection of

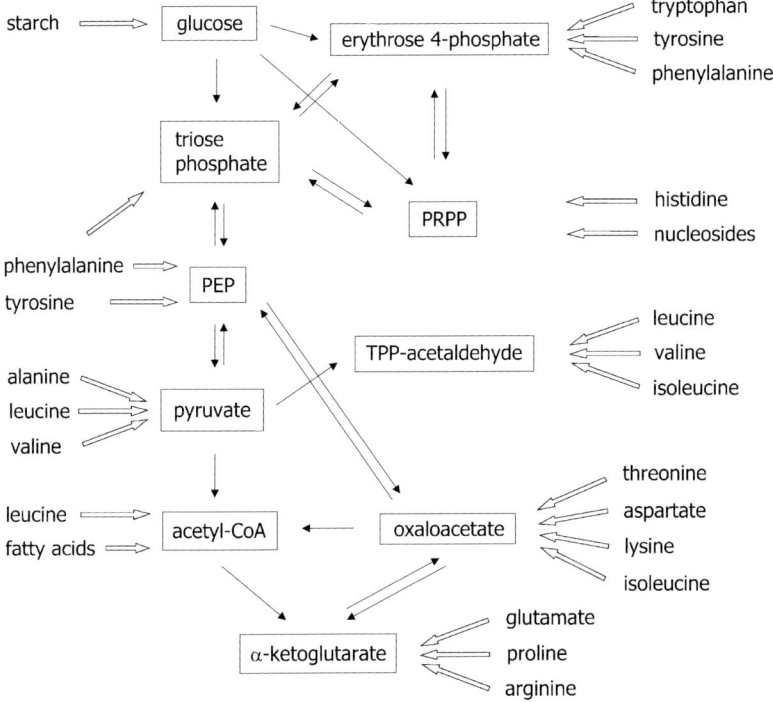

Figure 8. Scheme for the reconstruction of the labeling profiles in central metabolic intermediates ('hubs', shown in boxes) from the labeling patterns of amino acids, nucleosides, starch and fatty acids. Similar to the retrosynthesis approach for dissecting the precursors of a target compound in the organic synthesis, the retro-arrow indicates the retrobiosynthetic approach. The labeling patterns of metabolic hubs provide information about the flux through the metabolic network (schematically indicated by standard reaction arrows).

metabolite diversion as opposed to the qualitative description of net label transfer into one given metabolite that had been the source of errors in many of the earlier studies. It should be emphasized that the metabolites that can be easily used for a quantitative analysis of isotope patterns (e.g., amino acids, nucleosides, starch) provide the isotopolog profiles of approximately ten central intermediates ('hubs') in the metabolic network (Fig. 8). Since most of the basic building blocks of natural products are recruited from that cohort, the experimental approach is not limited to the question of terpene origin in plants but can be generally used to evaluate the biosynthetic history of natural products in a wide range of biological systems. However, for the complete delineation of metabolic flux in a given plant, isotopic equilibrium is one of the prerequisites. In light of the very long labeling times typically used in retrobiosynthetic studies, this assumption appears to be correct.

Conclusion

Labeling methods using isotopic tracers are in use since about 60 years and have contributed to the elucidation of most, if not all, metabolic pathways. Their power and complexity have been increased by the development of NMR methods for the analysis of enrichments and positional labeling, and of MS methods with high resolution and sensitivity for the detection of trace metabolites. In parallel, powerful softwares are being developed to handle the increasing amounts of data. In face of the considerable progress in the methods of analysis, classical limitations remain and require essential choices to be operated by the researcher. For instance, obtaining a rapid and uniform labeling of the tissues entirely depends on the structure of the plant material and is not always possible; supplying the labeled substrates by incubation in an aqueous medium requires special care to avoid disturbing the oxygenation of the tissues, which would dramatically affect their metabolism. Complementing the medium with specific nutrients or vitamins may also be necessary to reproduce physiological conditions [25]. A labeling method is defined by the substrates, the labeling time the analytical equipment and the labeling parameters analyzed, i.e., amounts of label, enrichments or positional of labeling. The present chapter emphasizes the choice of the labeling duration and its adequation with the model used for the qualitative and quantitative interpretation of the data as essential conditions for the success of labeling experiments.

While a given labeling method may appear as the most suitable for a particular material, pathway or question, more information is obtained when different methods are used in combination. The examples presented indicate that the interpretation of the labeling data depend essentially on the modeling of the pathways which is established from both labeling data and previous knowledge of either enzyme activities and their cellular localization, or genes with established or putative functions. In turn, as explained in [64] the labeling methods provide unique information on the dynamics of metabolism, which could not have been deduced from enzyme activities or gene expression data.

Short time labeling is the method of choice for the study of a particular metabolic pathway. It can also give access to the identification of rate limiting steps when coupled with models that include kinetic parameters [11]. Conversely, long-term labeling, in conditions of both metabolic and isotopic steady-state, leads to the calculation of a large number of fluxes in central metabolism. Recent studies have lead to the view of a central metabolism, from sucrose to PEP, with high rates of intermediate interconversion as compared to the fluxes towards the tricarboxylic acid cycle or the biosynthetic pathways. These results extend the concept of readily reversible reactions that was elaborated around sucrose metabolism [65] and may account for the flexibility and robustness of plant central metabolism [21, 66], at least in sugar fed sink tissues.

From the small number of detailed studies, some features, like the cycling of triose-P to hexose-P, appear to be general, while others are more variable. For example, the labeling of cytosolic and plastidial metabolites, may be similar [21, 25] or different [20] according to plant tissues, which may reflect different exchange rates between the cytosol and plastids. The significance of these differences is not clear at the moment, since relationships between features of central metabolism and developmental conditions of the tissues have been proposed in only few particular cases. The role of Rubisco in developing green embryo was clearly related with the accumulation of triglycerides [26]. Minor differences in flux patterns during the development of maize kernels were hypothetically related with changes in the demand for certain amino acids [29]. A profound reorganization of the metabolism with increased catabolism of proteins and lipids [34, 67], and impairment of growth [68] was related with a limitation of sugar supply. A general understanding of specific patterns in the plant central metabolism could be quickly obtained through an intensive exploitation of the labeling data obtained in steady-state condition (fluxomics). Data would be provided on the enrichments and isotopolog profiles of each of the 'central' metabolites presented in Figure 8, and probably a few others. They would be made available through database, and different models could be compared in the interpretation of these data.

As illustrated here through the example of isoprenoids, the use of positional labeling in the retrobiosynthetic analysis of steady-state labeling data makes it possible to establish the contribution of distinct pathways to the formation of stored compounds where the amounts of intermediates are too low to be analyzed. The incredible diversity of plant secondary metabolites has been revealed by MS-based metabolomics [69]. This diversity is probably sensitive to growth conditions and developmental stages [70]. For metabolites of interest, the aim will be to improve their production or accumulation in plants. The task would be relatively easy if they were end-products of linear pathway supplied with non-limiting substrates. More probably, some of the precursors may be limiting; and the metabolites of interest exposed to further conversion. The way of increasing their production will therefore be not obvious. Establishing the metabolic architecture leading to these metabolites (as in [11]), through short time label transfer or retrobiosynthetic analyses may be of great help. Associating this information, obtained in selected genotypes, to gene expression and metabolomic data would make a useful contribution to systems biology.

References

1. Roßmann A, Butzenlechner M, Schmidt H-L (1991) Evidence for a nonstatistical carbon isotope distribution in natural glucose. *Plant Physiol* 96: 609–614
2. Klumpp K, Schäufele R, Lötscher M, Lattanzi FA, Feneis W, Schnyder H (2005) C-isotope composition of CO2 respired by shoots and roots: fractionation during dark respiration? *Plant, Cell & Env* 28: 241–250
3. Kruger NJ, Ratcliffe RG, Roscher A (2003) Quantitative approaches for analysing fluxes through plant metabolic networks using NMR and stable isotope labelling. *Phytochem Rev* 2: 17–30
4. Roessner-Tunali U, Liu J, Leisse A, Balbo I, Perez-Melis A, Willmitzer L, Fernie AR (2004) Kinetics of labelling of organic and amino acids in potato tubers by gas chromatography-mass spectrometry following incubation in 13C labelled isotopes. *Plant J* 39: 668–679

5. Reiner J (1953) The study of metabolic turnover rates by means of isotopic tracers. I. fundamental relations. *Arch Biochem Biophys* 46: 53–81
6. Katz J, Wood H (1963) The use of $C^{14}O_2$ yields from Glucose-1- and -6-C^{14} for the evaluation of the pathways of glucose metabolism. *J Biol Chem* 238: 517–524
7. Katz K, Grunnet N (1979) Estimation of metabolic pathways in steady state *in vitro*. Rates of tricarboxylic acid and pentose cycle. In: H Kornberg (ed): *Techniques in metabolic research*, Elsevier Scientific Publishing Co, New York
8. ap Rees T (1980) Assessment of the contributions of metabolic pathways to plant respiration. In: D Davies (ed): *Metabolism and respiration*, Academic Press, New York, 1–29
9. Schuster S, Fell DA, Dandekar T (2000) A general definition of metabolic pathways useful for systematic organization and analysis of complex metabolic networks. *Nat Biotechnol* 18: 326–332
10. van Winden W, Verheijen P, Heijnen S (2001) Possible pitfalls of flux calculations based on C-13-labeling. *Metab Eng* 3: 151–162
11. McNeil SD, Rhodes D, Russell BL, Nuccio ML, Shachar-Hill Y, Hanson AD (2000) Metabolic modeling identifies key constraints on an engineered glycine betaine synthesis pathway in tobacco. *Plant Physiol* 124: 153–162
12. Rhodes D, McNeil S, Nuccio M, Hanson A (2004) Metabolic engineering and flux analysis of glycine betaine synthesis in plants: progress and prospects. In: B Kholodenko, HV Westerhoff (eds): *Metabolic engineering in the post genomic era*, Horizon Bioscience, Wymondham, UK
13. Kelleher JK (2004) Probing metabolic pathways with isotopic tracers: insights from mammalian metabolic physiology. *Metab Eng* 6: 1–5
14. Roscher A, Kruger NJ, Ratcliffe RG (2000) Strategies for metabolic flux analysis in plants using isotope labelling. *J Biotechnol* 77: 81–102
15. Hargreaves JA, ap Rees T (1988) Turnover of starch and sucrose in roots of *Pisum sativum*. *Phytochem* 27: 1627–1629
16. Dancer J, David M, Stitt M (1990) Water stress leads to a change of partitioning in favour of sucrose in heterotrophic cell suspension cultures of *Chenopodium rubrum*. *Plant Cell Environ* 13: 957–963
17. Hill ST, ap Rees T (1994) Fluxes of carbohydrate metabolism in ripening bananas. *Planta* 192: 52–60
18. Geigenberger P, Reimholz R, Geiger M, Merlo L, Canale V, Stitt M (1997) Regulation of sucrose and starch metabolism in potato tubers in response to short-term water deficit. *Planta* 201: 502–518
19. N'tchobo H, Dali N, NguyenQuoc B, Foyer CH, Yelle S (1999) Starch synthesis in tomato remains constant throughout fruit development and is dependent on sucrose supply and sucrose synthase activity. *J Exp Bot* 50: 1457–1463
20. Dieuaide-Noubhani M, Raffard G, Canioni P, Pradet A, Raymond P (1995) Quantification of compartmented metabolic fluxes in maize root tips using isotope distribution from (13C) or (14C) labeled glucose. *J Biol Chem* 270: 13147–13159
21. Rontein D, Dieuaide-Noubhani M, Dufourc Erick J, Raymond P, Rolin D (2002) The metabolic architecture of plant cells. Stability of central metabolism and flexibility of anabolic pathway during the growth cycle of tomato cells. *J Biol Chem* 277: 43948–43960
22. Alonso AP, Vigeolas H, Raymond P, Rolin D, Dieuaide-Noubhani M (2005) A new substrate cycle in plants: evidence for a high glucose-phosphate-to-glucose turnover from *in vivo* steady-state and pulse-labeling experiments with [C-13] glucose and [C-14] glucose. *Plant Physiol* 138: 2220–2232

23. Trethewey RN, Riesmeier JW, Willmitzer L, Stitt M, Geigenberger P (1999) Tuber-specific expression of a yeast invertase and a bacterial glucokinase in potato leads to an activation of sucrose phosphate synthase and the creation of a sucrose futile cycle. *Planta* 208: 227–238
24. Garlick AP, Moore C, Kruger NJ (2002) Monitoring flux through the oxidative pentose phosphate pathway using [1-14C]gluconate. *Planta* 216: 265–272
25. Schwender J, Ohlrogge JB, Shachar-Hill Y (2003) A flux model of glycolysis and the oxidative pentosephosphate pathway in developing *Brassica napus* embryos. *J Biol Chem* 278: 29442–29453
26. Schwender J, Goffman F, Ohlrogge JB, Shachar-Hill Y (2004) Rubisco without the Calvin cycle improves the carbon efficiency of developing green seeds. *Nature* 432: 779–782
27. Glawischnig E, Gierl A, Tomas A, Bacher A, Eisenreich W (2001) Retrobiosynthetic nuclear magnetic resonance analysis of amino acid biosynthesis and intermediary metabolism. Metabolic flux in developing maize kernels. *Plant Physiol* 125: 1178–1186
28. Glawischnig E, Gierl A, Tomas A, Bacher A, Eisenreich W (2003) Starch biosynthesis and intermediary metabolism in maize kernels. Quantitative analysis of metabolite flux by NMR. *Plant Physiol* 130: 1717–1727
29. Ettenhuber C, Spielbauer G, Margl L, Hannah L, Gierl A, Bacher A, Genschel U, Eisenreich W (2005) Changes in flux pattern of the central carbohydrate metabolism during kernel development in maize. *Phytochem* 66: 2632–2642
30. Eisenreich W, Ettenhuber C, Laupitz R, Theus C, Bacher A (2004) Isotopolog perturbation techniques for metabolic networks. Metabolic recycling of nutritional glucose in *Drosophila melanogaster*. *Proc Natl Acad Sci USA* 101: 6764–6769
31. Ettenhuber C, Radykewicz T, Kofer W, Koop H-U, Bacher A, Eisenreich W (2005) Metabolic flux analysis in complex isotopologous space. Recycling of glucose in tobacco plants. *Phytochem* 66: 323–335
32. Salon C, Raymond P, Pradet A (1988) Quantification of carbon fluxes through the tricarboxylic acid cycle in early germinating lettuce embryos. *J Biol Chem* 263: 12278–12287
33. Ferrario-Mery S, Hodges M, Hirel B, Foyer CH (2002) Photorespiration-dependent increases in phosphoenolpyruvate carboxylase, isocitrate dehydrogenase and glutamate dehydrogenase in transformed tobacco plants deficient in ferredoxin-dependent glutamine-alpha-ketoglutarate aminotransferase. *Planta* 214: 877–886
34. Dieuaide Noubhani M, Canioni P, Raymond P (1997) Sugar-starvation-induced changes of carbon metabolism in excised maize root tips. *Plant Physiol* 115: 1505–1513
35. Macnicol PK, Raymond P (1998) Role of phosphoenolpyruvate carboxylase in malate production by the developing barley aleurone layer. *Physiol Plant* 103: 132–138
36. Edwards S, Nguyen BT, Do B, Roberts JKM (1998) Contribution of malic enzyme, pyruvate kinase, phosphoenolpyruvate carboxylase, and the Krebs cycle to respiration and biosynthesis and to intracellular pH regulation during hypoxia in maize root tips observed by nuclear magnetic resonance imaging and gas chromatography-mass spectrometry. *Plant Physiol* 116: 1073–1081
37. Raymond P, Al-Ani A, Pradet A (1985) ATP production by respiration and fermentation, and energy charge during aerobiosis and anaerobiosis in twelve fatty and starchy germinating seeds. *Plant Physiol* 79: 879–884
38. Canvin D, Beevers H (1961) Sucrose synthesis from acetate in the germinating castor bean: kinetics and pathways. *J Biol Chem* 236: 988–995
39. Dieuaide M, Brouquisse R, Pradet A, Raymond P (1992) Increased fatty acid beta-oxidation after glucose starvation in maize root tips. *Plant Physiol* 99: 595–600
40. Pracharoenwattana I, Cornah J, Smith S (2005) *Arabidopsis* peroxisomal citrate synthase is required for Fatty Acid respiration and seed germination. *Plant Cell* 17: 2037–2048

41. Wiechert W (2001) C-13 metabolic flux analysis. *Metab Eng* 3: 195–206
42. Wiechert W, Mollney M, Petersen S, de Graaf AA (2001) A universal framework for C-13 metabolic flux analysis. *Metab Eng* 3: 265–283
43. Eisenreich W, Strauß G, Werz U, Bacher A, Fuchs G (1993) Retrobiosynthetic analysis of carbon fixation in the phototrophic eubacterium *Chloroflexus aurantiacus*. *Eur J Biochem* 215: 619–632
44. Sacchettini J, Poulter C (1997) Creating isoprenoid diversity. *Science* 277: 1788–1789
45. Bochar D, Friesen J, Stauffacher C, Rodwell V (1999) Biosynthesis of mevalonic acid from acetyl-CoA. In: D Cane (ed.): *Comprehensive natural product chemistry*, Pergamon, Oxford, 15–44
46. Eisenreich W, Rohdich F, Bacher A (2001) Deoxyxylulose phosphate pathway to terpenoids. *Trends Plant Sci* 6: 78–84
47. Eisenreich W, Bacher A, Arigoni D, Rohdich F (2004) Biosynthesis of isoprenoids via the non-mevalonate pathway. *Cell Mol Life Sci* 61: 1401–1426
48. Disch A, Hemmerlin A, Bach TJ, Rohmer M (1998) Mevalonate-derived isopentenyl diphosphate is the biosynthetic precursor of ubiquinone prenyl side chain in tobacco BY-2 cells. *Biochem J* 331: 615–621
49. Eisenreich W, Menhard B, Hylands PJ, Zenk MH, Bacher A (1996) Studies on the biosynthesis of taxol: the taxane carbon skeleton is not of mevalonoid origin. *Proc Natl Acad Sci USA* 93: 6431–6436
50. Eisenreich W, Sagner S, Zenk MH, Bacher A (1997) Monoterpenoid essential oils are not of mevalonoid origin. *Tetrahedron Letters* 38: 3889–3892
51. Lichtenthaler HK, Schwender J, Disch A, Rohmer M (1997) Biosynthesis of isoprenoids in higher plant chloroplasts proceeds via a mevalonate-independent pathway. *FEBS Lett* 400: 271–274
52. Eichinger D, Bacher A, Zenk MH, Eisenreich W (1999) Analysis of metabolic pathways via quantitative prediction of isotope labeling patterns: a retrobiosynthetic 13C NMR study on the monoterpene loganin. *Phytochem* 51: 223–236
53. Eisenreich W, Rieder C, Grammes C, Hessler G, Adam KP, Becker H, Arigoni D, Bacher A (1999) Biosynthesis of a Neo-epi-verrucosane diterpene in the liverwort *Fossombronia alaskana* – A retrobiosynthetic NMR study. *J Biol Chem* 274: 36312–36320
54. Eichinger D, Bacher A, Zenk MH, Eisenreich W (1999) Quantitative assessment of metabolic flux by C-13 NMR analysis. Biosynthesis of anthraquinones in *Rubia tinctorum*. *J Am Chem Soc* 121: 7475
55. Margl L, Ettenhuber C, Istvan G, Zenk MH, Bacher A, Eisenreich W (2005) Biosynthesis of benzofuran derivatives in root cultures of *Tagetes patula* via phenylalanine and 1-deoxy-D-xylulose 5-phosphate. *Phytochem* 66: 887–899
56. Fellermeier M, Eisenreich W, Bacher A, Zenk MH (2001) Biosynthesis of cannabinoids: incorporation experiments with 13C-labeled glucoses. *Eur J Biochem* 268: 1596–1604
57. Goese M, Kammhuber K, Bacher A, Zenk MH, Eisenreich W (1999) Biosynthesis of bitter acids in hops. A 13C-NMR and 2H-NMR study on the building blocks of humulone. *Eur J Biochem* 263: 447–454
58. Adam P, Arigoni D, Bacher A, Eisenreich W (2002) Biosynthesis of hyperforin in *Hypericum perforatum*. *J Med Chem* 45: 4793
59. Wang CZ, Maier UH, Eisenreich W, Adam P, Obersteiner I, Keil M, Bacher A, Zenk MH (2001) Unexpected biosynthetic precursors of amarogentin – a retrobiosynthetic 13C NMR study. *Eur J Org Chem* 1459–1465
60. Schuhr C, Radykewicz T, Sagner S, Latzel C, Zenk M, Arigoni D, Bacher A, Rohdich F, Eisenreich W (2003) Quantitative assessment of metabolite flux by NMR spectroscopy.

Crosstalk between the two isoprenoid biosynthesis pathways in plants. *Phytochem Rev* 2: 3–16

61. Adam KP, Zapp J (1998) Biosynthesis of the isoprene units of chamomile sesquiterpenes. *Phytochem* 48: 953–959
62. Itoh D, Karunagoda RP, Fushie T, Katoh K, Nabeta K (2000) Nonequivalent labeling of the phytyl side chain of chlorophyll a in callus of the hornwort *Anthoceros punctatus*. *J Nat Prod* 63: 1090–1093
63. Yang JW, Orihara Y (2002) Biosynthesis of abietane diterpenoids in cultured cells of *Torreya nucifera* var. *radicans*: biosynthetic inequality of the FPP part and the terminal IPP. *Tetrahedron* 58: 1265–1270
64. Fernie AR, Geigenberger P, Stitt M (2005) Flux an important, but neglected, component of functional glenomics. *Curr Opin Plant Biol* 8: 174–182
65. Geigenberger P, Stitt M (1993) Sucrose synthase catalyses a readily reversible reaction *in vivo* in developing potato tubers and other plant tissues. *Planta* 189: 329–339
66. Spielbauer G, Margl L, Hannah LC, Römisch W, Ettenhuber C, Bacher A, Gierl A, Eisenreich W, Genschel U (2006) Robustness of central carbohydrate metabolism in developing maize kernels. *Phytochem* 67: 1460–1475
67. Brouquisse R, Gaudillere JP, Raymond P (1998) Induction of a carbon-starvation-related proteolysis in whole maize plants submitted to light/dark cycles and to extended darkness. *Plant Physiol* 117: 1281–1291
68. Gibon Y, Blasing OE, Palacios-Rojas N, Pankovic D, Hendriks JHM, Fisahn J, Hohne M, Gunther M, Stitt M (2004) Adjustment of diurnal starch turnover to short days: depletion of sugar during the night leads to a temporary inhibition of carbohydrate utilization, accumulation of sugars and post-translational activation of ADP-glucose pyrophosphorylase in the following light period. *Plant J* 39: 847–862
69. Keurentjes JJB, Fu J, de Vos CHR, Lommen A, Hall RD, Bino RJ, van der Plas LHW, Jansen RC, Vreugdenhil D, Koornneef M (2006) The genetics of plant metabolism. *Nat Genet* 38: 842–849
70. Baxter I, Borevitz J (2006) Mapping a plant's chemical vocabulary. *Nat Genet* 38: 737–738

Plant Systems Biology
Edited by Sacha Baginsky and Alisdair R. Fernie
© 2007 Birkhäuser Verlag/Switzerland

Network visualization and network analysis

Victoria J. Nikiforova and Lothar Willmitzer

Max-Planck-Institut für Molekulare Pflanzenphysiologie, Am Mühlenberg 1, 14476 Potsdam-Golm, Germany

Abstract

Network analysis of living systems is an essential component of contemporary systems biology. It is targeted at assemblance of mutual dependences between interacting systems elements into an integrated view of whole-system functioning. In the following chapter we describe the existing classification of what is referred to as biological networks and show how complex interdependencies in biological systems can be represented in a simpler form of network graphs. Further structural analysis of the assembled biological network allows getting knowledge on the functioning of the entire biological system. Such aspects of network structure as connectivity of network elements and connectivity degree distribution, degree of node centralities, clustering coefficient, network diameter and average path length are touched. Networks are analyzed as static entities, or the dynamical behavior of underlying biological systems may be considered. The description of mathematical and computational approaches for determining the dynamics of regulatory networks is provided. Causality as another characteristic feature of a dynamically functioning biosystem can be also accessed in the reconstruction of biological networks; we give the examples of how this integration is accomplished. Further questions about network dynamics and evolution can be approached by means of network comparison. Network analysis gives rise to new global hypotheses on systems functionality and reductionist findings of novel molecular interactions, based on the reliability of network reconstructions, which has to be tested in the subsequent experiments. We provide a collection of useful links to be used for the analysis of biological networks.

Introduction

A living organism consists of a lot of elements (e.g., genes, proteins, metabolites, etc.) organized in a functional structure capable simultaneously to maintain its homeostasis and to develop. In addition, this structure must be able to react to the changes in both external and internal environment. This reaction itself constitutes a chain of consecutive events starting from signal perception through signal transduction and various subsequent transformations towards an endpoint response reaction. These events need to be integrated in a proper spatial and temporal context. The events in such chains are changes in a state of elements, and information concerning these changes propagates along the chain. From this explanation, the answer to the biological questions why and how a particular response to a given signal develops seems to be relatively straightforward. However, the complexity of living systems is so high, that to date hardly any such chains of reactions have been elucidated. Actually, for the vast majority of reactions our knowledge is at a rudimentary 'black box' stage: we know the initial signal (the exciter) and a response endpoint, but how spatio-temporal aspects of responses are executed remains largely unknown. A further complexity is introduced by the fact that a single exciter generally influences more than one physiological reaction. For the above-described simplified concept of information exchange this suggests that the chains of consecutive events occurring in response to the exciter must branch, and change in a state of each element within a chain can result in multiple downstream effects. This response plurality can nowadays be easily illustrated with the use of transcript profiles, which are rapidly accumulating in public repositories and hence available for the research community. In most underlying physiological experiments a single environmental parameter is altered, and in response expression of a large number of genes is changed. For example, in experiments in which sulfur was depleted from the *Arabidopsis* growth medium, up to 5% of all genes and 11.5% of measured metabolites exhibited significantly different levels [1]. These multiple changes in response to a single initial exciter have to be extrapolated to the whole system of response development. Each new change in a chain (being in turn an exciter for the downstream changes) is also potentially able to cause multiple changes downstream in the network. Thus, information on the initial exciter spreads in multiple downstream directions, forming a dense causally directed network of interactions. Studying the network of interacting elements within living systems is facilitating efforts to fill the 'response black box' – a task that represents a major challenge for network analysis as a component of contemporary systems biology.

Types of recognized biological networks

According to the Webster's dictionary, a network is an intricately connected system of things or people. A type of a biological network is defined by what these 'things' are (nodes, vertices, etc.), what the nature of their connections (edges) is, and ideally why these things are connected. Below we give the examples of the most common types of biological (often termed also cellular or molecular) networks, with comments on what knowledge is usually gained from networks of these types. It is worth mentioning that in this relatively new research area the terminology is not yet well-established. Table 1 illustrates the frequency of different terms used for biological networks in the related literature as of January 2006.

In what are currently termed metabolic networks, or biochemical reaction networks, vertices are represented by metabolites (substances), and metabolic reactions are represented by directed edges, which interconnect substrates and products of these reactions. Metabolic networks describe the potential pathways that may be used by a cell to accomplish metabolic processes. These are probably the first cellular networks, which biologists started to reconstruct as schematic representations

Table 1. Terminology of biological networks

	Term in Ovid Database Server (http://ovid.gwdg.de/)	Frequency
0	biological network(s)	235
0	cellular network(s) (used also in, e.g., Telecommunication Systems)	1,089
0	molecular network(s)	400
0	biomolecular network(s)	9
0	bioregulatory network(s)	4
1	metabolic network(s)	626
1	biochemical reaction network(s)	45
2	transcription network(s)	47
2	network(s) of transcription interactions	1
2	gene regulation network(s)	26
2	gene-regulatory network(s) (used broader)	234
2	transcriptional regulation network(s)	14
2	regulatory network(s) (used very broad)	1,666
3	protein interaction network(s)	218
3	protein–protein interaction network(s)	62
3	interactome	101
4	correlation network(s) (not only biological networks)	64
4	co-expression network(s)	5
4	coexpression network(s)	9
4	expression network(s)	71
5	signaling network(s)	1,249
	signaling network(s)	1,030
	signaling network(s)	223
6	gene network(s)	552
6	genetic regulatory network(s)	113

of a sum of biosynthetic pathways deduced from biochemical studies. Nowadays the vast biochemical information is compiled in specialized databases, and metabolic networks on top of these data serve as a visualization tool for multiple interconnections between their elements. As an example of such repositories, BioCyc [2] is a collection of 205 (as of January 2006) Pathway/Genome Databases, each of which describes the genome and metabolic pathways of a single organism. Among these organisms plant biologists will find a comprehensive *Arabidopsis* Pathway/Genome Database called AraCyc [3]. Connected to the BioCyc repository is the MetaCyc database, which, in distinction to the organism-specific databases, is a reference source on metabolic pathways from many organisms [4]. Another example is the KEGG PATHWAY [5], a collection of manually drawn pathway maps representing our up-to-date knowledge on the molecular interaction and reaction networks. Although very rich, this database may be less recommended for plant biologists, as the reference metabolic networks represent non-plant metabolism. The enzymes known for plants can be mapped on these networks, but the reactions which are known not to occur in plants will still stay in the networks as connecting links. However, keeping in mind that, contrary to conventional wisdom, our current knowledge of the structure of plant cellular metabolism is far from complete [6], expansion and integration of the knowledge of metabolism in well characterized 'post-genome' organisms into plant biology will facilitate faster progress in plant systems.

In transcription networks (termed also: networks of transcription interactions, gene regulation networks, gene-regulatory networks, transcriptional regulation networks or simply regulatory networks) directed edges reflect interactions between transcription factors and the genes they regulate or the DNA sites to which they bind, with the direction from the transcription factor to the regulated gene. These networks describe potential pathways cells can use to regulate global gene expression programs. This is a newer type of cellular network which started to develop with the accumulating knowledge on protein factors regulating transcription of target genes by means of binding to the regulatory elements contained in their promoters. As with biochemical repositories, the information on experimentally verified interactions is also collected in major electronically accessible data bases. Here analysis at the network level is essential, because each transcription factor generally regulates the expression of more than one gene, the expression of each gene is often regulated by more than one transcription factor, and furthermore, the expression of transcription factors themselves can be regulated by the other transcription factors in a cascade-like manner. Thus, this type of information exchange also forms a dense network of interactions.

For many model systems the complete arrays of transcription factors and their target genes have been deciphered and compiled into electronic repositories. The major data repository for gene regulation in *Escherichia coli* is stored in RegulonDB [7], while the GRID database compiles information on physical interactions for three organisms whose genomes have been deciphered: yeast *Saccharomyces cerevisiae*, fly *Drosophila melanogaster* and worm *Caenorhabditis elegans*. Among plant-specific databases, the major ones which collect information on transcription factors and cis-regulatory elements are AGRIS, DATF, PlantCare and Place. Data on identified molecular interactions are also collected within the more general databases (such as BIND [8]), which are organism- and interaction-type unspecific. The analysis of genome-scale transcription networks is exemplified by the papers [9] for *E. coli* and [10] for yeast, but no comprehensive survey of this type exists yet for plants.

In the other type of cellular graphs – protein interaction networks – the nodes are proteins, and two nodes are connected by a non-directed edge if the two proteins bind to each other. In parallel with the rapid development of modern molecular techniques for determining protein–protein interactions, such as high-throughput yeast two-hybrid strategies [11], proteome-scale reconstructions of global protein interaction networks have been carried out for some model organisms. An organism's total set of protein–protein interactions is often termed as its interactome [12, 13]. Similarly to the data on metabolic and transcriptional interactions, that concerning protein–protein interactions is stored in electronic repositories and often utilized to construct interactome networks of model organisms, such as yeast [14],

Drosophila [15], *Bacillus subtilis* [16], *Caenorhabditis elegans* [17], the malaria parasite *Plasmodium falciparum* [18] and even humans [19, 20]. Among plants, the interactome of *Arabidopsis* will most probably be the first described. To date, the first *Arabidopsis* interactome fragments have been recently reconstructed, e.g., de Folter and colleagues [21] presented a plant interactome map of proteins from the *Arabidopsis thaliana* MADS box transcription factor family. This network fragment adds data on plants to a growing collection of available interaction maps for a number of different organisms.

Besides organism-specific databases on protein–protein interactions, several large repositories collect information on protein interactions in different organisms, or even more general, on all known biomolecular interactions of different types. One such major collection for data on experimentally verified protein interactions is the Database of Interacting Proteins (DIP [22]), which stores the information on more than 55,000 protein interactions in 110 different organisms (as of January 2006). The above-mentioned BIND compiles published information on more than 200,000 biomolecular interactions in 1,528 different organisms, including 1,537 interactions described for *Arabidopsis thaliana* (as of January 2006). Although the plant-related part of the BIND database remains relatively small (in BIND only 0.76% of all interaction records refer to plants) cataloguing and networking protein interactions is a rapidly expanding area with high gene function discovery potential. The success of such approaches depends on combined efforts of large scientific consortia and mapping of the *Arabidopsis* interactome has been included as an integrated component of the 2010 Project, aimed at determining the function of all genes in *Arabidopsis thaliana*.

In correlation networks nodes are genes (these networks are often termed also as gene coexpression networks, or just expression networks) or/and metabolites; two nodes are connected with non-directed edges, if patterns of changes in their expression/concentration correlate significantly to each other. Unlike in the previously described types of cellular networks, in correlation networks connections do not directly represent a physical interaction between nodes, but coexpression or co-behavior, under applied conditions. The items with similar patterns of co-behavior are usually considered to be more likely functionally associated, due to a variety of different biological reasons. These functional associations imply an exchange of information between items. The whole correlation network represents a sum of such associations, with the branching paths, along which the information is processed in order to finally accomplish endpoint biological reactions. Building of such correlation networks attempts to reconstruct real dynamic interacting networks of genes in the genetic regulatory circuitry. The approach seems to be adequate, as these real networks result *in vivo* in complex gene expression and metabolite concentration patterns.

The initial datasets for reconstruction of correlation networks are 'omics'-scale profiles of gene expression and metabolite concentrations (what is often termed as transcriptome and metabolome, correspondingly). Current approaches to attain transcript and metabolic profiles are described in the previous chapters. Available collections of transcript profiles are already large and continue to grow rapidly and the necessity of such repositories for metabolic profiles is widely recognized. Major repositories of genome-scale transcript profiles are compiled in Table 2. Some of these, for example M-CHiPS, NASCArrays or Genevestigator, provide convenient tools for data mining, acting as data warehouses rather than mere repositories. In several of these databases there exists the possibility for pair-wise correlation analysis. For example, utilizing NASCArrays one can build two gene scatter plots to compare expression patterns of two genes, or with another tool, Gene Correlator of Genevestigator repository, coexpression of two genes over a set of array chips can be visualized.

The potential for the analysis of coexpression for functional genetics has been already recognized in pre-genomic era [29, 30], tested experimentally and proved to be useful for decisions on functions of examined genes (e.g., [31, 32]). Later, when 'omics'-scale gene expression/metabolic concentration profiles became available, global analysis of pattern similarities began to be applied [33–35]. Approximately at the same time the first studies on functional genomics based on transcriptional correlations were carried out [36]. Since these pioneering studies systematic approaches for identifying the biological functions of novel genes have been widely applied, signifying an era of genome-wide functional analysis. Finally, matrices of pair-wise correlations across genome-scale arrays have been computed and global correlation networks were built from these correlation matrices. For example, Kim and co-workers assembled data from *Caenorhabditis elegans* DNA microarray experiments [37] involving multiple growth conditions, developmental stages, and varieties of mutants. In this study co-regulated genes were grouped together and visualized in an expression map that displayed correlations of gene expression profiles. Already in this early study of one of the first correlation networks their high potential in gene discovery was visualized demonstrating that it is possible to assign functions through identification of genes that are co-regulated with known sets of genes or even to uncover previously unknown genetic functions. Correlation network analysis has subsequently been applied to yeast, worm, fly and human, and combined analysis of all four allowed identification of global coexpression relationships and their evolutionary conservation [38]. Subsequent demonstrations of the high level of co-regulation conservation in the evolution of prokaryotes and eukaryotes [39] implies that functional relationships predicted from coexpression network analysis in one species can be transferred to another species.

As the next cognitive step alterations in coexpression relationships in two distinct coexpression networks have been studied [40]. With this approach it was possible to show, that functional changes such as alteration in energy metabolism, promotion of cell growth and enhanced immune activity were accompanied with coexpression changes. We shall discuss this approach in more detail below in a chapter devoted to network comparison.

Metabolite correlation networks can be exemplified by the studies of Weckwerth, Fiehn and colleagues [41, 42]. Unlike gene expression correlation networks, most metabolite correlation networks concern plant systems. Recently, given the availability of both metabolite and gene expression profiles, the use of cross-correlation analysis in search for functional gene-metabolite associations became possible. It has been demonstrated by fungal and plant biologists, that the integration of transcript

Table 2. Major repositories for genome-scale transcript profiles

Repository name [Reference #]	Web link	No. of Experiments (sample series)[a]	No. of hybridizations (samples, arrays, slides)[a]	Organisms	Comments
GEO – Gene Expression Omnibus [23]	http://www.ncbi.nlm.nih.gov/geo/	2,967	67,837	variable	367 experiments on *Arabidopsis*
ArrayExpress [24]	http://www.ebi.ac.uk/arrayexpress/	1,226	34,486	variable	no Arabidopsis data
SMD – Stanford Microarray Database	http://genome-www5.stanford.edu//		10,516	variable	630 hybridizations for *Arabidopsis*
M-CHiPS [25]	http://www.mchips.org/	40	316	variable	no Arabidopsis data
yMGV – yeast microarray global viewer [26]	http://transcriptome.ens.fr/ymgv/index.php	1,544		2 yeasts	
Expression Connection	http://db.yeastgenome.org/cgi-bin/expression/expressionConnection.pl			yeast	
Webminer	http://genome-www.stanford.edu/cgi-bin/webminer/mkjavascript			yeast	
NASCArrays	http://affymetrix.arabidopsis.info/narrays/experimentbrowse.pl	234		*Arabidopsis* and other species	
Genevestigator Database [27]	http://web.uni-frankfurt.de/fb15/botanik/mcb/AFGN/atgenex.htm	183	2,317	Arabidopsis	
AtGenExpress	http://web.uni-frankfurt.de/fb15/botanik/mcb/AFGN/atgenex.htm	47	1,387	Arabidopsis	
MAEDA/RARGE at RIKEN [28]	http://rarge.gsc.riken.go.jp/microarray/microarray.pl			Arabidopsis	
SGED – Solanaceae Gene Expression Database	http://www.tigr.org/tdb/potato/SGED_index2.shtml	33	1,072	Solanaceae	

[a] as of January 2006

and metabolic profiles can facilitate the identification of candidate genes for biotechnology [43, 44]. In subsequent studies, combined metabolomics and transcriptomics data were mined and clusters of co-regulated genes and metabolites were determined that displayed coordinated behavior under given experimental conditions [45, 46]. Finally, the entire network of gene-metabolite correlations has been reconstructed from combined sets of transcript and metabolic profiles [47]. From such reconstructions, a global network of information exchange in a living organism is revealed allowing prediction of master controllers of homeostasis. Weckwerth and Morgenthal [48] recently summarized what biologists can gain when analyzing metabolite correlation networks. From studies on network topology putative regulators of underlying processes can be identified as highly connected nodes, or hubs. Metabolic correlation networks can be further superimposed on biochemical reaction networks; through this analysis unexpected pleiotropic changes in genetically modified plants can be identified and assigned to those parts of metabolism which are influenced by genetic manipulation [49]. Knowledge gained from the analysis of gene expression correlation networks is based on the underlying assumption that identified clusters of co-expressed genes are co-regulated. Gene expression at the level of transcription is regulated by transcription factors which bind to specific regulatory sequences in the promoter regions of regulated genes. That many genes are co-regulated suggests the presence of common regulatory sequences in the promoters of clustered genes and makes their analysis a priority in network studies. The validity of such promoter analysis was realized in early studies on correlations of patterns of gene expression [50]. To understand combinatorial control of gene expression, hierarchical and modular organization of regulatory DNA sequence elements in the promoters of co-expressed genes has been examined [34]. For such studies global gene expression correlation networks can be of extreme use, as they intrinsically contain and process the information encoded by transcription networks. Modern research on transcriptomics coupled to promoter analysis has allowed the identification of novel transcription factor target genes [51] and putative regulatory motifs [52], elucidation and prediction of complex regulatory events [53].

Signaling networks are often distinguished as another type of molecular network [54, 55]. These networks represent signal transduction pathways, where nodes are proteins or small molecules, and directed links are signal transduction events. The basic knowledge for reconstruction of such networks comes from low-throughput experiments on individual molecules. Resulting signaling networks are usually assembled around a single signaling cascade, as, for example, the signaling network of bacterial chemotaxis [56] or multiple studies on cancer signaling (reviewed by [57, 58]). In this sense such signaling pathways may be regarded as subnetworks, or network fragments of a global signaling network. Nevertheless their complexity is high due to a big number of the involved elements, branching, feedforward and feedback regulations and cross-talk with other signaling cascades [59, 60].

In plant biology several signaling networks have also been resolved at the molecular level, for example the signaling network of the plant immune system [61] or hydrogen peroxide signaling network that mediates plant programmed cell death [62]. Such studies can be concentrated also on signaling molecules, which may be common for several signaling pathways. For example, nitric oxide and hydrogen peroxide are key signaling molecules produced in response to various stimuli and involved in a diverse range of plant signal transduction processes. One such process is stomatal closure controlled by guard cell signal transduction. By the combined efforts of several laboratories the whole signaling network which controls stomatal closure is being assembled molecule by molecule. Through the analysis of this network in its spatial and temporal resolution a close interrelationship between the involved molecules have been identified [63–66].

In spite of the fact that common signaling molecules have been identified, the present state of knowledge cannot say how molecular information is processed through a network of interlacing signal transduction pathways. Reconstruction of a whole network of interlacing signaling cascades remains a challenging task. In this direction, there are attempts to assemble the whole signaling network, although still limited to single processes. For example, Janes and co-workers [67] constructed a systems model of 7,980 intracellular signaling events that links response outputs associated with apoptosis. Due to globality of the model, it was possible to predict multiple responses induced by a combination of factors.

In what are often called gene networks (or genetic regulatory networks), nodes are genes that are connected with arrowed links directed from gene A to gene B, if for example a mutation (perturbed expression level) in gene A leads to changed expression of gene B. Thus, gene networks show the phenomenological interactions between gene activities. Although in this approach only the transcriptome is considered, gene relationships are basically mediated by proteins and metabolites, and in this way all biochemistry underlying gene–gene interactions is implicitly present in gene networks. Besides network connectivity, regulatory strengths of gene–gene interactions can be quantified from experimental data and represented by, e.g., a thickness of a connecting edge (for example, by an approach suggested by [68]), introducing quantitative aspects to gene networks. Gene networks can be reconstructed from single gene perturbations, as was done, for example, by modulating activin in mice [69], human fibroblast response [70], or by perturbing the action of a key regulator of floral asymmetry in *Arabidopsis* [71]. If perturbations were applied to all genes in a genome, the global gene network of an organism would be uncovered. On the way to such globalization, the repositories of compiled information of single-gene mutations of 'post-genome' organisms and resulting databases of essential genes, like DEG [72] could be used.

As summarized by Chan with colleagues [70], reconstruction of gene networks from gene expression data is useful for:

1. identifying important genes in relation to a disease or a biological function
2. gaining an understanding on the dynamic interaction between genes
3. predicting gene expression values at future time points
4. predicting drug effects over time.

Currently less utilized are protein sequence similarity-based networks. In patterns of protein domains the latter are connected if appearing in genome sequences in combinations [73]. Protein domain universe graphs (PDUG) are constructed by representing the nonredundant set of protein structural domains as nodes and using the structural similarity between those domains to define the edges on the graph [74].

Other types of biological networks usually represent an integration of the above-described network types in different combinations based on multiple datasets, representing any relationship between a set of genes, mRNAs, metabolites or proteins. New types of network can be generated by an enrichment of any of these networks with data from diverse genetic sources. For example, Garten and colleagues [75] superimposed transcription network and gene expression correlation network of yeast to filter out false positive associations from so-called location data on transcription factor proteins with their spectrum of promoter-binding sites determined *in vivo*. In yeast cellular network modelled by Yu and Li [76], data on transcription factor, gene relationships, microarray data and prior biological knowledge are integrated. As distinguishing features resulting from this integration, the combinatorial nature of transcription regulation, an estimate of transcription factor activity and condition specificity of the relationships are considered. Lu and co-workers [77] integrated initial yeast protein interaction network with diverse sources of genomic evidence, ranging from coexpression relationships to similar phylogenetic profiles. As a result, they observed measurable improvement in prediction performance of protein networks. In another approach undertaken by Patil and Nielsen [78] integration of genome-scale metabolic network and gene expression data enabled systematic identification of so-called reporter metabolites, important in metabolic regulation. It was possible to identify also the significantly correlated metabolic subnetworks after direct or indirect perturbations of the metabolism. de Lichtenberg and colleagues [79] used gene expression data from different stages of the yeast cell cycle, integrated it with a protein network and discovered that most of the protein complexes are comprized of both periodically and constitutively expressed proteins, which suggests that the former control complex activity by a mechanism of just-in-time assembly. Ihmels and co-workers [80] integrated large-scale expression data with the structural description of yeast metabolic network and found that only distinct branches at metabolic branchpoints are coexpressed and that individual isozymes were often separately co-regulated with distinct processes. Ideker and co-workers [81] inferred models of transcriptional regulation through integrating the data on protein–protein and protein–DNA interactions, the directionality of signal transduction in protein–protein interactions, as well as signs of the immediate effects of these interactions in what they call physical networks.

Obviously, the list of integrated networks has increased dramatically in the last two years alone and may be continued with almost any combination of data.

Types of representations of biological networks

With the use of high-throughput methods of modern biology the information on molecular interactions or co-behavior, cell regulation and signal transduction is rapidly accumulating. Although very complex by its nature, this data can be assembled in a simpler form of network graphs of interconnected elements. The information contained in such graphs can be of varying precision, depending on the availability of underlying knowledge. For example, in the networks describing interactome edges are usually unambiguous: connection between two proteins represents the possibility of direct binding which has been experimentally proven. However, the symbols used in other network types may lack strict definitions (often reflecting a lack of exact knowledge). To illustrate this, Kitano and colleagues [82] give an example of a typical signal transduction diagram, in which an arrow symbol could be interpreted four different ways: activation, translocation, dissociation of protein complex and residue modification.

To be able to share and to exchange knowledge gained from network analysis, systems biologists need to 'speak the same language', i.e., apply similar sets of formalization rules in the process of building such networks. While, to date, no consensus has yet been reached several approaches such as that of Pirson and colleagues [83], who elaborated a simple symbolic representation set of 18 controls for signal transduction networks, have been attempted. This set of formalization rules was further extended by KW Kohn [84] to additionally cover protein interaction and transcription networks. The elaborated graphical method could deal with both 'heuristic' and 'explicit' diagrams. Heuristic diagrams are important to build networks, when detailed knowledge of all possible reaction paths is not available, while 'explicit' means that the diagrams are totally unambiguous and suitable for computer simulation. This work was a step forward in information standardization from human- to machine-readable form of representing and communicating biological networks. The innovation in this direction was the development of the Systems Biology Markup Language (SBML), an open XML-based format for representing biochemical reaction networks. With the help of SBML models common to research in many areas of computational biology, including cell signaling pathways, metabolic pathways, gene regulation networks and others can be described [85].

Network topologies

After it has become possible to assemble information around a biological system in the form of a network of molecular interactions, it's time now to get the knowledge on how the functioning of the entire biological system is accomplished by means of the analysis of assembled network. To make it clear why biologists need to study an assembly to understand a biosystem, an analogy with a comprehensive technical system consisting of a lot of pieces is often exploited. Indeed, to understand functioning of the entire biosystem from a sum of studies on functionality of individual molecules is similar to studying the ship components to obtain knowledge on how a ship retains buoyancy and moves in a desired direction. For conceiving the entire functioning of both systems, knowledge on functionality of separate components, although being absolutely necessary is not sufficient, it is rather a matter of assembling and interaction of the component parts. For biosystems these properties are indicated by the structure, or topology, of an assembled network.

Early topological studies of cellular networks revealed several common characteristic features. Assemblies of molecular interactions usually represent complex heterogeneous networks, with nests of more dense connections. These nests are recognized as network modules, allowing network fragmentation into functional subnetworks. Network structure often involves a hierarchy of levels.

Aspects of structure can be deduced from statistical analysis of several parameters of network topology, in particular a number of connections (connectivity) for network elements and connectivity degree distribution, the degree of node centralities, clustering coefficient, network diameter and average path length.

Connectivities

In a biological network representation two nodes are connected to each other by edges, if an information exchange between these nodes occurs. Each node may be connected to distinct numbers of other nodes. From multiple analyses of biological network topologies, it is well established that connectivities are distributed among nodes with high inhomogeneity: the majority of nodes have a small number of connections, while a minority have a big number of connections. In large networks, the probability function $P(k)$ for the connectivity degree k may follow a behavior, described by the formula $P(k) = Ak^{-\gamma}$, called a power law. In a logarithmic scale this function takes a shape of a line, with the slope reflected by γ. Such distribution of a connectivity degree means that none of the nodes can be chosen as a scale representative from connectivity degree of which the judgement on connectivities of the other nodes may be drawn. That is why the networks with such connectivity degree distribution are often referred to as scale-free networks. Scale-free property of large networks was first distinguished by Barabasi and Albert [86]. After that, numerous large networks were described as being scale-free. Among biological networks, approximate scale-freeness was detected for many systems including, among others, metabolic networks of 43 different organisms [87], a pattern of protein domain combinations occurring in 40 genomes [73] further expanded to a protein domain universe graph [74] and gene-metabolite correlation network of *Arabidopsis* [47].

Scale-free networks possess a set of universal properties. First, paths by which information from any node can reach any other node, are relatively short. This feature was called a 'small-word' property [88]. The consequence of this feature for topology of scale-free networks is their high density and relatively small diameter. This in turn, taken together with a vast number of weakly connected nodes, brings us to the next consequence that is high redundancy of network paths. This property is very important for network stability. Indeed, if information from one node can reach another node by many redundant paths, then the probability to break information exchange by disturbance of any casual node from these paths is low. This means that scale-free networks are very robust against casual disturbances [89]. High stress tolerance of biological systems can be deduced also from robustness of a scale-free network of stress information processing. However, this property has an evident underside. The network integrity can be easily disrupted by the disturbance of highly connected nodes, called hubs. This determines the potential importance of elements with high numbers of connections in maintaining homeostasis of a biosystem. For biotechnology and biomedicine, such hubs represent target elements to influence system functioning. However, it has to be mentioned here that the latest well-defined studies on topologies of technological and biological networks clarify the relationship between scale-freeness and power law distribution and suggest that the connectivity degree distribution of many biological networks is often better described by distributions other than the popular power law. Affirmative conclusions, which are often deduced from scale-freeness of biological networks, have to be assessed critically for the quantitative understanding of complex biological processes [90, 91].

Centralities

The ranking of system elements (nodes) using centralities is another tool for estimating the importance, or influence strength, of a node. Such tools are mainly used in the analysis of social networks, where centrality measures are commonly described as indices of prestige, prominence, importance, and power – the four Ps [92]. Centrality is considered to weight indispensability of a node for information processing between distant nodes. A classical illustration implies a network of two clusters connected to each other with one node. This node is considered to be centrally positioned, or central. Although in a minimal case it may bear only two connections, one to each of the clusters (and thus is of low connectivity), it is nevertheless crucially important for keeping the integrity of the whole network. In terms of informational processing, information (a parcel) cannot be delivered from any node of one cluster to any other node of another cluster, bypassing the node which connects two clusters. Being central for information processing through the network, this node therefore is able to influence a lot of other nodes and consequently is of high importance for system functionality.

In network topology analysis, several centrality measures are utilized [93]. The degree centrality [94, 95] is interpreted as a measure of immediate influence. As opposed to connectivity, the degree centrality of a node considers not only a number of direct connections of this node, but also connectivities of its direct neighbors. Indeed, if a node has just a few connections, but through these connections is bound to a highly connected hub, then the probability of the information to be processed through this node is still high. The eigenvector centrality [96] can be considered as an extended degree centrality which is proportional to the sum of the centralities of the node's neighbors [93]. Another centrality measure, betweenness centrality [97], gives an estimation of how often a node appears on the way of an informational parcel between any two other nodes, and by this defines the control influence strength of the node whose centrality is being measured. Congenerous to this measure is the closeness centrality [94, 95], which in social networks is most frequently used to measure relative access to network resources and information, and can also be interpreted as measuring the degree of independence from others in the network

[98]. The subgraph centrality [93] characterizes the participation of each node in all subgraphs in a network, with smaller subgraphs having higher importance. To describe the centers of biological networks, further methods for geometric centrality measures were considered, namely excentricity, status, and centroid value that were originally used in the context of resource placement problems [99].

In biological networks the most important nodes are traditionally searched among those highest connected (hubs). However, this approach is not always successful, for example in the analysis of yeast protein interaction network the essentiality of a gene was poorly related to the number of interactors of the corresponding protein [100]. Centrality measures as an alternative to connectivity are increasingly attempted for this means. For example in the yeast protein interaction network, centrality of the genes was associated with the essential functions of the genes [101], and when compared with node connectivities, the ranking introduced by the subgraph centrality was more highly correlated with the lethality of individual yeast proteins [93]. Ma and Zeng [102] have identified the most central metabolites in a metabolic network by measuring the closeness centrality of the nodes, which correlated with the average path length. By the analysis of the betweenness centrality of protein domains in the graph of protein domain structures a gatekeeper protein domain, removal of which partitions the largest cluster into two large sub-clusters, was found. As was suggested, the loss of such gatekeeper protein domains in the course of evolution may be responsible for the creation of new fold families [103]. The centrality measure was recently also applied in biomedicine, where it helped to estimate, e.g., the importance of differentially expressed genes in lung cancer tissues [104], or the relevance of different mediators in the human immune cell network [105]. As was shown by a comparative study of protein interaction networks of three evolutionary distant eukaryotes: yeast, worm, and fly, the centrality of proteins had similar distributions; proteins that had a more central position in all three networks, regardless of the number of direct interactors, evolve more slowly and are more likely to be essential for survival [106].

By analogy with the connectivity degree distribution, which follows a power law in most large biological networks, Goh and co-workers [107] found that the betweenness centrality in biological scale-free networks also displays a power law distribution, and an exponent of this distribution can be used as a discriminating factor to classify the scale-free networks. Power law distribution was demonstrated also for the betweenness centrality values of protein domains in the graph of protein domain structures [103].

Clustering coefficient

The clustering coefficient is another statistical measure to characterize large networks. It quantifies the cohesiveness of the neighborhood of a node, in other words, how well connected the neighbors of a vertex in a graph are. In real networks it decreases with the vertex degree connectivity [108]. The clustering coefficient of a node is defined as the ratio between the number of edges linking nodes adjacent to this node and the total possible number of edges among them [88]. In other words, the clustering coefficient quantifies how close the local neighborhood of a node is to being part of a clique, a region of the graph (a subgraph) where every node is connected to every other node [109].

Real networks are generally characterized by a high clustering coefficient [88, 110]. For biological networks, a high average clustering coefficient was found, for example, in protein interaction and metabolic networks [111, 112], indicating a high level of redundancy and cohesiveness [109]. In gene expression networks generated from large model-organism expression datasets the average clustering coefficient was also several orders of magnitude higher than would be expected for similarly sized scale-free networks [113].

The diversity of cohesiveness of local neighborhoods is characterized by averaging the clustering coefficients of nodes that have the same connectivity degree. The function resulting from this procedure was decreasing in metabolic networks [114] and protein interaction networks [112]. This suggests that low-degree nodes tend to belong to highly cohesive neighborhoods whereas higher-degree nodes tend to have neighbors that are less connected to each other [109].

As an example application, in the recent study by Wei and colleagues [115] clustering coefficient was used to find out the superior one of the two possible mechanisms of the tRNA sequences evolution, namely point mutation and complementary duplication. From comparison of clustering coefficients in two alternative networks, which were constructed, based on these two possible mechanisms it was concluded that modern tRNA sequences evolved primarily by the mechanism of complementary method, and point mutation is an important and indispensable auxiliary mechanism during the evolutionary event.

Network diameter

In a graph theory, a network diameter is a global metric of its structure. It is defined as the average path length among all nodes. Together with average path lengths, the network diameter is considered as a measure of systems functionality, like, for example, in a study of robustness and vulnerability of the p53 protein interaction network [116]. In another example, using the path of shortest length, Said and coworkers [117] identified that the toxicity-modulating proteins in yeast have more interactions with other proteins, leading to a greater degree of metabolic adaptation upon modulating the functioning of these proteins.

Considering dynamics in biological networks

As a biological system is alive and ever-changing, it functions in time, or dynamically. Dynamical behavior is its intrinsic property and implies dynamical behavior of its constituting elements. Networks, now widely applied for systems biology, may be analyzed statically, or may consider this dynamical behavior, depending on the network type and on the nature of the datasets underlying network reconstruction. For metabolic, transcription and protein interaction networks, usual representation as graphs reflects the static properties of a system. The standard approach to model network dynamics is through sets of coupled differential equations, describing how the concentrations of the various products evolve over time [118]. However, such a model requires knowledge of the various reaction rates and rate-order kinetics. To overcome this drawback temporal data can be integrated into these networks. For example de Lichtenberg and colleagues analyzed the dynamics of protein complexes during the yeast cell cycle by means of integration of temporal data on protein interactions and gene expression [79], revealing previously unknown components and modules. In modeling the dynamics of another type of initially static network, a metabolic network, large-scale biochemical systems approaches, such as the network thermodynamics theory, biochemical systems theory, metabolic control analysis, and flux balance analysis are used. P Ao [119] modeled dynamics of a metabolic network by adding four dynamical structure elements: potential function, translocation matrix, degradation matrix, and stochastic force. Network dynamics was determined by these four elements being in balance, which gave rise to a special stochastic differential equation. This allowed experimental data being displayed stochasticity which carried important biological information.

As opposed to the above-mentioned networks, which are static by the nature of underlying data utilized, correlation networks are built from temporal (or sometimes concentrational) series of transcript or/and metabolite profiles. This defines the dynamical property of a resulting correlation network, which can be analyzed by cluster analysis and the systematic search for characteristic patterns of gene expression associated with a state of interest [120–123]. The dynamical property can also be implemented into the analysis of static networks by integrating with dynamical network types, as was demonstrated, for example, by Guthke and co-workers [124] in studies of the kinetics of the immune response to bacterial infection. In another study on yeast transcriptional regulatory network, molecular interactions in the cellular transcription, translation, and degradation machineries were incorporated into dynamic mathematical models of the biochemical system by finding the most changed parameters from yeast oligonucleotide microarray expression patterns in cases where a phenotype difference existed between two samples [125]. On a genomic scale, the dynamics of a biological network was analyzed for multiple conditions in yeast by integrating transcriptional regulatory information and gene expression data [126]. In another approach, which we would call vertical integration, dynamics is implemented into a biological network by combining different levels of system description. Applicability and limitations of modeling the dynamics of cellular networks with this approach were demonstrated by Vilar and colleagues [127] on the lac operon of *Escherichia coli* as a prototype system. Here, three levels (molecular, cellular, and that of cell population) were integrated into a single model, and by this dynamical aspects of the system were captured.

Several mathematical and computational approaches have been suggested for determining the dynamics of regulatory networks: including linear [128] and nonlinear [129] models, time-series analysis [130, 131] and Bayesian networks of dependencies [132, 133]. The dynamics of a biological system can be investigated by computing kinetic curves for molecular components (RNA, proteins) using the method of generalized threshold models [134]. A dynamic network model can also be deduced from a simple discrete model by postulating logical rules that formally summarize legacy data, as was demonstrated by plant biologists for interaction of the so-called ABC homeotic floral genes in *Arabidopsis* floral organ determination [135].

Generally, the highly nonlinear dynamics exhibited by genetic regulatory systems can be predicted by either of two important theoretical approaches: the continuous approach, based on reaction-kinetics differential equations, and the Boolean approach, based on difference equations and discrete logical rules [136, 137]. With these approaches biological systems can be characterized into an ordered regime where the system is robust against perturbations, and a chaotic regime where the system is extremely sensitive to perturbations. In a case study of HeLa cells its underlying genetic network appeared to operate either in the ordered regime or at the border between order and chaos but did not appear to be chaotic [138].

Causal directionality in biological networks

Causality is another characteristic feature of a dynamically functioning system. Depending on the nature of underlying type of an informational exchange biological networks can be either directed or undirected. Causal directionality in the biological networks is subject for reconstruction, when cause-and-effect relationship of the interactions between two components is well defined, e.g., the direction of metabolic flow from substrates to products in metabolic networks, the information flow from transcription factors to the genes that they regulate in transcription networks, propagation of signal transduction events in signaling networks, or influence on gene expression in gene networks. Such networks are causally directed. In undirected networks, such as protein interaction networks or protein sequence similarity-based networks, the relationships are mutually equidirectional. Some biological networks, although possessing intrinsic causal directionality, stay as undirected graphs, because edge directions are difficult or even not possible to identify. This applies to a great extent to networks reconstructed from high-throughput metabolic, proteomic or genomic analysis. As can be illustrated by gene coexpression networks, although genes with similar expression profiles are likely to regulate each other or be regulated by another common gene, from co-response analysis it is impossible to infer any notion of causality – which gene is regulated and which gene is regulating. However, if such networks are built from dynamic measurements of responses, which yield hierarchical information about causal relations in the underlying system, then causal relationships in these networks can be inferred. This approach was probed, for example, on hormone and insulin signaling using tyrosine residues phosphorylation data [139]. Similarly, response dynamics elucidates causality, when the information is used regarding the time lag between species at which the highest correlation was found [122]. In the new multiscale fuzzy clustering method fuzzy cluster centers can be used to discover causal relationships between groups of co-regulated genes. With this method applied to gene expression data, a new regulatory relationship concerning trehalose regulation of carbohydrate metabolism in *Arabidopsis* was found [140]. In another example, causal directionality was implemented to gene-metabolite correlation network with the use of *a priori* knowledge on the molecule, which excites the systems response and can thus be considered as a 'cause'. In such network propagation of the information flow from the exciter to physiological endpoints can be followed [47]. To derive causal influences in cellular signaling networks, machine learning was applied to the simultaneous measurement of multiple phosphorylated protein and phospholipid components in thousands of individual primary human immune system cells. Perturbing these cells with molecular interventions drove the ordering of connections between pathway components [141].

The problem of causality in biological networks can be accessed also by means of integrating with directed networks. In a causal inference approach transcriptional regulatory networks of yeast were constructed using gene expression data, promoter sequences and information on transcription factor binding sites [142]. In this method identified active transcription factors provide the causal effect as 'treatments' measured quantitatively, and gene expression levels are viewed as 'responses'. In a study of the pheromone response in yeast, causal relationships were implemented into the non-directed network of protein–protein interactions by integrating with the directed networks of protein–DNA interactions and signal transduction [81].

Comparative network analysis

Now, as enormous amounts of data are available on molecular interaction networks, the next cognition step for system biologists implies new questions about network dynamics and evolution. These questions can be approached by means of network comparison. In such analysis communication networks for steady state and perturbation, or for organisms of different evolutionary distance in normal growth and in response to the same perturbing agent, can be compared. By comparing topologies of the resulting alternative communication networks constitutive and exciter-specific communication paths can be revealed, as well as hubs as specific controllers of the response development. Moreover, network comparisons can be used systematically to catalog conserved network regions, each representing a functionally homologous mechanism or pathway [143]. This approach also helps to resolve some technical aspects of network analysis. One of the major such problems is generally the high noise component in biological networks. This problem can be approached, for example, by comparing a network reconstructed from real data with a network built from the same dataset, subjected to shuffling procedure and thus assumed to be information-free. As a result of such comparison, noise component can be subtracted from the real data-based network. Comparative analysis of real networks also helps to address the problem of noise. Thus, by comparing networks drawn from different species or conditions [144–146], it was possible to reinforce the common signal present in both networks while reducing the noise component. Network comparison was helpful also in separating true protein–protein and protein–DNA interactions from false positives [147], annotating interactions with functional roles and, ultimately, organizing large-scale interaction data into models of cellular signaling and regulatory machinery [148].

In biological applications, network comparison is becoming increasingly fruitful. We shall illustrate this with several examples. As was shown by the analysis of metabolic networks, comparison of network topologies for 43 organisms revealed hierarchical modularity in the network organization [114]. Pairwise comparison of protein interaction networks of bacteria and yeast allowed detection of evolutionarily conserved pathways [149] and significantly conserved protein complexes [150]. Further cross-species study of protein–protein interaction networks, now of worm, fly and yeast, revealed remarkable similarities in network structures [106], and identified previously not described protein functions and interactions [151]. Network comparison was applied also to gene coexpression networks. In cancer research, studies on two distinct coexpression networks: a tumor network and normal network showed that cancer affected many coexpression relationships accompanied with functional changes [40]. These case studies demonstrate that network comparisons provide essential biological information beyond what is gained from the analysis of separate networks.

A growing demand for statistical techniques and tools applicable for network comparison meets with a growing response by bioinformaticians. In this vein a technique for finding branching structure shared by a set of phylogenetic networks was recently introduced [152]. Kelley and co-workers [149] implemented a strategy for aligning two protein–protein interaction networks that combines interaction topology and protein sequence similarity, which was further developed into a PathBLAST tool for alignment of protein interaction networks [143]. Another tool called Cfinder allows finding overlapping dense groups of nodes in networks [153]. The reader will find the collection of corresponding links in Table 3 below.

Testing biological networks

Analysis of biological networks gives life to new global hypotheses on systems functionality and reductionist findings of novel molecular interactions. The reliability of these hypotheses will be based on the general reliability of the network reconstruction procedure. If among numerous findings revealed through network analysis a significant number matches with prior experimental knowledge, this can generally serve as a validation of the network analysis methodology employed. However this approach evidently cannot validate each individual finding and as such cannot substitute for wet-laboratory experimentation.

The use of *a priori* knowledge is best illustrated by the studies on the yeast integrated regulatory network. Its reliability was tested on datasets related to the pheromone response pathway, and the resulting model showed consistence with previous studies on the pathway [81]. Similarly, in the network model of bacteria and yeast protein complexes several of these complexes matched well with prior ex-

Table 3. Networking tools

Tool name [Reference #]	Designation	Web link
Pajek [157]	analysis of large networks	http://vlado.fmf.uni-lj.si/pub/networks/pajek/
Cytoscape [158]	visualizing molecular interaction networks and integrating these interactions with gene expression profiles and other state data	http://www.cytoscape.org/
VANTED [159]	Visualization and Analysis of Networks containing Experimental Data	http://vanted.ipk-gatersleben.de/
VisANT [160]	visualizing and analyzing many types of biological networks	http://visant.bu.edu/
BiNGO [161]	assessing overrepresentation of gene ontology categories in biological networks	http://www.psb.ugent.be/cbd/papers/BiNGO/
Centibin [162]	calculation and visualization of centralities for biological networks	http://centibin.ipk-gatersleben.de/
Cfinder [153]	finding overlapping dense groups of nodes in networks	http://angel.elte.hu/%7Evicsek/
PathBLAST [143]	alignment of protein interaction networks	http://www.pathblast.org/
TopNet [163]	comparing network topologies	http://networks.gersteinlab.org/genome/interactions/networks/core.html
CellDesigner [82]	diagrammatic network editing software	http://www.celldesigner.org/

perimental knowledge on complexes in yeast only and thus served for validation of the methodology [150]. In biomedical studies, the importance of identified hubs for network function was supported by the severe phenotypes exhibited by human patients and animal models when these genes were mutated [154].

Similarly the use of direct experimentation for validation of biological networks has also been applied to the yeast integrated regulatory network: whereby the knockout of genes and subsequent phenotyping confirmed the effects which were predicted by the network model [81]. Mutation has also been used strategically in cancer research in order to test the significance of the results drawn from the network analysis [116]. In the same study another method of experimental testing was tried, namely the effects of tumor inducing viruses were compared with those derived from network analysis. Protein interaction networks were tested by two hybrid experiments in which approximately half of 60 inferred interaction predictions were confirmed [151]. However, in spite of the general acceptance of the reductionist methods of experimental confirmation in biology, the problem of testing the reliability of a reconstructed biological network cannot be fully approached by such methods for all network types. Where it is possible, network construction as the method for analysis of the entire system's functionality by means of assembling coherence between the elements in complex systems can be reliably tested by the assembly of an alternative network. The expected experiments on this may imply, e.g., analysis of information conductivity in a network reconstructed from the similar data source, but obtained on a system with a hub gene/protein knocked out, and therefore will lay in a field of network comparison. Here, matching of the predicted information conductivity to that one in an alternative network will work for confirmation of the reliability of the reconstructed network.

Intrinsic properties of biological networks – are there any?

Recent advances in networking studies allow a comparative analysis of many large networks of biological, social and technological nature (e.g., [153, 155, 156]). In these studies a question is asked on the existence of common properties for these large networks and systems they describe. It was found, that, while on the one hand, complex systems, indeed, share several common properties, on the other hand, each system is characterized by unique parameters. Identification of regularities being specific for biosystems may lead to better understanding of the uniqueness of life phenomenon and may imply also a practical interest in developing the new information technologies of complex systems management.

Software solutions for network visualization and analysis with useful links

Modern software networking tools can handle multiple data types from distinct technologies. Some of these tools are multifunctional developments for general networking studies like Pajek, Cytoscape, and VANTED. The others represent more specialized tools created for the analysis of separate network properties, like network centralities (Centibin), or overrepresented gene ontologies (BiNGO). Network comparison studies can be approached with PathBLAST and TopNet. The reader will find short descriptions of functionality and applicability for the major networking tools with the corresponding links and references in Table 3.

Furthermore, the set of software tools helpful in networking studies, which has been developed for pathway analysis, is given in Table 4. Among these tools, AraCyc, a collection of biochemical pathways described in *Arabidopsis*, is designated for the networking of plant biosystems.

Among the other useful software developments, in Table 5 we provide the list of those, which are the most commonly used as data sources for network reconstructions. The last two links are devoted to the universal networking language SBML and the data integration tool Pointillist.

Table 4. Pathways: databases and analysis tools

Tool/Database name [Reference #]	Designation	Link
KEGG PATHWAY [5]	collection of manually drawn pathway maps for the molecular interaction and reaction networks	http://www.genome.ad.jp/kegg/pathway.html
BioCyc [2]	collection of pathway/genome databases plus the BioCyc open chemical database	http://www.biocyc.org/
AraCyc [3]	biochemical pathway database for *Arabidopsis*	http://www.arabidopsis.org/tools/aracyc/
MetaCyc [4]	database of nonredundant, experimentally elucidated metabolic pathways	http://metacyc.org/
PaVESy [164]	Pathway Visualization Editing System	http://pavesy.mpimp-golm.mpg.de/PaVESy.htm
KnowledgeEditor [165]	interactive modeling and analyzing biological pathways based on microarray data	

Table 5. Databases of molecular interactions and other

Name [Reference #]	Designation	Web link
RegulonDB [7]	database on mechanisms of transcription regulation and operon organization in *Escherichia coli*	http://regulondb.ccg.unam.mx
GRID [166]	database of genetic and physical interactions in yeast, fly and worm	http://biodata.mshri.on.ca/grid
Ospray [167]	visualization of complex interaction networks	http://biodata.mshri.on.ca/osprey
BIND [8]	Biomolecular Interaction Network Database	http://www.bind.ca/Action
DIP [22]	Database of Interacting Proteins	http://dip.doe-mbi.ucla.edu/
PPI [19]	Human protein–protein interaction network database	http://141.80.164.19/neuroprot/ppi_search.php
KEGG [168]	Kyoto Encyclopedia of Genes and Genomes	http://www.genome.ad.jp/kegg/
DEG [72]	Database of Essential Genes	http://tubic.tju.edu.cn/deg/
SBML [85]	Systems Biology Markup Language	http://www.sbml.org/
Pointillist [169]	inferring the set of elements affected by a perturbation of a biological system	http://magnet.systemsbiology.net/software/Pointillist/

References

1. Nikiforova VJ, Kopka J, Tolstikov V, Fiehn O, Hopkins L, Hawkesford MJ, Hesse H, Hoefgen R (2005) Systems re-balancing of metabolism in response to sulfur deprivation, as revealed by metabolome analysis of *Arabidopsis* plants. *Plant Physiol* 138: 304–318
2. Karp PD, Ouzounis CA, Moore-Kochlacs C, Goldovsky L, Kaipa P, Ahren D, Tsoka S, Darzentas N, Kunin V, Lopez-Bigas N (2005) Expansion of the BioCyc collection of pathway/genome databases to 160 genomes. *Nucleic Acids Res* 33: 6083–6089
3. Zhang PF, Foerster H, Tissier CP, Mueller L, Paley S, Karp PD, Rhee SY (2005) MetaCyc and AraCyc. Metabolic pathway databases for plant research. *Plant Physiol* 138: 27–37
4. Krieger CJ, Zhang PF, Mueller LA, Wang A, Paley S, Arnaud M, Pick J, Rhee SY, Karp PD (2004) MetaCyc: a multiorganism database of metabolic pathways and enzymes. *Nucleic Acids Res* 32: D438–D442
5. Ogata H, Goto S, Sato K, Fujibuchi W, Bono H, Kanehisa M (1999) KEGG: Kyoto Encyclopedia of Genes and Genomes. *Nucleic Acids Res* 27: 29–34
6. Sweetlove L, Fernie AR (2005) Tansley Review: Regulation of metabolic networks. Understanding metabolic complexity in the systems biology era. *New Phytol* 168: 9–23
7. Salgado H, Gama-Castro S, Martinez-Antonio A, Diaz-Peredo E, Sanchez-Solano F, Peralta-Gil M, Garcia-Alonso D, Jimenez-Jacinto V, Santos-Zavaleta A, Bonavides-Martinez C et al. (2004) RegulonDB (version 4.0): transcriptional regulation, operon organization and growth conditions in *Escherichia coli* K-12. *Nucleic Acids Res* 32: D303–D306
8. Alfarano C, Andrade CE, Anthony K, Bahroos N, Bajec M, Bantoft K, Betel D, Bobechko B, Boutilier K, Burgess E et al. (2005) The Biomolecular Interaction Network Database and related tools 2005 update. *Nucleic Acids Res* 33: D418–D424
9. Shen-Orr SS, Milo RM, Alon U (2002) Network motifs in the transcriptional regulation network of *Escherichia coli*. *Nature Genet* 31: 64–68
10. Lee TI, Rinaldi NJ, Robert F, Odom DT, Bar-Joseph Z, Gerber GK, Hannett NM, Harbison CT, Thompson CM, Simon I et al. (2002) Transcriptional regulatory networks in *Saccharomyces cerevisiae*. *Science* 298: 799–804
11. Zhong JH, Zhang HM, Stanyon CA, Tromp G, Finley RL (2003) A strategy for constructing large protein interaction maps using the yeast two-hybrid system: Regulated arrays and two-phase mating. *Genome Res* 13: 2691–2699
12. Cusick ME, Klitgord N, Vidal M, Hill DE (2005) Interactome: gateway into systems biology. *Hum Mol Gen* 14: R171–R181
13. Skipper M (2005) A protein network of one's own proteins. *Nature Rev Mol Cell Biol* 6: 824–825
14. Uetz P, Giot L, Cagney G, Mansfield TA, Judson RS, Knight JR, Lockshon D, Narayan V, Srinivasan M, Pochart P et al. (2000) A comprehensive analysis of protein–protein interactions in *Saccharomyces cerevisiae*. *Nature* 403: 623–627
15. Giot L, Bader JS, Brouwer C, Chaudhuri A, Kuang B, Li Y, Hao YL, Ooi CE, Godwin B, Vitols E et al. (2003) A protein interaction map of *Drosophila melanogaster*. *Science* 302: 1727–1736
16. Hoebeke M, Chiapello H, Noirot P, Bessieres P (2001) SPiD: a subtilis protein interaction database. *Bioinformatics 17: 1209–1212*
17. Li S, Armstrong CM, Bertin N, Ge H, Milstein S, Boxem M, Vidalain PO, Han JD, Chesneau A, Hao T et al. (2004) A map of the interactome network of the metazoan *C. elegans*. *Science* 303: 540–543

18. LaCount DJ, Vignali M, Chettier R, Phansalkar A, Bell R, Hesselberth JR, Schoenfeld LW, Ota I, Sahasrabudhe S, Kurschner C et al. (2005) A protein interaction network of the malaria parasite *Plasmodium falciparum*. *Nature* 438: 103–107
19. Stelzl U, Worm U, Lalowski M, Haenig C, Brembeck FH, Goehler H, Stroedicke M, Zenkner M, Schoenherr A, Koeppen S et al. (2005) A human protein–protein interaction network: a resource for annotating the proteome. *Cell* 122: 957–968
20. Rual J-F, Venkatesan K, Hao T, Hirozane-Kishikawa T, Dricot A, Li N, Berriz GF, Gibbons FD, Dreze M, Ayivi-Guedehoussou N et al. (2005) Towards a proteome-scale map of the human protein–protein interaction network. *Nature* 437: 1173–1178
21. de Folter S, Immink RGH, Kieffer M, Parenicova L, Henz SR, Weigel D, Busscher M, Kooiker M, Colombo L, Kater MM et al. (2005) Comprehensive interaction map of the Arabidopsis MADS box transcription factors. *Plant Cell 17: 1424–1433*
22. Salwinski L, Miller CS, Smith AJ, Pettit FK, Bowie JU, Eisenberg D (2004) The database of interacting proteins: 2004 update. *Nucleic Acids Res* 32: D449–451
23. Barrett T, Suzek TO, Troup DB, Wilhite SE, Ngau WC, Ledoux P, Rudnev D, Lash AE, Fujibuchi W, Edgar R (2005) NCBI GEO: mining millions of expression profiles – database and tools. *Nucleic Acids Res* 33: D562–566
24. Parkinson H, Sarkans U, Shojatalab M, Abeygunawardena N, Contrino S, Coulson R, Farne A, Garcia Lara G, Holloway E, Kapushesky M et al. (2005) ArrayExpress – a public repository for microarray gene expression data at the EBI. *Nucleic Acids Res* 33: D553–D555
25. Fellenberg K, Hauser NC, Brors B, Hoheisel JD, Vingron M. (2002) Microarray data warehouse allowing for inclusion of experiment annotations in statistical analysis. *Bioinformatics* 18: 423–433
26. Le Crom S, Devaux F, Jacq C, Marc P (2002) yMGV: helping biologists for yeast microarray data mining. *Nucleic Acid Res* 30: 76–79
27. Zimmermann F, Hirsch-Hoffmann M, Hennig L, Gruissem W (2004) GENEVESTIGATOR. *Arabidopsis* microarray database and analysis toolbox. *Plant Physiol* 136: 2621–2632
28. Seki M, Narusaka M, Ishida J, Nanjo T, Fujita M, Oono Y, Kamiya A, Nakajima M, Enju A, Sakurai T et al. (2002) Monitoring the expression profiles of 7000 *Arabidopsis* genes under drought, cold, and high-salinity stresses using a full-length cDNA microarray. *Plant J* 31: 279–292
29. Oliver S (1996) A network approach to the systematic analysis of yeast gene function. *Trends in Genetics* 12: 241–242
30. Hodgman TC (2000) A historical perspective on gene/protein functional assignment. *Bioinformatics* 16: 10–15
31. Blochzupan A, Decimo D, Loriot M, Mark MP, Ruch JV (1994) Expression of nuclear retinoic acid receptors during mouse odontogenesis. *Differentiation* 57: 195–203
32. Yamazaki M, Majeska RJ, Yoshioka H, Moriya H, Einhorn TA (1997) Spatial and temporal expression of fibril-forming minor collagen genes (types V and XI) during fracture healing. *J Orthopaedic Res* 15: 757–764
33. Cho RJ, Campbell MJ, Winzeler EA, Steinmetz L, Conway A, Wodicka L, Wolfsberg TG, Gabrielian AE, Landsman, Lockhart DJ et al. (1998) A genome-wide transcriptional analysis of the mitotic cell cycle. *Mol Cell* 2: 65–73
34. Zhang MQ (1999) Promoter analysis of co-regulated genes in the yeast genome. *Comput Chem* 23: 233–250
35. Eisen MB, Spellman PT, Brown PO, Botstein D (1998) Cluster analysis and display of genome-wide expression patterns. *Proc Natl Acad Sci USA* 95: 14863

36. Chu S, DeRisi J, Eisen M, Mulholland J, Botstein D, Brown PO, Herskowitz I (1998) The transcriptional program of sporulation in budding yeast. *Science* 282: 699–705
37. Kim SK, Lund J, Kiraly M, Duke K, Jiang M, Stuart JM, Eizinger A, Wylie BN, Davidson GS (2001) A gene expression map for *Caenorhabditis elegans*. *Science* 293: 2087
38. Stuart JM, Segal E, Koller D, Kim SK (2003) A gene-coexpression network for global discovery of conserved genetic modules. *Science* 302: 249–255
39. Snel B, van Noort V, Huynen MA (2004) Gene co-regulation is highly conserved in the evolution of eukaryotes and prokaryotes. *Nucleic Acids Res* 32: 4725–4731
40. Choi JK, Yu US, Yoo OJ, Kim S (2005) Differential coexpression analysis using microarray data and its application to human cancer. *Bioinformatics* 21: 4348–4355
41. Kose F, Weckwerth W, Linke T, Fiehn O (2001) Visualizing plant metabolomic correlation networks using clique-metabolite matrices. *Bioinformatics* 17: 1198–1208
42. Steuer R, Kurths J, Fiehn O, Weckwerth W (2003) Observing and interpreting correlations in metabolomic networks. *Bioinformatics* 19: 1019–1026
43. Askenazi M, Driggers EM, Holtzman DA, Norman TC, Iverson S, Zimmer DP, Boers ME, Blomquist PR, Martinez EJ, Monreal AW et al. (2003) Integrating transcriptional and metabolite profiles to direct the engineering of lovastatin-producing fungal strains. *Nature Biotechnol* 21: 150–156
44. Urbanczyk-Wochniak E, Luedemann A, Kopka J, Selbig J, Roessner-Tunali U, Willmitzer L, Fernie AR (2003) Parallel analysis of transcript and metabolic profiles: a new approach in systems biology. *EMBO Rep* 4: 989–993
45. Hirai MY, Yano M, Goodenowe DB, Kanaya S, Kimura T, Awazuhara M, Arita M, Fujiwara T, Saito K (2004) Integration of transcriptomics and metabolomics for understanding of global responses to nutritional stresses in *Arabidopsis thaliana*. *Proc Natl Acad Sci USA* 101: 10205–10210
46. Hirai MY, Klein M, Fujikawa Y, Yano M, Goodenowe DB, Yamazaki Y, Kanaya S, Nakamura Y, Kitayama M, Suzuki H et al. (2005) Elucidation of gene-to-gene and metabolite-to-gene networks in *Arabidopsis* by integration of metabolomics and transcriptomics. *J Biol Chem* 280: 25590–25595
47. Nikiforova VJ, Daub CO, Hesse H, Willmitzer L, Hoefgen R (2005) Integrative gene-metabolite network with implemented causality deciphers informational fluxes of sulphur stress response. *J Exp Bot* 56: 1887–1896
48. Weckwerth W, Morgenthal K (2005) Metabolomics: from pattern recognition to biological interpretation. *Drug Discov Today* 10: 1551–1558
49. Weckwerth W, Loureiro ME, Wenzel K, Fiehn O (2004) Differential metabolic networks unravel the effects of silent plant phenotypes. *Proc Natl Acad Sci USA* 101: 7809–7814
50. Spellman PT, Sherlock G, Zhang MQ, Iyer VR, Anders K, Eisen MB, Brown PO, Botstein D, Futcher B (1998) Comprehensive identification of cell cycle-regulated gene of the yeast *Saccharomyces cerevisiae* by microarray hybridization. *Mol Biol Cell* 9: 3273–3297
51. Vlieghe K, Vuylsteke M, Florquin K, Rombauts S, Maes S, Ormenese S, Van Hummelen P, Van de Peer Y, Inze D, De Veylder L (2003) Microarray analysis of E2Fa-DPa-overexpressing plants uncovers a cross-talking genetic network between DNA replication and nitrogen assimilation. *J Cell Sci* 116: 4249–4259
52. Liu FL, VanToai T, Moy LP, Bock G, Linford LD, Quackenbush J (2005) Global transcription profiling reveals comprehensive insights into hypoxic response in *Arabidopsis*. *Plant Physiol* 137: 1115–1129
53. Venter M, Botha FC (2004) Promoter analysis and transcription profiling: Integration of genetic data enhances understanding of gene expression. *Physiol Plant* 120: 74–83

54. Xia Y, Yu HY, Jansen R, Seringhaus M, Baxter S, Greenbaum D, Zhao HY, Gerstein M (2004) Analyzing cellular biochemistry in terms of molecular networks. *Annu Rev Biochem* 73: 1051–1087
55. Schlitt T, Brazma A (2005) Modelling gene networks at different organisational levels. *FEBS Letters* 579: 1859–1866
56. Kollmann M, Lovdok L, Bartholome K, Timmer J, Sourjik V (2005) Design principles of a bacterial signalling network. *Nature* 438: 504–507
57. Bagnato A, Spinella F, Rosano L (2005) Emerging role of the endothelin axis in ovarian tumor progression. *Endocr Relat Cancer* 12: 761–772
58. Kundu JK, Surh YJ (2005) Breaking the relay in deregulated cellular signal transduction as a rationale for chemoprevention with anti-inflammatory phytochemicals. *Mutat Res – Fund Mol Mech Mut* 591: 123–146
59. Katagiri F (2004) A global view of defense gene expression regulation – a highly interconnected signaling network. *Curr Opin Plant Biol* 7: 506–511
60. Zhao J, Davis LC, Verpoorte R (2005) Elicitor signal transduction leading to production of plant secondary metabolites. *Biotech Adv* 23: 283–333
61. Feechan A, Kwon E, Yun BW, Wang YQ, Pallas JA, Loake GJ (2005) A central role for S-nitrosothiols in plant disease resistance. *Proc Natl Acad Sci USA* 102: 8054–8059
62. Gechev TS, Minkov IN, Hille J (2005) Hydrogen peroxide-induced cell death in *Arabidopsis*: Transcriptional and mutant analysis reveals a role of an oxoglutarate-dependent dioxygenase gene in the cell death process. *IUBMB Life* 57: 181–188
63. Murata Y, Pei ZM, Mori IC, Schroeder J (2001) Abscisic acid activation of plasma membrane Ca2+ channels in guard cells requires cytosolic NAD(P)H and is differentially disrupted upstream and downstream of reactive oxygen species production in abi1-1 and abi2-1 protein phosphatase 2C mutants. *Plant Cell* 13: 2513–2523
64. MacRobbie EAC (2002) Evidence for a role for protein tyrosine phosphatase in the control of ion release from the guard cell vacuole in stomatal closure. *Proc Natl Acad Sci USA* 99: 11963–11968
65. Mustilli AC, Merlot S, Vavasseur A, Fenzi F, Giraudat J (2002) *Arabidopsis* OST1 protein kinase mediates the regulation of stomatal aperture by abscisic acid and acts upstream of reactive oxygen species production. *Plant Cell* 14: 3089–3099
66. Bright J, Desikan R, Hancock JT, Weir IS, Neill SJ (2006) ABA-induced NO generation and stomatal closure in *Arabidopsis* are dependent on H2O2 synthesis. *Plant J* 45: 113–122
67. Janes KA, Albeck JG, Gaudet S, Sorger PK, Lauffenburger DA, Yaffe MB (2005) Systems model of signaling identifies a molecular basis set for cytokine-induced apoptosis. *Science* 310: 1646–1653
68. de la Fuente A, Brazhnik P, Mendes P (2002) Linking the genes: inferring quantitative gene networks from microarray data. *Trends Genetics* 18: 395–398
69. Mazhawidza W, Winters SJ, Kaiser UB, Kakar SS (2006) Identification of gene networks modulated by activin in L beta T2 cells using DNA microarray analysis. *Histol Histopathol* 21: 167–178
70. Chan ZSH, Kasabov N, Collins L (2006) A two-stage methodology for gene regulatory network extraction from time-course gene expression data. *Expert Systems with Applications* 30: 59–63
71. Costa MMR, Fox S, Hanna AI, Baxter C, Coen E (2005) Evolution of regulatory interactions controlling floral asymmetry. *Development* 132: 5093–5101
72. Zhang R, Ou HY, Zhang CT (2004) DEG, a Database of Essential Genes. *Nucleic Acids Res* 32: D271–D272

73. Apic G, Gough J, Teichmann SA (2001) Domain combinations in archaeal, eubacterial and eukaryotic proteomes. *J Mol Biol* 310: 311–325
74. Dokholyan NV, Shakhnovich B, Shakhnovich EI (2002) Expanding protein universe andits origin from the biological Big Bang. *Proc Natl Acad Sci USA* 99: 14132–14136
75. Garten Y, Kaplan S, Pilpel Y (2005) Extraction of transcription regulatory signals from genome-wide DNA – protein interaction data. *Nucleic Acids Res* 33: 605–615
76. Yu T, Li K-C (2005) Inference of transcriptional regulatory network by two-stage constrained space factor analysis. *Bioinformatics* 21: 4033–4038
77. Lu LJ, Xia Y, Paccanaro A, Yu H, Gerstein M (2005) Assessing the limits of genomic data integration for predicting protein networks. *Genome Res* 15: 945–953
78. Patil KR, Nielsen J (2005) Uncovering transcriptional regulation of metabolism by using metabolic network topology. *Proc Natl Acad Sci USA* 102: 2685–2689
79. de Lichtenberg U, Jensen LJ, Brunak S, Bork P (2005) Dynamic complex formation during the yeast cell cycle. *Science* 307: 724–727
80. Ihmels J, Levy R, Barkai N (2004) Principles of transcriptional control in the metabolic network of *Saccharomyces cerevisiae*. *Nat Biotechnol* 22: 86–92
81. Yeang CH, Ideker T, Jaakkola T (2004) Physical network models. *J Comput Biol* 11: 243–262
82. Kitano H, Funahashi A, Matsuoka Y, Oda K (2005) Using process diagrams for the graphical representation of biological networks. *Nat Biotechnol* 23: 961–966
83. Pirson I, Fortemaison N, Jacobs C, Dremier S, Dumont JE, Maenhaut C (2000) The visual display of regulatory information and networks. *Trends Cell Biol* 10: 404–408
84. Kohn KW (2001) Molecular interaction maps as information organizers and simulation guides. *Chaos* 11: 84–97
85. Hucka M, Finney A, Sauro HM, Bolouri H, Doyle JC, Kitano H, Arkin AP, Bornstein BJ, Bray D, Cornish-Bowden A et al. (2003) The systems biology markup language (SBML): a medium for representation and exchange of biochemical network models. *Bioinformatics* 19: 524–531
86. Barabasi AL, Albert R (1999) Emergence of scaling in random networks. *Science* 286: 509–512
87. Jeong H, Tombor B, Albert R, Oltvai ZN, Barabasi A-L (2000) The large-scale organization of metabolic networks. *Nature* 407: 651–654
88. Watts DJ, Strogatz SH (1998) Collective dynamics of 'small-world' networks. *Nature* 393: 440–442
89. Albert R, Jeong H, Barabási AL (2000) Error and attack tolerance of complex networks. *Nature* 406: 378–382
90. Stumpf MPH, Ingram PJ (2005) Probability models for degree distributions of protein interaction networks. *Europhysics Letters* 71: 152–158
91. Arita M (2005) Scale-freeness and biological networks. *J Biochem* 138: 1–4
92. Borgatti SP (1995) Centrality and AIDS. *Connections* 18: 112–115
93. Estrada E, Rodriguez-Velazquez JA (2005) Subgraph centrality in complex networks. *Physical Review E* 7105: 6103
94. Freeman LC (1979) Centrality in social networks conceptual clarification. *Soc Networks* 1: 215–239
95. Albert R, Jeong H, Barabasi AL (1999) Internet – Diameter of the World-Wide Web. *Nature* 401: 130–131
96. Bonacich P (1972) Factoring and weighting approaches to status scores and clique identification. *J Math Sociol* 2: 113–120

97. Freeman LC (1977) A set of measures of centrality based on betweenness. *Sociometry* 40: 35–41
98. Hagen G, Killinger DK, Streeter RB (1997) An analysis of communication networks among Tampa Bay economic development organizations. *Connections* 20: 13–22
99. Wuchty S, Stadler PF (2003) Centers of complex networks. *J Theor Biol* 223: 45–53
100. Coulomb S, Bauer M, Bernard D, Marsolier-Kergoat MC (2005) Gene essentiality and the topology of protein interaction networks. *Proc R Soc Lond [Biol]* 272: 1721–1725
101. Jeong H, Mason SP, Barabasi AL, Oltvai ZN (2001) Lethality and centrality in protein networks. *Nature* 411: 41–42
102. Ma HW, Zeng AP (2003) The connectivity structure, giant strong component and centrality of metabolic networks. *Bioinformatics* 19: 1423–1430
103. Dokholyan NV (2005) The architecture of the protein domain universe. *Gene* 347: 199–206
104. Wachi S, Yoneda K, Wu R (2005) Interactome-transcriptome analysis reveals the high centrality of genes differentially expressed in lung cancer tissues. *Bioinformatics* 21: 4205–4208
105. Tieri P, Valensin S, Latora V, Castellani GC, Marchiori M, Remondini D, Franceschi C (2005) Quantifying the relevance of different mediators in the human immune cell network. *Bioinformatics* 21: 1639–1643
106. Hahn MW, Kern AD (2005) Comparative genomics of centrality and essentiality in three eukaryotic protein-interaction networks. *Mol Biol Evol* 22: 803–806
107. Goh KI, Oh E, Jeong H, Kahng B, Kim D (2002) Classification of scale-free networks. *Proc Natl Acad Sci USA* 99: 12583–12588
108. Soffer SN, Vázquez A (2005) Network clustering coefficient without degree-correlation biases. *Physical Review E* 71: 057101
109. Albert R (2005) Scale-free networks in cell biology. *J Cell Sci* 118: 4947–4957
110. Albert R, Barabasi A-L (2002) Statistical mechanics of complex networks. *Rev Mod Phys* 74: 47–97
111. Wagner A, Fell DA (2001) The small world inside large metabolic networks. *Proc R Soc Lond [Biol]* 268: 1803–1810
112. Yook SH, Oltvai ZN, Barabási AL (2004) Functional and topological characterization of protein interaction networks. *Proteomics* 4: 928–942
113. Carter SL, Brechbuhler CM, Griffin M, Bond AT (2004) Gene co-expression network topology provides a framework for molecular characterization of cellular state. *Bioinformatics* 20: 2242–2250
114. Ravasz E, Somera AL, Mongru DA, Oltvai ZN, Barabasi AL (2002) Hierarchical organization of modularity in metabolic networks. *Science* 297: 1551–1555
115. Wei FP, Meng M, Li S, Ma HR (2006) Comparing two evolutionary mechanisms of modern tRNAs. *Mol Phylogenet Evol* 38: 1–11
116. Dartnell L, Simeonidis E, Hubank M, Tsoka S, Bogle IDL, Papageorgiou LG (2005) Robustness of the p53 network and biological hackers. *FEBS Letters* 579: 3037–3042
117. Said MR, Begley TJ, Oppenheim AV, Lauffenburger DA, Samson LD (2004) Global network analysis of phenotypic effects: protein networks and toxicity modulation in Saccharomyces cerevisiae. *Proc Natl Acad Sci USA* 101: 18006–18011
118. Voit E (2000) *Computational Analysis of Biochemical Systems*. Cambridge University Press, Cambridge
119. Ao P (2005) Metabolic network modelling: Including stochastic effects. *Computers & Chem Eng* 29: 2297–2303

120. Holstege FC, Jennings EG, Wyrick JJ, Lee TI, Hengartner CJ, Green MR, Golub TR, Lander ES, Young R (1998) Dissecting the regulatory circuitry of a eukaryotic genome. *Cell* 95: 717–728
121. Tavazoie S, Hughes JD, Campbell MJ, Cho RJ, Church GM (1999) Systematic determination of genetic network architecture. *Nat Genet* 22: 281–285
122. D'haeseleer P, Liang S, Somogyi R (2000) Genetic network inference: from co-expression clustering to reverse engineering. *Bioinformatics* 16: 707–726
123. Wagner A (2001) How to reconstruct a large genetic network from n gene perturbations in fewer than n^2 easy steps. *Bioinformatics* 17: 1183–1197
124. Guthke R, Moller U, Hoffmann M, Thies F, Topfer S (2005) Dynamic network reconstruction from gene expression data applied to immune response during bacterial infection. *Bioinformatics* 21: 1626–1634
125. Cavelier G, Anastassiou D (2005) Phenotype analysis using network motifs derived from changes in regulatory network dynamics. *Proteins* 60: 525–546
126. Luscombe NM, Babu MM, Yu HY, Snyder M, Teichmann SA, Gerstein M (2004) Genomic analysis of regulatory network dynamics reveals large topological changes. *Nature* 431: 308–312
127. Vilar JMG, Guet CC, Leibler S (2003) Modeling network dynamics: the lac operon, a case study. *J Cell Biol* 161: 471–476
128. Tegner J, Yeung MKS, Hasty J, Collins JJ (2003) Reverse engineering gene networks: Integrating genetic perturbations with dynamical modeling. *Proc Natl Acad Sci USA* 100: 5944–5949
129. Wahde M, Hertz J (2000) Coarse-grained reverse engineering of genetic regulatory networks. *Biosystems* 55: 129–136
130. Arkin A, Shen P, Ross J (1997) A test case of correlation metric construction of a reaction pathway from measurements. *Science* 277: 1275–1279
131. Remondini D, O'Connell B, Intrator N, Sedivy JM, Neretti N, Castellani GC, Cooper LN (2005) Targeting c-Myc-activated genes with a correlation method: Detection of global changes in large gene expression network dynamics. *Proc Natl Acad Sci USA* 102: 6902–6906
132. Friedman N, Linial M, Nachman I, Pe'er D (2000) Using Bayesian networks to analyze expression data. *J Comput Biol* 7: 601–620
133. Tamada Y, Kim S, Bannai H, Imoto S, Tashiro K, Kuhara S, Miyano S (2003) Estimating gene networks from gene expression data by combining Bayesian network model with promoter element detection. *Bioinformatics* 19: II227–II236
134. Tchuraev RN, Galimzyanov AV (2001) Modeling of actual eukaryotic control gene subnetworks based on the method of generalized threshold models. *Mol Biol* 35: 933–939
135. Espinosa-soto C, Padilla-Longoria P, Alvarez-Buylla ER (2004) A gene regulatory network model for cell-fate determination during *Arabidopsis* thalianal flower development that is robust and recovers experimental gene expression profiles. *Plant Cell* 16: 2923–2939
136. Shmulevich I, Dougherty ER, Kim S, Zhang W (2002) Probabilistic Boolean networks: a rule-based uncertainty model for gene regulatory networks. *Bioinformatics* 18: 261–274
137. Ramo P, Kesseli J, Yli-Harja O (2005) Stability of functions in Boolean models of gene regulatory networks. *Chaos* 15: 34101
138. Shmulevich I, Kauffman SA, Aldana M (2005) Eukaryotic cells are dynamically ordered or critical but not chaotic. *Proc Natl Acad Sci USA* 102: 13439–13444
139. Kam Z (2002) Generalized analysis of experimental data for interrelated biological measurements. *Bull Math Biol* 64: 133–145

140. Du P, Gong H, Wurtele ES, Dickerson JA (2005) Modeling gene expression networks using fuzzy logic. *IEEE T Syst Man Cy B* 35: 1351–1359
141. Sachs K, Perez O, Pe'er D, Lauffenburger DA, Nolan GP (2005) Causal protein-signaling networks derived from multiparameter single-cell data. *Science* 308: 523–529
142. Xing B, van der Laan MJ (2005) A causal inference approach for constructing transcriptional regulatory networks. *Bioinformatics* 21: 4007–4013
143. Kelley BP, Yuan BB, Lewitter F, Sharan R, Stockwell BR, Ideker T (2004) PathBLAST: a tool for alignment of protein interaction networks. *Nucleic Acids Res* 32: W83–W88
144. Forst CV, Schulten K (1999) Evolution of metabolisms: a new method for the comparison of metabolic pathways using genomics information. *J Comput Biol* 6: 343–360
145. Ogata H, Fujibuchi W, Goto S, Kanehisa M (2000) A heuristic graph comparison algorithm and its application to detect functionally related enzyme clusters. *Nucleic Acids Res* 28: 4021–4028
146. Dandekar T, Schuster S, Snel B, Huynen M, Bork P (1999) Pathway alignment: application to the comparative analysis of glycolytic enzymes. *Biochem J* 343: 115–124
147. von Mering C, Krause R, Snel B, Cornell M, Oliver SG, Fields S, Bork P (2002) Comparative assessment of large-scale data sets of protein–protein interactions. *Nature* 417: 399–403
148. Ideker T, Lauffenburger DA (2003) Building with a scaffold: emerging strategies for high- to low-level cellular modeling. *Trends Biotechnol* 21: 255–262
149. Kelley BP, Sharan R, Karp RM, Sittler T, Root DE, Stockwell BR, Ideker T (2003) Conserved pathways within bacteria and yeast as revealed by global protein network alignment. *Proc Natl Acad Sci USA* 100: 11394–11399
150. Sharan R, Ideker T, Kelley B, Shamir R, Karp RM (2005) Identification of protein complexes by comparative analysis of yeast and bacterial protein interaction data. *J Comput Biol* 12: 835–846
151. Sharan R, Suthram S, Kelley RM, Kuhn T, McCuine S, Uetz P, Sittler T, Karp RM, Ideker T (2005) Conserved patterns of protein interaction in multiple species. *Proc Natl Acad Sci USA* 102: 1974–1979
152. Choy C, Jansson J, Sadakane K, Sung WK (2005) Computing the maximum agreement of phylogenetic networks. *Theor Comput Sci* 335: 93–107
153. Palla G, Derenyi I, Farkas I, Vicsek T (2005) Uncovering the overlapping community structure of complex networks in nature and society. *Nature* 435: 814–818
154. Clipsham R, Zhang YH, Huang BL, McCabe ERB (2002) Genetic network identification by high density, multiplexed reversed transcriptional (HD-MRT) analysis in steroidogenic axis model cell lines. *Mol Genet Metab* 77: 159–178
155. Girvan M, Newman MEJ (2002) Community structure in social and biological networks. *Proc Natl Acad Sci USA* 99: 7821–7826
156. Barabasi AL, de Menezes MA, Balensiefer S, Brockman J (2004) Hot spots and universality in network dynamics. *Eur Physical J B* 38: 169–175
157. Batagelj V, Mrvar A (2003) Pajek – Analysis and visualization of large networks. In: M Jünger, P Mutzel (eds): *Graph Drawing Software*. Springer, Berlin, 77–103
158. Shannon P, Markiel A, Ozier O, Baliga NS, Wang JT, Ramage D, Amin N, Schwikowski B, Ideker T (2003) Cytoscape: A software environment for integrated models of biomolecular interaction networks. *Genome Res* 13: 2498–2504
159. Junker BH, Klukas C, Schreiber F (2006) VANTED: A system for advanced data analysis and visualization in the context of biological networks. *BMC Bioinformatics* 7: 109
160. Hu Z, Mellor J, Wu J, Yamada T, Holloway D, DeLisi C (2005) VisANT: data-integrating visual framework for biological networks and modules. *Nucleic Acids Res* 33: W352–W357

161. Maere S, Heymans K, Kuiper M (2005) BiNGO: a Cytoscape plugin to assess overrepresentation of gene ontology categories in biological networks. *Bioinformatics* 21: 3448–3449
162. Koschützki D, Lehmann KA, Peeters L, Richter S, Tenfelde-Podehl D, Zlotowski O (2005) Centrality Indices. In: U Brandes, T Erlebach (eds): *Network Analysis. LNCS Tutorial 3418.* Springer, 16–61
163. Yu HY, Zhu XW, Greenbaum D, Karro J, Gerstein M (2004) TopNet: a tool for comparing biological sub-networks, correlating protein properties with topological statistics. *Nucleic Acids Res* 32: 328–337
164. Ludemann A, Weicht D, Selbig J, Kopka J (2004) PaVESy: Pathway visualization and editing system. *Bioinformatics* 20: 2841–2844
165. Toyoda T, Konagaya A (2003) KnowledgeEditor: a new tool for interactive modeling and analyzing biological pathways based on microarray data. *Bioinformatics 19: 433–434*
166. Breitkreutz BJ, Stark C, Tyers M (2003) The GRID: the General Repository for Interaction Datasets. *Genome Biol* 4: R23
167. Breitkreutz BJ, Stark C, Tyers M (2003) Osprey: A network visualization system. *Genome Biol* 4: R22
168. Kanehisa M, Goto S, Kawashima S, Okuno Y, Hattori M (2004) The KEGG resources for deciphering the genome. *Nucleic Acids Res* 32: D277–D280
169. Hwang D, Rust AG, Ramsey S, Smith JJ, Leslie DM, Weston AD, Atauri PD, Aitchison JD, Hood L, Siegel AF et al. (2005) A data integration methodology for systems biology. *Proc Natl Acad Sci USA* 102: 17296–17301

Plant Systems Biology
Edited by Sacha Baginsky and Alisdair R. Fernie
© 2007 Birkhäuser Verlag/Switzerland

Current challenges and approaches for the synergistic use of systems biology data in the scientific community

Christian H. Ahrens*, Ulrich Wagner*, Hubert K. Rehrauer, Can Türker and Ralph Schlapbach

Functional Genomics Center Zurich, Winterthurerstrasse 190, Y32H66, CH-8057 Zurich, Switzerland
* equal contribution

Abstract

Today's rapid development and broad application of high-throughput analytical technologies are transforming biological research and provide an amount of data and analytical opportunities to understand the fundamentals of biological processes undreamt of in past years. To fully exploit the potential of the large amount of data, scientists must be able to understand and interpret the information in an integrative manner. While the sheer data volume and heterogeneity of technical platforms within each discipline already poses a significant challenge, the heterogeneity of platforms and data formats across disciplines makes the integrative management, analysis, and interpretation of data a significantly more difficult task. This challenge thus lies at the heart of systems biology, which aims at a quantitative understanding of biological systems to the extent that systemic features can be predicted. In this chapter, we discuss several key issues that need to be addressed in order to put an integrated systems biology data analysis and mining within reach.

Introduction

Today's rapid development and broad application of high-throughput analytical technologies are transforming biological research. The reductionist approach of studying individual or a few genes or gene products and their function in intricate detail, as it has been practiced for most of the last century, is shifting towards a global approach, where a large portion or even all molecular elements of an organism (genes, proteins, metabolites, and other molecular species) can be studied in parallel. This paradigm shift has been catalyzed by the availability of an increasing number of complete genome sequences. As a consequence, enormous amounts of data are being generated. Genomics technologies like DNA sequencing and gene expression analysis have led the way and can be considered established standard in many research projects. In contrast, large scale proteomics and metabolomics technologies have matured only recently. However, already at this stage they often produce several-fold greater amounts of analytical data than genomics technologies, even though a concurrent analysis of all elements is not yet feasible. Information thus has become both the bounty and the bane of life science laboratories. This flood of data gives scientists opportunities undreamt of in past years to understand the fundamentals of biological processes, to study the regulation of individual components or entire pathways in health *versus* disease, and to explore the effects of compounds with potential as therapeutic drugs. Yet, to realize those opportunities, scientists must be able to understand and interpret the information in an integrative manner. While the sheer data volume and heterogeneity of technical platforms within each discipline already pose a significant challenge, the heterogeneity of platforms and data formats across disciplines makes the integrative management, analysis, and interpretation of data a significantly more difficult task.

This challenge thus lies at the heart of systems biology, which aims at a quantitative understanding of biological systems to the extent that systemic features can be predicted. A typical systems biology workflow includes:

1. standardized qualitative and quantitative data collection and management
2. proper data integration allowing comparative evaluation
3. modeling of the experimental situation, and
4. perturbation of the systems and prediction of the outcome [1].

These steps have also been referred to by Douglas Lauffenburger as *"the four M's of systems biology: measurement, mining, modeling, manipulation"* [2]. Today, the large scale experimental datasets required for step 1 frequently include genomic information, single nucleotide polymorphism (SNP) data, gene expression, protein expression and protein interaction data, as well as metabolite data. In addition, integration of data other than of analytical origin including scientific literature or pathway information, sub-cellular localization or other image data can greatly increase the significance of a finding and put it into its proper biological context.

In this chapter, we discuss several key issues that need to be addressed in order to put a systems biological data analysis and mining within reach. We focus on the large scale experimental data collection and management and data integration steps. A detailed depiction of modeling approaches that use the acquired data for the prediction of systems behavior will follow in the next two chapters. In order to provide widely usable information, we describe in the following general concepts that are relevant to all scientific fields, but will stress plant systems biology specific efforts wherever suitable.

The need for data standards

The collection of large scale experimental datasets from several functional genomics technologies and their subsequent integration is an essential initial step towards a systems biology perspective. It promises to provide novel insights that cannot be gained through analysis of datasets originating from any specific technology platform alone. Since these datasets are typically quite heterogeneous in terms of data type, comprehensiveness, quality and semantics and are usually stored in a multitude of heterogeneous and autonomous repositories, adequate strategies for standardization at various levels are critical.

Already when focusing on one specific functional genomics technology platform, data heterogeneity can pose a significant challenge for research institutions. Among the first to experience the pitfalls of an unstructured, non-standardized accumulation of large amounts of data were the molecular biologists that set out to exploit large DNA sequence databases. The example of Genbank, an important data repository for storage and exchange of DNA sequence information, illustrates how the few restrictions that were initially made with respect to data format and, more importantly, to annotation and description vocabulary, had massive influence on the usefulness and quality of the resource [3]. The sequence information originated from only relatively few massive sequencing projects and many independent, loosely organized but highly focused sequencing approaches carried out by different labs all over the world. As a consequence, the sequence data in Genbank is highly redundant and lacks standardized schemes for gene names and descriptions. By concentrating on a few single genes or gene families researchers in a specific field could still keep track of their data. Eliminating redundancies and agreeing on naming schemes or at least mapping redundant gene names to each other has been tedious but was feasible in a limited context. For a systematic approach however, manual curation and analysis of data is impossible. Unstructured and non-standardized sequence information is hardly exploitable by computer algorithms. Therefore, the first projects that brought a minimal degree of standardization in sequence information were of great importance for the plethora of later developments in the field. The RefSeq initiative of the National Center for Biotechnology Information (NCBI), the host institution of Genbank [4], is one of these DNA sequence curation efforts. Even prior to this effort, the establishment of Swiss-Prot [5], represented a cornerstone in bioinformatics. Swiss-Prot is a non-redundant, highly curated database for protein sequence information that is structured according to its own standards for format and content. Examples of the annotation fields that are provided for each record include protein name (including aliases and synonyms), protein descriptions, literature annotations, sequence features like functional domains and protein modifications, and database cross-references.

The generation and content of genome sequence data is relatively easy to describe at a technical level. The resulting data are uniform in structure and the principles of the analytical techniques do not differ significantly. The remaining molecular categories (genes, transcripts and proteins) in functional genomics however are more complex. For example, a wide range of platforms can be used to determine gene expression on the transcriptome level. As a consequence, a wealth of publications exists that describe the comparability, or the lack thereof, of data obtained from different platforms with quite heterogeneous results and conclusions. Interestingly, Jarvinen et al. report that already differences in the annotation of target sequences and respective probes can lead to a lack of comparability between the experimental results from different transcriptomics platforms [6]. The use of standardized gene sequence annotations would help to dramatically reduce this source of error.

The challenge of data heterogeneity becomes more pronounced when moving towards an integration of different technologies in order to enable a global systematic view [7]. To allow the integration and comparison of data originating from both geographically and technologically disparate technology platforms, standardization of data needs to be employed at different levels:

- Using controlled vocabularies, which involve definition of the naming schemes of entities and their relationships: In computer science, particular hierarchically structured concepts, called ontologies, have been widely used to define controlled vocabularies. Ontologies have been increasingly popular in the field of functional genomics.
- Capturing of the detailed experimental information and instrument settings: This is an often overlooked key factor for understanding and reproducing an experiment, and for comparison of data from the same experimental source and platform. This set of information is an absolute requirement for comparisons across platforms and technologies.
- Structuring of the data must be done according to generally accepted standards (e.g., following the MAGE-OM database scheme): As different projects have very different needs, restrictions and resources, it is unlikely that this type of standardization will ever be fulfilled to the perfect end. Nonetheless, standardized data has to be provided in order to allow computer programs to carry out mappings, or translations, between the different formats in which the data are stored and structured.
- Defining easily exchangeable data formats: Today, data formats including the XML file formats MAGE-ML (for microarray data) or mz-XML (for proteomics data) have led the way to large standardization efforts.

The higher the level of standardization, the more analytical data becomes amenable to computational approaches for data management, analysis and knowledge discovery. In parallel, standardization is also a prerequisite for the establishment of infrastructures that allow the scientific community to store, share, and exploit the massive data content. To realize the discussed benefits, standardization requires from the research community a willingness to find and adhere to a consensus. In addition, the dialog between instrument manufacturers and researchers with the aim to enable standardized exchange of the data formats will greatly enhance the ability to integrate and subsequently mine data coming from different instrument platforms in a given technology field [7].

Ontologies

The complex nature and classification of biological data

Several factors complicate the effective handling, exchange and modeling of biological data. Firstly, biological data very often are described using ambiguous terms and even researchers working in the same field rarely agree completely on how to define objects (like gene names) and relationships between them. Therefore, many synonymous expressions exist for identical objects and relationships. On the other hand, objects and relationships with different or partially different meaning are often named in a homonymous way [8]. Secondly, biological information is typically not complete at the time when categories for its classification and hierarchical organization are first established. The incomplete and dynamic nature of biological data, requires that categories and concepts need to be changed and terminologies have to be revised continuously [9] in order to combine both well-established knowledge and the latest findings in an integrated manner.

Integrating biological knowledge with molecular data from functional genomics experiments in a standardized way should allow researchers to efficiently exchange and explore their findings [10]. The more comprehensive such a standardization, the better integrated datasets can be further exploited and interpreted by computational means, e.g., for data mining or inference calculations. One promising and widely used method to achieve such standardization is to set up an ontology. An ontology represents a collection of common terms, the meaning of the terms and the formal relations between the terms as agreed upon by a group of experts in a respective field. In other words, an ontology embodies a field of formalized knowledge that should be as explicit and complete as possible. Within an ontology, individual terms and concepts are defined by a set of statements that connect them to other terms and concepts with structuring rules using 'description logic' [11]. This description logic can be implemented and visualized as a Directed Acyclic Graph (DAG), as shown in Figure 1. DAGs are well suited to represent multiple hierarchical relationships as each 'child' term can have more than one 'parent' term. In the example of Figure 1 the term 'transport' is part and child of the terms 'cellular physiological process' and 'establishment of localization'.

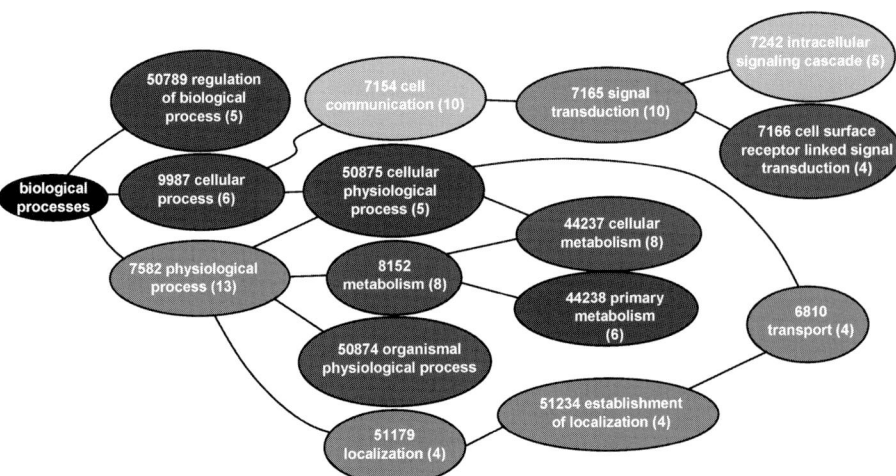

Figure 1. Example of a NetAffx Gene Ontology Mining Tool output [93]. The example displays the terms and relationships for a subset of an ontology, represented as a Directed Acyclic Graph.

More elaborate definitions for an ontology exist in the field of computer science, where their use is well-established [12]. However, for the purpose of this article, we will emphasize their use rather than technical aspects.

Examples of biological ontologies

One major factor that determines the usability of an ontology is a sociological one: it has to be generally accepted by a respective user community, since the quality of the ontology depends on a strong commitment of the community during the tedious setup, as well as the subsequent further development and curation stages of the ontology. The Open Biomedical Ontologies (OBO) consortium represents an important step towards an international repository of biological standards and ontologies [13, 14]. OBO serves as an inventory of and a link to well-structured controlled vocabularies for shared use across different biological and medical domains.

The effort of the Plant Ontology (PO) consortium to develop and curate ontologies that describe plant structures and growth/developmental stages is in close alignment with the OBO [15]. Rather than setting up a large collection of vocabularies, the main interest of the PO consortium is to describe the denotation and the relationship of the terms, and thus to integrate diverse vocabularies used in plant anatomy, morphology and growth and developmental stages. At their website (www.plantontology.org), any node (i.e., term within the ontology) can be selected with the ontology browser, which will reveal the associated plant structures or growth and developmental stages (Fig. 2) in the DAG structure.

The members of the PO consortium adopted and extended the main concepts set up by the Gene Ontology consortium [15], which is the implementation of one of the most widely used and best established concepts of a bio-ontology. The Gene Ontology aims to provide standardized and controlled vocabularies for genome annotation. These vocabularies are organized in three main categories which are structured as mutually independent hierarchies [16, 17]. These categories include (i) biological process and the genes involved therein, (ii) molecular function of the genes and their products, and (iii) sub-cellular localization of the gene products. In both prokaryotes and eukaryotes many biological functions are carried out by homologous genes. It is therefore possible to provide consistent descriptors for gene products in different databases and to standardize classifications for sequences and sequence features throughout the set of prokaryotic and eukaryotic organisms. By the end of 2005, the GO covered around 20,000 terms and their relationships with almost 170,000 genes mapped to them.

Applications of bio-ontologies

If an ontology is linked to OBO, an identifier (ID) is attributed to each entry in the ontology. This ID can be used to connect entries in a database to an ontology or to connect entries in two or several databases. Such links can be exploited to standardize entries in a database, which in return helps to more easily and efficiently exchange,

Current challenges and approaches for the synergistic use of systems biology data   283

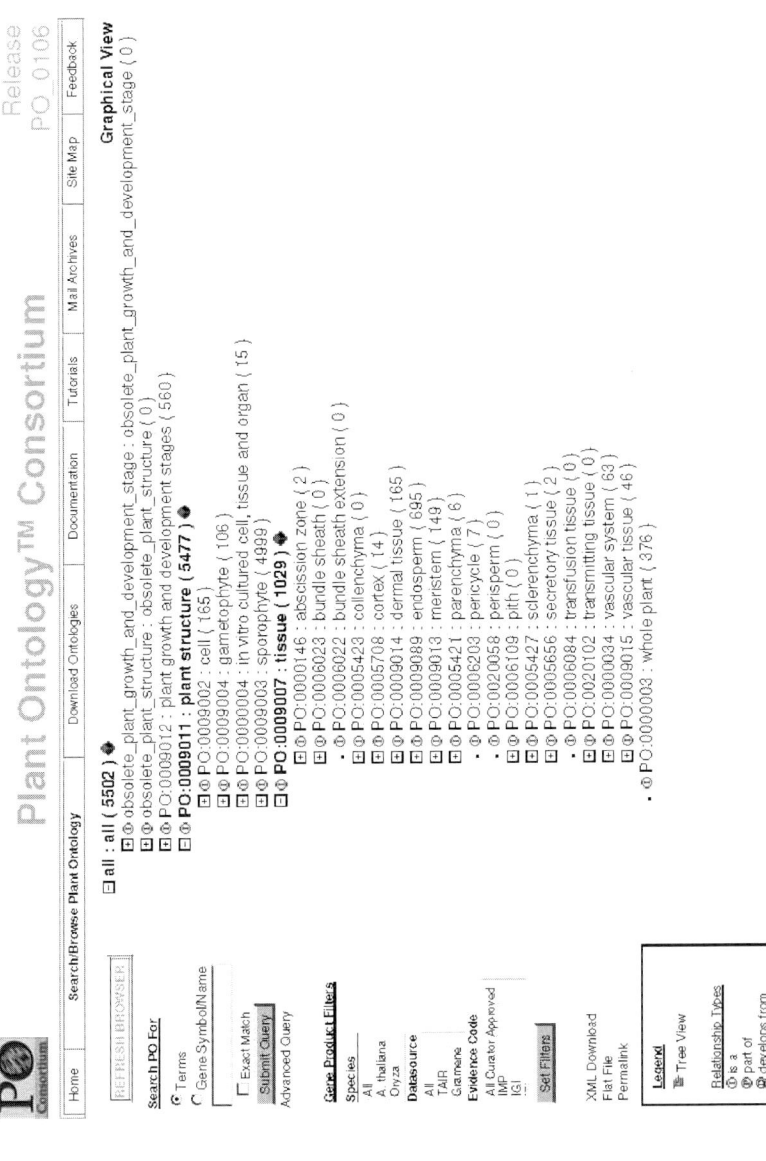

Figure 2. Example of a plant structure ontology available through the Plant Ontology browser (www.plantontology.org). For the term tissue, all associated and defined subcategories with their respective identifier are displayed.

reproduce and interpret data. As computer-interpretability is key for the setup of ontologies, cross-database queries can be carried out effectively when the ontologies are mapped to each other. In the example of Plant Ontology and the possible integration with Gene Ontology, one of the potential aims could be to combine aspects of evolution and taxonomy with information about the genetic equipment of certain species or with information about gene expression. A first and promising approach of such an integration effort in the field of plant biology is the Genevestigator project [18].

Combining GO annotations with gene expression data has recently become very popular and has developed into a standard element of gene expression data analysis. If we follow for illustrative reasons the data analysis of a typical microarray experiment in a workflow-based manner, a first result is often a list of genes that are significantly up- or down-regulated between different conditions. Determining significantly overrepresented GO categories within this gene list can give a fairly refined picture about which functions, biological roles and sub-cellular locations are mainly affected by a change of the experimental conditions [19]. A panel of web-based bioinformatics tools, such as Onto-Express [20, 21] or GOTM [22], and standalone tools, such as ermineJ [23] or BiNGO [24], have been developed to provide this type of information and are available to the scientific community. Very recent developments show that use of Gene Ontology vocabulary can improve established methods for microarray data analysis. As an example, the goCluster program allows to integrate annotation information, clustering algorithms and visualization tools [25]. Lottaz and Spang implemented an algorithm that exploits functional annotations from the Gene Ontology database to build biologically focused classifiers [26]. These classifiers are used to uncover potential molecular disease sub-entities and associate them to biological processes without compromising overall prediction accuracy.

Controlled vocabularies as defined by Gene Ontology are increasingly used for the functional classification of genes and gene products. For example, efforts have been undertaken to associate, as comprehensively as possible, all *Arabidopsis* genes with GO terms [27]. This type of information can help to establish the functional annotation of genomes of newly sequenced organisms. Comparative genomics approaches can be further facilitated by quantitatively assessing similarities and dissimilarities of sets of genes or even genomes [28]. The database Gramene [29] is a good example of how such a comparative genome analysis can be successfully conducted in different grass species after different types of information (genetic and physical mapping data, gene localization and descriptions of phenotypic characters and mutations, genomic and EST data, protein structure and function analysis, interpretation of biochemical pathways) have been integrated with controlled GO vocabularies [30, 31].

Summary

Ontologies have been successfully applied to many different areas of biological research. Data that is annotated in a standardized and commonly accepted scheme facilitates a better understanding of even large datasets by researchers, and makes them amenable to computational exploitation. This is critical for experimental approaches that employ 'Omics' techniques, which produce data at a large scale. It can be foreseen that data repositories such as public databases and publishers will require the association of data with controlled vocabularies [10]. However, from a technical point of view, improvements in the field of bio-ontologies are still needed, as many of them apparently do not conform to the international standards for ontology design and description [32].

Current standardization approaches in transcriptomics

Quite early after the advent and increasing use of microarrays for gene expression measurements, researchers recognized the benefit of sharing and synergistically using expression data generated all over the world [33]. This triggered the foundation of the Microarray Gene Expression Data society (MGED). Since then, MGED has done pioneering work in establishing standards for the description and the exchange of microarray data, thereby laying the foundation for the big popularity of the public microarray data repositories ArrayExpress and Gene Expression Omnibus.

MGED's two main contributions to the scientific community in the years 1999–2002 were the establishment of guidelines for experiment annotations, the Minimum Information About Microarray Experiments (MIAME) standard, and the Microarray Gene Expression Object Model (MAGE-OM) together with the associated Microarray Gene Expression Markup Language (MAGE-ML). With these guidelines, MGED tackled several inherent problems when sharing microarray gene expression data:

- How to describe biological samples, their conditions, treatments, and analysis in a standardized way?
- How to characterize microarray measurements?
- How to exchange microarray data between different technological platforms and software packages such that the results can be compared?

MIAME: Minimal Information About a Microarray Experiment

The most challenging and most critical guideline is the set of MIAME requirements. These describe the data and associated meta-information that is necessary to unambiguously analyze a gene expression dataset. Obviously this includes the ability to reproduce and verify results that have been derived from an expression dataset. MGED has deliberately chosen to keep the MIAME requirements informal, i.e., MIAME does neither provide a controlled vocabulary for the meta-information nor does it include any definition of ontologies for sample, experiment, etc. Instead, the MIAME checklist [34] is specified in simple text form. It groups the required information into six parts:

1. Experimental design
2. Array design
3. Samples
4. Hybridizations
5. Measurements, and
6. Normalization controls.

It also includes a description of the kind of information the respective categories have to contain.

The general formulation of the MIAME requirements makes them applicable to a wide range of microarray applications and to many different organisms. A disadvantage is that it does not give enough detailed guidance for some specific fields. In order to address this shortcoming, the MIAME has been extended to be more specific for plants [35]. This extension includes ontology terms for an accurate characterization of growth stages and plant organs. Other extensions have been made to better support toxicogenomics applications [36].

While MIAME is a mere requirement list, a solution for how to deal with microarray data in a way that satisfies the MIAME requirements is specified by the MAGE-OM together with the MGED Ontology (MO) and the MAGE Markup Language (MAGE-ML).

MAGE-OM: The MAGE Object Model

The MAGE-OM models microarray experiments in a MIAME-compliant way using the Unified Modeling Language (UML). The MAGE-OM defines objects like Array, Hybridization and HybridizationProtocol and how they are related to each other. Figure 3 shows a sub diagram of the MAGE-OM. Basically, MAGE-OM provides a framework for the structured, machine-interpretable representation of microarray experiments.

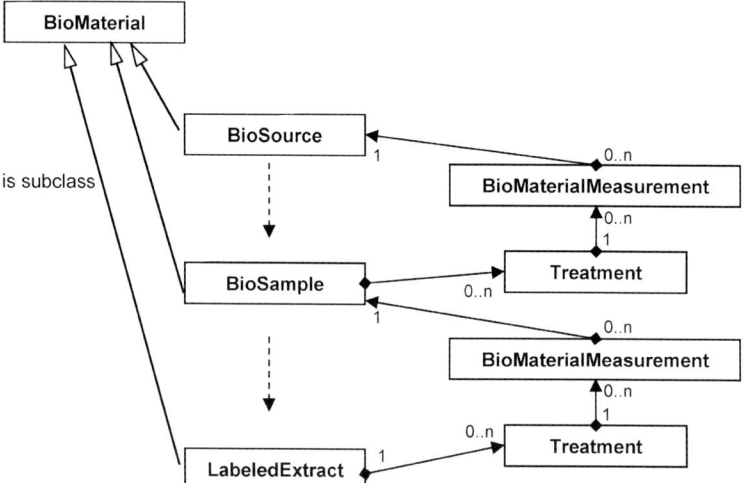

Figure 3. Excerpt of the MAGE-OM with the objects BioSource, BioSample, and LabeledExtract as subclasses of the object BioMaterial. These objects describe the respective organism from which a tissue sample is drawn and from which labeled RNA is generated. The arrows show the relationships between the objects. Numbers at either end of the arrows indicate how many instances of one object type can have a direct relationship to how many instances of the other object type.

MAGE-ML: The MAGE Markup Language

MAGE-ML [37] defines an XML format for the storage and transmission of microarray expression data. An XML-file is a text-file where content elements are encapsulated by tags, just like in HTML (which is actually a specific implementation of XML). In XML, the set of allowed tags and their meaning can be specified in a DTD file or in an XML schema. For the representation of microarray experiments, the set of tags is given by the MAGE-ML.DTD which is provided by the Objects Management Group [38]. Presently, most of the microarray analysis and storage systems can handle microarray expression data in MAGE-ML format. For people who want to implement their own microarray software, MGED provides a toolkit (MAGE-STK) for reading and writing of MAGE-ML files.

MO: The MGED Ontology

The MO finally provides the standard terms for the annotation of microarray experiments. For example, it tries to comprehensively define all allowed terms for the category sex of an individual: F, F$^-$, Hfr, female, hermaphrodite, male, mating type a, mating type alpha, mating type h$^-$, mating type h$^+$, mixed sex, unknown sex. By strictly adhering to these terms to characterize the sex, this annotation can easily be computer-processed. Unfortunately not all terms of the MO are fully elaborated within the MAGE-OM. However, the overall structure of MO is consistent with MAGE-OM. The definitions within MO enable the unambiguous annotation as well as structured queries of microarray data using the ontology annotation. They guarantee that data semantics remain unchanged when exchanging expression data between different systems. It is one of the open biomedical ontologies (OBO).

Summary

MGED has established scientific community standards and resources for describing, sharing, and integrating microarray data. These standards do not only cover the actual measurement data, but also the annotation of biological samples as well as information about the respective probe sequences. The incorporation of these three domains and their respective standards made the MGED initiative so useful and successful. The MGED standards are followed and implemented by the major data repositories as well as the major commercial and public domain software systems. A microarray experiment that is represented as a MAGE-OM cannot only be analyzed automatically; it can also be integrated with other microarray experiments that are represented according to the MAGE-OM.

Despite all the benefits that MGED has created there are also some drawbacks. These are due to the fact that the MGED initiative was the first of its kind and touched unknown territory. Retrospectively, one realizes that many of the definitions and standards are useful, but suffer from insufficient rigor [32]. This is also recognized by the MGED society, who state on their web-site (www.mged.org) that the "*boundaries between MIAME concepts, the MIAME-compliant MAGE-OM and the MGED ontology, that try to define and structure the MIAME concepts, are neither well defined nor easy to understand*" [39].

The success of MGED's work on the establishment of data standards has triggered similar initiatives in proteomics and plant metabolomics [40]. Both fields benefit from the experience gained within the transcriptomics field. The MGED society continues the work on improving and extending the established standards. Currently, the focus is on extension of the concepts to toxicogenomics, *in situ* hybridizations, and immunochemistry experiments and on incorporation of the respective ontologies.

Current standardization approaches in proteomics

After description of the well-established standardization initiatives in transcriptomics, we provide a concise summary of similar, however less advanced efforts in proteomics. We also briefly touch upon some of the additional challenges in this field. The emerging standardization efforts in the plant metabolomics field [40] will not be described here.

PSI: The Proteomics Standards Initiatives

The proteomics community also responded to the urgent need for standardized approaches. The Human Proteome Organization (HUPO), founded in 2001 with the aim to unify national and regional proteomics societies and to work on common guidelines, laid out a first set of initiatives [41], among them the Proteomics Standards Initiative (PSI). The PSI workgroup was formed to define and set up proteomics standards in order to enable an accurate description of data, centralized data storage and exchange of data between researchers and centralized repositories [42]. The efforts are focused mainly on three areas:

(i) capture of general proteomics standards (GPS), including a broad proteomics data model, Minimum Information About a Proteomics Experiment (MIAPE), and an ontology (PSI-Ont)
(ii) molecular interaction standard (PSI-MI)
(iii) mass spectrometry standards (PSI-MS)

General proteomics standards
While PSI could benefit greatly from the work of MGED, the proteomics field faces additional challenges. In particular, the definition of the MIAPE requires to capture a larger set of metadata [43] since proteomics data is much more context-dependent than transcriptomics data. Protein levels change rapidly and do not necessarily correlate with gene expression levels, and the roughly 300 different posttranslational modifications can occur in various combinations [44]. Thus proteomics experiments provide a snapshot in time of the biological sample under study [45]. The much

more heterogeneous biochemical properties of proteins compared to nucleic acids, the much higher complexity of the proteome and the several orders of magnitude greater dynamic range of protein expression [46] make proteomics a challenging enterprize.

MIAPE is designed as a broad data model that can accommodate both 2-D gel based and multi-dimensional liquid chromatography tandem mass spectrometry (LC-MS/MS) based approaches. To make the task more manageable, the PSI decided to focus on development of PSI-MI and PSI-MS, while developing GPS alongside [43]. The PSI plans to develop an ontology (PSI-Ont) that supports standard data formats like mzData (see below). Ultimately the PSI ontology will form a part of the Functional Genomics Ontology (FuGO).

PSI-MI: The PSI Molecular Interaction Standard
The function of protein complexes is context-dependent, and can change depending on the associated proteins, and even in a temporally regulated fashion [47]. Protein–protein interaction data thus can add significant value to systems biology studies. The molecular interaction standard that describes these interactions is the most advanced of the PSI efforts and has been published [48]. A consortium of major public interaction database providers that include DIP [49], MINT [50], MPact [51] and IntAct [52] has agreed to adopt the PSI-MI standard and to enable researchers to download data from their website [53] in this format. These repositories provide access to interaction data from several model organisms. The amount of interaction data for plant species, however, is so far only minimal.

It is envisioned that the PSI-MI standard will be extended to include other types of interacting molecules, such as RNA, DNA and small molecules [48].

PSI-MS: The PSI Mass Spectrometry Standard
The mass spectrometry standard is being actively developed with major contributions coming both from the HUPO-PSI group, and a separate consortium of academic and commercial labs. Two standards have been implemented so far: PSI's mzData and mzXML of the second group [54]. MzData is a data format that aims at uniting the large number of current formats (pkl, dta, mgf, etc.) into one. Importantly, since it captures processed data in the form of peak lists, it is not a substitute for the raw file formats of the respective instrument vendors. It is supported by many instrument vendors (for the conversion of raw files to mzData) and database search engine vendors.

The mzXML data standard was designed to capture a more detailed set of information, including the raw data, and draws on XML's advantages of portability and extendability [54]. Importantly, it was designed to allow to execute some limited analyses on the acquired mass spectrometry data as well, and thus satisfies additional requirements such as speed of access to individual scans. To enable fast access, an index of all MS/MS scans is included.

While initially designed to address different issues, the standards have come closer to each other and both groups have agreed to work on one common future standard.

Summary

Much work remains to be done in the establishment of standards in the proteomics field. Only recently, additional standardization initiatives were started by the PSI to address the description of posttranslational modifications and the 2-D gel electrophoresis workflow [55]. Proteomics technologies are developing at a rapid rate, and the more traditional 2-D gel-based proteomics approaches that have been practiced since the mid 70s are more and more complemented and replaced by multi-dimensional liquid chromatography tandem mass spectrometry (LC-MS/MS) based approaches, also called shotgun proteomics. These hold exceptional potential to dig deeper into the complex proteome and to overcome some of the drawbacks of 2-D gel-based analysis, which include identification of both low-abundance proteins and membrane proteins, the protein class of highest interest for the pharmaceutical industry. However, since the complex protein samples are enzymatically digested into peptides prior to mass spectrometric analysis, especially the last step of assigning identified peptides back to the protein sequences is computationally challenging and not unambiguous [56]. The need to establish standards for these approaches and guidelines on how proteomics results should be published has been identified [57]. Key contributions in this area are tools like Peptide and Protein Prophet that assign probability values to peptide or protein identifications [58, 59], and which have helped the significant increase and growing impact of shotgun proteomics studies. A more detailed description of the bioinformatics aspects of shotgun proteomics have been reviewed recently [60].

The latest technological approach to large scale protein identification and characterization – top down proteomics using high accuracy Fourier transform mass spectrometers for the identification of complete proteins – features even less data standards but bears the promise to reduce data complexity as it abolishes the need to computationally assign peptide sequences to the corresponding proteins after the analysis. Irrespective of the analytical approach used, standardization will be a prerequisite for future data integration and data exchange within the scientific community.

Data management, distribution and repositories

A true systems biology approach aims to integrate data from transcriptomics, proteomics and additional functional genomics platforms. Standardization efforts for the production and management of transcriptomics or proteomics data will be beneficial for such an integrated view. However, for the integration of data from different platforms several additional issues have to be solved. Furthermore, strategies for the integrated storage and efficient querying of the data, such as a federated database or a central data warehouse approach have to be chosen. We present one solution for this kind of data integration as it is currently being implemented by the Functional Genomics Center Zurich. We also describe selected publicly available systems for storage of transcriptomics and proteomics data along with associated analysis capabilities, respectively.

Current challenges and approaches for the synergistic use of systems biology data   291

Data integration as basis for the synergistic usage of data

To provide the technical basis for a synergistic usage of system biology data, various forms of heterogeneity have to be overcome. With respect to data storage the following issues have to be considered:

- Since scientific data is generated, processed, and analyzed by different (kinds of) instrument PCs, data is inherently distributed.
- Huge amounts of generated data (often more than several Terabytes) must be stored and retrieved efficiently.
- Large parts of the data are unstructured and undocumented.
- Redundant data is processed on several instrument PCs or servers, usually relying on heterogeneous (instrument-specific) data formats.
- Scientific data is often only available in form of proprietary data files, complicating the post-processing of the data.
- Biological data is inherently context-sensitive. The conditions that existed at the time of data generation have to be captured.

In life sciences research environments, a physical (tight) integration of the required data types into one global database is demanding due to various reasons. First of all, the data is inherently distributed over a number of data resources. These data sources are maintained by autonomous organizations, applying their own rules how to access and treat 'their' data. In addition, these data sources are usually heterogeneous with respect to data representation forms (structured database entries, XML documents, object graphs, flat files, etc.) and/or data access interfaces (Web forms, interactive SQL, simple file operations, etc.). Above that, the data itself may be represented completely differently, for instance, with respect to its naming or structuring [61]. Moreover, the local data sources usually are dynamic. They evolve both in terms of datasets (e.g., new rows in database tables) as well as data schemata (e.g., new or altered table definitions and ontologies, respectively). To correctly reflect such local changes, the integrated global database must be updated permanently. This however assumes that the local data sources can always be monitored – which is often not the case in practice. Despite these challenges, a number of companies have successfully applied the database warehouse approach to integrate disparate datasets. This was facilitated by the fact that they control a number of proprietary heterogeneous data sources and that they have the necessary manpower for the tedious database schema development, data collection, and continuous update tasks.

Clearly, a synergistic usage of system biology data must rely on a uniform access to such heterogeneous resources. System-crossing queries shall be supported in a user-friendly way by transparently identifying and resolving relationships and conflicts (synonyms, homonyms, etc.) between the various data resources. An example for such a system-crossing query might be: "*Query for a given protein or gene, show all relevant evidence from the literature that implies this protein/gene in a biological process (indexing gene names, symbols, aliases and extraction of information from literature, combination of search-terms), and display the respective gene and protein expression levels.*" Such queries require a semantical linking

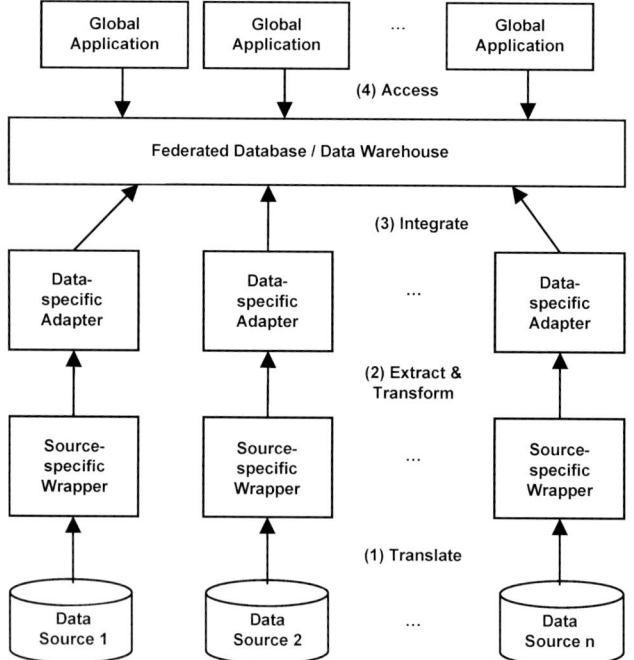

Figure 4. Mapping layers common to a federated database and data warehouse approach. The integration layer allows global applications, such as data analysis and visualization tools, to transparently access integrated data without any knowledge of the detailed structure of the underlying data sources.

among the various data resources. Standardization efforts such as MGED or the Systems Biology Markup Language (SBML) [62] are essential to solve the problem of semantically linking different data resources.

Queries over system biology data are often formulated in an *ad hoc* style, exploiting query refinement as means to incrementally get closer to the desired query results. Since these queries can also become complex and computationally expensive, efficient combinations of information retrieval (text-based search) and database search techniques (structured querying) have to be combined [63]. The latter requirement is essential in the field of life sciences research because large parts of the data (e.g., annotations of experiments or samples) are only available in unstructured text format. Currently, to perform a complex query, a scientist needs to break down the query into sub-queries targeted to the appropriate sources and integrate the results retrieved. This is very demanding since the scientist need not only be able to (technically) access the various data sources but also to correctly interpret the individual query results.

Federated databases and data warehousing approaches allow to hide this complexity from the users [64, 65]. As sketched in Figure 4, they provide transparent access to data from different resources. Without such an integration layer, all global applications would need to know the detailed structure of the corresponding local data resources.

The following layers in the presented architecture provide the required mappings:

1. The wrappers are programs or scripts that are used to overcome the syntactic and conceptual heterogeneity of the various data sources. The wrappers translate the data into a common language, i.e., the language of the federated schema or data warehouse. This language might be for example SQL, XML (eXtensible Markup Language), or RDF (Resource Description Framework). Ideally, these wrappers are provided by the owner of the data sources, which however is usually not the case.
2. The adapters extract and transform the part of the local data sources that are relevant for the global applications.
3. The federated database or the data warehouse integrates the extracted and transformed data. While in the federated database approach this integration is only virtually on schema basis, the data warehousing approach performs a physical integration. In the federated database approach, the data stays in the local repositories and are brought together at run-time depending on the queries. In the data warehousing approach, the corresponding data is loaded (copied) into a central data warehouse. Additional mechanisms are required to maintain the consistency of the data warehouse, and make sure that the data is up to date. The critical task in both approaches is to define the global schema and to resolve conflicts among the various data resources. One example for such conflicts is a naming conflict, e.g., the same gene is named differently in two data sources. Languages like SQL do not provide explicit constructs to state and resolve such conflicts. 'Same as' relationships between data objects can only be formulated intricately by introducing additional tables managing the corresponding information. RDF provides a more promising framework to better capture the semantics of local data resources. All data is represented as graphs. The data of different resources can easily be put together by simple union of graphs. Conflict resolution can then be performed by introducing additional edges between the nodes of the united graph, e.g., an edge 'same as' between two nodes representing the same gene.
4. Global applications such as data analysis or visualization tools are built on top of the federated database or data warehouse, respectively. Thus, they do not have to know the structure of the various local data sources.

Current commercial and research prototype systems tackling the problem of federated data storage and search require much handwork to write wrappers and adapters to access and integrate the various data sources. As briefly sketched in the next sections, a number of systems exist that support specific applications in the area of transcriptomics and proteomics, especially with respect to the storage of scientific data. These systems mainly rely on an integrated database solution with some added data warehousing functionality.

As an example for a system that supports federated data storage and search, we briefly describe the architecture of the system being built at the Functional Genomics

Center Zurich (FGCZ). In our scenario, there are a number of technology platforms and instruments generating huge amounts of raw data. Furthermore, there are internal as well as external data sources providing (partly) structured or semi-structured data. The architecture depicted in Figure 5 shall provide a framework to capture all relevant data that accumulate during the process of an experiment, starting with the preparation of the sample and ending with the data analysis. A central goal is to maintain the heterogeneous and often undocumented data generated by the different instruments in a searchable fashion. For that, the integration layer includes a workflow engine that supports the scientists with appropriate workflows. An example workflow may be composed of the following steps: The scientist (1) prepares the experiment sample, (2) performs the experiment (possibly repetitions thereof), and (3) uses a data analysis tool to investigate the results. All these steps produce data that might be relevant for a later search. In the first step, all data related to the biological sample must be captured (sample organism, cell line or tissue, protein or nucleic acid concentration and quality, protocols, etc.). In particular, unique sample IDs must be given to the samples in order to make them distinguishable. In the second step, large datasets are generated, which often are transformed into different formats. In the last step, the data analysis tools may produce summary data describing the experiment results. At the end, the scientist has to store and annotate all data relevant for a federated search in a global data store. In this way, data from different experiments, possibly processed on different instruments and technology platforms, become globally available and thus searchable through a common portal. Besides, the detailed experiment data is stored in specific marts, e.g., for transcriptomics and proteomics that provide additional technology-specific data analysis and search functionalities.

As indicated in Figure 5, the integration framework can also be used to connect external data sources such as literature or ontology databases. The main problem here is to write wrappers that transform the vocabulary of the external data sources to that of the global data store.

Current solutions for data repositories in transcriptomics

Today there are many microarray gene expression databases available in the internet (see Tab. 1 for a short, non-exhaustive list). These databases serve different purposes including:

1. repository of raw data
2. online analysis of expression data
3. visualization of individual expression profiles
4. functional expression analysis, and
5. comprehensive coverage of expression data related to a specific species or development stage.

Among these, the ArrayExpress and Gene Expression Omnibus (GEO) databases are of general interest, since the major journals accept them as public repositories

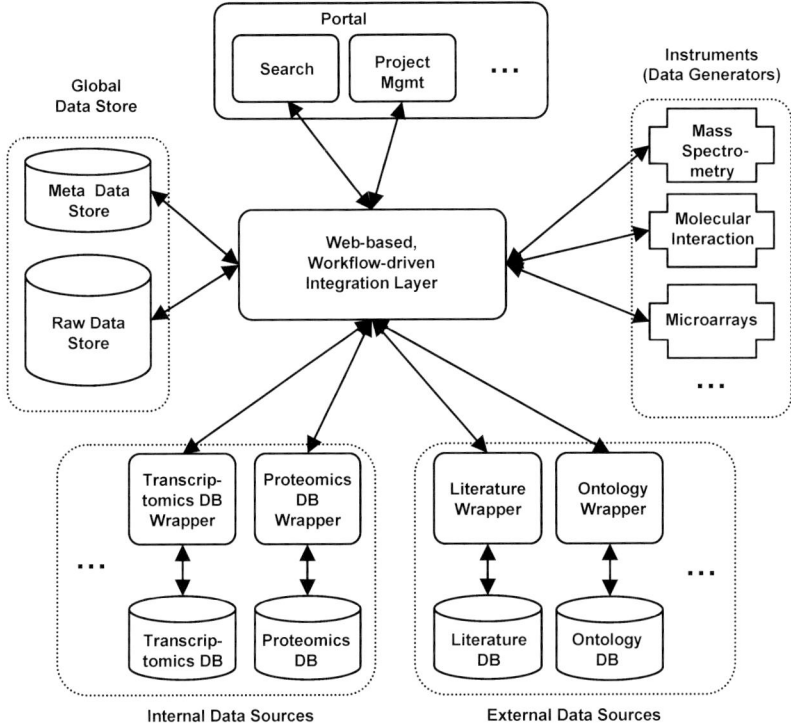

Figure 5. Architecture of the FGCZ data integration concept. The various internal as well as external data sources and data generators (instruments) available at FGCZ are integrated using a workflow-driven layer. Specific workflows capture raw as well as meta data into the global data store. Based on this global data store, a Web portal supports federated queries.

for array data that accompanies a publication. Both databases store array data and associated annotations in a standardized way that is compliant to the MIAME requirements and the MAGE Object model. Through these data repositories, researchers have access to a wealth of gene expression data that is ready to be loaded into their favorite expression analysis software, where it subsequently can be analyzed together with their own expression data.

In the following subsections we outline the features and coverage of ArrayExpress and GEO in more detail.

ArrayExpress
The major goals of EBI's ArrayExpress repository [66, 67] are to provide an archive for microarray data generated within research projects, especially those related to scientific publications, and to grant access to microarray data from disparate sources in a standardized form. ArrayExpress was the first microarray repository that adhered to the standards put forward by the MGED society. The repository, which went online in 2002, has attracted an increasing number of submissions. By Decem-

Table 1. Overview over selected gene expression databases

Name	Description and URL
ArrayExpress	Only slightly smaller than GEO and also accepted by journals as repository for data accompanying publications. Restricted to microarray data, big emphasis on data standardization. http://www.ebi.ac.uk/arrayexpress/
GeneVestigator	A curated database for *Arabidopsis* data from the GeneChip platform with a web-interface for functional expression analysis. https://www.genevestigator.ethz.ch/
GEO	The largest repository for microarray and other – omics data. Journals recognize this DB as repository for data accompanying publications. http://www.ncbi.nlm.nih.gov/geo/
GermOnline	A cross-species community knowledgebase on germ cell growth and development that organizes biological knowledge and microarray data. http://www.germonline.org/
NASCarray	All the *Arabidopsis* expression data produced at the microarray facility of the Nottingham *Arabidopsis* Stock Centre together with MIAME-compliant annotation. Focuses on the Affymetrix platform but also hosts two-color slides. http://affymetrix.arabidopsis.info
PEPR	A database with the primary goal of providing intuitive and user-friendly web-access to Affymetrix microarray data. http://pepr.cnmcresearch.org/home.do
PlasmoDB	A datawarehouse providing systems biology information of the malaria pathogen *Plasmodium*. Contains sequence, gene, protein, expression, pathway, phylogeny, 3D structure and other – omics data. http://plasmodb.org/
RAD	A public gene expression database containing data from array-based and nonarray-based (SAGE) experiments supporting MIAME compliant data submissions and data browsing, query and retrieval http://www.cbil.upenn.edu/RAD/php/index.php
SMD	A platform-independent, MIAME-compliant repository for gene expression/molecular abundance data. It is separated in a private and a public part. http://genome-www.stanford.edu/microarray

ber 2005, it contained data from ~1,200 expression studies comprising ~44,000 samples.

The ArrayExpress repository is literally built using the MAGE-OM as blueprint. With the help of a code generation tool that was developed by the EBI, the following elements of ArrayExpress were directly generated from the MAGE-OM: the database schema, functions for MAGE-ML import/export, data retrieval, and default visualizations. Through this approach, ArrayExpress is inherently compliant with the entire MAGE-OM and can easily be updated if the MAGE-OM is revised.

In Figure 6, we show the structure of the entire ArrayExpress suite with the repository as the central element. Data can be loaded through the MIAMExpress inter-

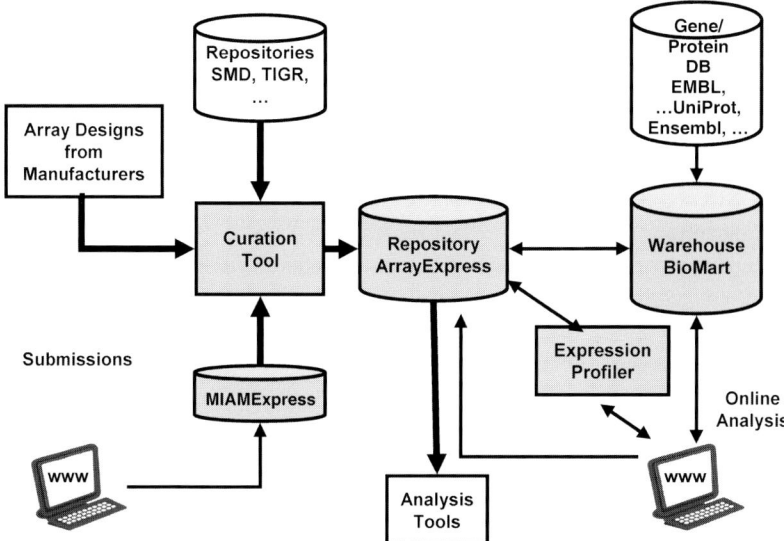

Figure 6. Setup of EBI's ArrayExpress database (grey elements) and the directly associated infrastructure. The thick arrows show all flows where data is transferred in MAGE-ML format. Users can download expression data for personal use or use EBI's ExpressionProfiler and the BioMart Data Warehouse for an online data analysis.

face, which facilitates submission of small datasets (<50 hybridizations) through an interactive web interface, and features a batch submission mechanism for larger studies. From MIAMExpress, the data is subsequently transferred to a curation tool where the data is automatically checked for completeness and formal correctness. In addition, EBI curators manually check the quality and meaningfulness of the annotations. In the case of problems, the curators ask the submitters to improve their data annotation. Through this curation step EBI guarantees that all data in the repository has an improved quality annotation, which is a prerequisite for subsequent meaningful automated analysis. The ExpressionProfiler is a web-based expression analysis tool, which allows users to directly analyze expression data in the repository. A warehouse solution for ArrayExpress is being set up, but is not yet completed. It is a gene-centric database as opposed to the repository, which is array-centric. It will enable gene-based queries, so that users can analyze expression values of individual genes across many studies. It will also provide integration with external sequence and protein databases so that a functional expression analysis will be possible. Data retrieval is easily achieved via a web interface that provides a simple query form where experimental data can be searched for species, experimental design, experimental factor, array design, etc. The data can be downloaded either in MAGE-ML or in raw format.

Gene Expression Omnibus
NCBI's Gene Expression Omnibus (GEO) is currently the largest public repository of microarray and other molecular abundance data. Similar to ArrayExpress, its goal

is to provide access to high-throughput data generated by the research community all over the world. GEO is explicitly open to all high-throughput data, not only microarray data, but also SAGE (Serial Analysis of Gene Expression), mass spectrometry peptide profiling data, and others. GEO's focus has been to generate a versatile and robust repository without rigorously enforcing data standards. As of December 2005, the repository holds 2,400 gene expression studies comprising ~64,000 samples.

The GEO repository was designed with the goal of being maximally flexible and open to accommodate data generated by future technologies. Therefore the NCBI abstained from rigid formal requirements and standards but chose a data representation that covers only the essential aspects that are expected to be common to all existing and upcoming high-throughput technologies. As a result, data from different technologies can indeed easily be hosted by GEO, but individual data types, as for example microarray data, cannot be modeled in much detail. Upload, online-usage, and download of data is extremely intuitive and can easily be performed by researchers that are neither familiar with the specific data-generating technologies nor the formal data standards.

The GEO repository has three key components: platform, sample and series. The *platform* covers the set of probes or reporters that is interrogated in a given experiment. For Affymetrix GeneChips, this is the list of probes contained on a given chip type. The *sample* holds the abundance data measured using a biological sample and a platform. Additionally, the sample has annotation fields that characterize the sample. A *series*, finally groups all samples that were analyzed in an experiment using a given platform. The user has to provide these three parts to complete a submission. Submissions can either be done interactively using web-forms, or directly by providing the information as tabular text following the GEO's SOFT (Simple Omnibus Format in Text) format, or in MAGE-ML format. Submitted data is curated for syntactic correctness and meaningfulness of the annotation.

In addition to the submission centric repository, GEO provides access to the data through two additional databases that are dataset and profile oriented. From the submissions, GEO curators group related samples into biologically meaningful datasets. To illustrate this we might think of a study of the gene expression response of *Arabidopsis thaliana* and *Triticum aestivum* to compound treatment with measurements for several doses and time points. The entire dataset from this study is uploaded as a single 'Gene Expression Series' holding the data for each sample of the study. The curators would then organize the data in two different datasets, one for *Arabidopsis thaliana* and one for *Triticum aestivum* and within each data set the replicates belonging to the same dose and time point would be grouped together in subsets. The manually curated datasets can be queried based on the annotation and GEO provides the online visualization and statistical analysis of these datasets. Orthogonal to the GEO datasets are the GEO profiles that provide visualizations of reporter signals across the samples contained in a dataset.

GEO has succeeded in establishing an easy-to-use and therefore popular repository that generates a significant benefit for the scientific community. However, the benefit of being versatile has been achieved at the cost of a lacking detailed data modeling capability. Consequently, an automated integration and analysis of data from different experiments is generally not possible. Further adding to this is the fact that GEO only encourages the submission of MIAME compliant data but does not enforce it. Thus, even for manual analysis, there may not be enough unambiguous annotation details available to analyze the data correctly. Finally, GEO does not require the submission of raw data as ArrayExpress does, so it does not provide the option to perform an alternative processing, which may be advisable in the light of updated sequence and gene information [68].

Plant-specific gene expression data repositories
By the end of 2005, ArrayExpress and Gene Expression Omnibus covered expression data for 35 [67] and over 100 different organisms [69], respectively. In ArrayExpress, around 25% of the data are plant-specific, the majority of which is from *Arabidopsis*. For the submissions and storage of *Arabidopsis* gene expression data the At-MIAMEexpress data submission tool was developed [70]. This tool facilitates annotation with a more precise description of plant-specific experimental conditions. The power of using a systematic annotation of microarray data is compellingly shown by a data warehouse approach such as Genevestigator [18]. Genevestigator provides researchers in the field of plant biology with a panel of software tools to mine expression information of *Arabidopsis* genes in conjunction with contextual information of about 2,000 Affymetrix GeneChip experiments. Of particular interest are the Gene Atlas tool, which combines tissue or organ annotation with gene expression data, and the Gene Chronologer tool, which combines different plant growth stages [71] with gene expression data.

Much of the experimental data in Genevestigator are gathered from public gene expression data repositories that are more or less specific for *Arabidopsis thaliana*, as for example the Nottingham *Arabidopsis* Stock Centre's (NASC, [72]) microarray database. It mostly contains Affymetrix GeneChip experiments and allows for text searching and data mining using a range of different software tools. The *Arabidopsis* Information Resource database for gene expression also hosts two-channel experiments [73]. Besides the expression values, this database hosts more detailed information concerning the experimental design and the source of the RNA used for hybridization, as well as the data analysis and protocols. Several other organism-specific gene expression databases exist for rice (Rice Expression Database (RED), [74]), soybean (soybean genomics and microarray database [75]), maize (Maize Oligonucleotide Array Project, [76]) or cereals (BarleyBase, [77]). Unfortunately, some of these do not yet contain an attractive amount of data. Since the data are often not mirrored in the bigger repositories, it is still worthwhile to exploit these data sources for a specific research question.

First steps towards creating a gene expression database for all plant species with additional web-based data mining software tools represent the DRASTIC database (Database Resource for the Analysis of Signal Transduction In Cells, [78]) and the INSIGHT (INference of cell SIGnaling HypoTheseS) software package. DRASTIC harbors almost 18,000 records from more than 500 peer-reviewed publications, which detail the results from around 300 treatments affecting more than 70 different plant species. So far, the data is somewhat confined to experiments that study gene expression changes of plants exposed to various pathogens, chemicals or other treatments such as drought, salt and low temperature. It is expected that such a setup will hold significant value for the plant community after expansion and coverage of more diverse research questions.

Current solutions for data repositories in proteomics

Since proteins ultimately carry out most cellular functions, the study of their interactions, sub-cellular localization, and posttranslational modifications provide important insights that cannot be gained by global transcriptome analysis. Thus they hold extremely valuable information for systems biology studies. A detailed description of proteomics databases that serve these distinct research areas is beyond the scope of this chapter. Since we already briefly touched upon selected repositories for protein interaction data and a number of databases for 2-D gel based results are well-known (for a list see [79]), we only describe SBEAMS, a publicly available integrated database and analysis solution for shotgun proteomics data, in more detail. An effort to establish a repository for peptide information that is mapped onto the respective genome sequence, PeptideAtlas [80], has recently been initiated. Since this project will help to integrate gene expression and protein expression data, to confirm gene models purely predicted by *in silico* methods and to identify genes missed by current gene prediction algorithms it will be of great interest for systems biology studies.

SBEAMS: Systems Biology Experiment Analysis Management System
SBEAMS [81] has been developed by the Institute for Systems Biology in Seattle as a framework for collecting, exploring and exporting the large amounts of data generated by several functional genomics technology platforms. At its core, SBEAMS is not a single program, but rather a set of software tools designed to work with data stored in various evolving relational databases. The major available modules include those for microarray data, for image data storage and processing, and finally a proteomics module for LC-MS/MS shotgun proteomics data, which we will describe in more detail below. SBEAMS combines a relational database with a web front-end for querying the database and providing integrated access to remote data sources; it is accessible from any platform and requires no client installations. The current version of SBEAMS uses the Microsoft SQL Server as back end database server, although portions of SBEAMS can also be run on the open source database systems MySQL and PostgreSQL.

In detail, the SBEAMS proteomics module allows users to store, manage, annotate, analyze, and compare data originating from shotgun proteomics experiments. It captures all relevant experimental parameters from the initial sample preparation up to the database search settings, the experimental data along with user-entered annotations, and the results. A cornerstone is the streamlined integration with a semi-automated pipeline that allows to process the raw data into peptide and protein assignments along with probabilistic accuracy estimates for the respective peptide or protein identifications [58, 59]. This feature provides an automated initial validation for large datasets. However, individual spectra can still be retrieved and manually inspected in an interactive manner.

In a proteomics workflow, users can enter their project, experiment, and sample information into the database. Subsequently, the data output of the analysis pipeline are loaded into SBEAMS. Other protein sequence information such as transmembrane domains, signal peptides, or a multitude of other protein parameters can be computed and integrated into the database schema. The proteomics module enables a flexible subsequent interactive exploration and analysis of the data with all features and annotations that are stored and integrated in the database, including the quality assignments. A variety of filtering tools and data visualization tools are critical to work with the large datasets and allow efficient prioritization. Analysis tools and custom scripts built on top of SBEAMS can be used to summarize the experimental results, to compare different experiments and to export selected datasets. As SBEAMS relies on a database, all data can be used for later analysis and comparison across experiments and even across platforms if the required modules are installed. Some of the key functionalities provided by the proteomics module are summarized below:

- Peptide Prophet and Protein Prophet output exploration, filtering and prioritization of the long lists of peptide or protein identifications, creation of summaries of identified peptides and proteins for one or more experiments
- browsing of search hits and the respective spectra with tools for quantitative protein expression analysis (XPRESS and ASAPRatio)
- search based on various features (e.g., different peptide probability scores or quantitation values)
- browsing all possible tryptic peptides for biosequence sets
- export of selected data into a number of data formats (xls, csv) for further analysis
- linking to Cytoscape's data visualization capabilities [82]

Since SBEAMS relies on a fully fledged relational database system, advanced users can enter complex SQL queries and implement any kind of program, e.g., data mining algorithms, on top of SBEAMS. While this flexibility can greatly enhance the ability to discover novel insights through integration of various information sources, it requires dedicated skilled staff to accomplish this.

Concluding remarks and outlook

Life science research in the last decade has started to move from a predominantly reductionistic approach to a more global, systematic analysis. This paradigm shift has been catalyzed by the availability of complete genome sequences for a number of eukaryotic model organisms. These include the first completely sequenced unicellular eukaryote *Saccharomyces cerevisiae* (baker's yeast) in 1996 [83], the round worm *Caenorhabditis elegans* (The *C. elegans* consortium, 1998) as first multicellular eukaryote, the fruitfly *Drosophila melanogaster* [84], and the thale cress *Arabidopsis thaliana* as the first plant genome sequence at the end of 2000 [85].

The plant community capitalized on the experiences and methods developed for other model organisms, and has made remarkable progress in the 5 years following publication of the *Arabidopsis* genome sequence [86]. Current gene prediction algorithms provide a rough estimate of the total number of genes in the respective organisms, and for *A. thaliana*, approximately 26,200 protein coding genes generate close to 28,000 distinct proteins by alternative splicing [86]. These gene predictions have provided the basis to study all genes in parallel using transcriptomics technologies. In addition, whole genome tiling arrays can be made and hybridized with cRNA in order to identify global areas of transcriptional activity. In *A. thaliana*, such an approach did not only corroborate *in silico* gene predictions that previously lacked any experimental evidence, but also led to the discovery of actively transcribed genome regions and genes that were missed by current prediction algorithms [87]. In analogy to other model organisms, application of high-throughput *RNAi* screens are expected to add significant value to studies that try to elucidate the molecular function of large numbers of gene products in parallel [88].

Since the level of gene expression does not correlate in all cases with the level of expressed protein, nor does it allow the study of sub-cellular localization, post-translational modifications, and protein–protein interactions, proteomics has a key role in functional genomics and systems biology studies. Technology in this area is progressing at a fast pace. While current methodologies allow for the quantitative analysis of several hundred of proteins in a single experiment [89], further technological advances are necessary to reliably study significantly more proteins at the same time [90].

Our review has focused on large scale data collection and data integration, the first two stages of a systems biology workflow. In particular, we have addressed the efforts in the transcriptomics and proteomics fields, and detailed initiatives to standardize data formats and to provide publicly available data repositories. Similar initiatives in the metabolomics field are not described. However, this field in particular will address the later phases of a systems biology approach, i.e., the description of a system by accurate models and prediction of a phenotype when manipulating various parameters of the model. While these later phases of a systems biology approach when performed on integrated data from all disciplines will undoubtedly gain importance in the future, such studies so far have been rare. Trey Ideker's landmark study in yeast, which integrated several large datasets of one focused part of the entire system, the galactose pathway, has clearly demonstrated that such an approach can unravel novel components and shed new light onto a system that has been extensively studied for more than 30 years [91]. More ambitious systems biology integration projects such as the Virtual Plant project [92], therefore name 2010 as a realistic time point to start the investigation of systems biology questions in the plant at a larger scale.

Acknowledgements

The authors would like to thank the Plant Gene Ontology Consortium for permission to reproduce Figure 2 and Alvis Brazma (EBI, Hinxton) for permission to reproduce

Figure 6. We also thank Patrick Pedrioli (ISMB, ETH Zurich) for reviewing the section on Standards in Proteomics, Erich Brunner and Sandra Lövenich (Institute of Zoology, University of Zurich) for reviewing the section on Proteomics Data Repositories, and Mike Scott for carefully proof-reading the manuscript.

References

1. Bork P, Serrano L (2005) Towards cellular systems in 4D. *Cell* 121:507–509
2. Lauffenburger D (2003) Systems biology. *Chem Eng News* 81: 45–55
3. Maglott DR, Katz KS, Sicotte H, Pruitt KD (2000) NCBI's LocusLink and RefSeq. *Nucleic Acids Res* 28: 126–128
4. Pruitt KD, Katz KS, Sicotte H, Maglott DR (2000) Introducing RefSeq and LocusLink: curated human genome resources at the NCBI. *Trends Genet* 16: 44–47
5. Bairoch A, Apweiler R (2000) The SWISS-PROT protein sequence database and its supplement TrEMBL in 2000. *Nucleic Acids Res* 28: 45–48
6. Jarvinen AK, Hautaniemi S, Edgren H, Auvinen P, Saarela J, Kallioniemi OP, Monni O (2004) Are data from different gene expression microarray platforms comparable? *Genomics* 83: 1164–1168
7. Hack CJ (2004) Integrated transcriptome and proteome data: the challenges ahead. *Brief Funct Genomic Proteomic* 3: 212–219
8. Schulze-Kremer S (2002) Ontologies for molecular biology and bioinformatics. *In Silico Biol* 2: 179–193
9. Rojas I, Ratsch E, Saric J, Wittig U (2004) Notes on the use of ontologies in the biochemical domain. *In Silico Biol* 4: 89–96
10. Blake J (2004) Bio-ontologies–fast and furious. *Nat Biotechnol* 22: 773–774
11. Bard JB, Rhee SY (2004) Ontologies in biology: design, applications and future challenges. *Nat Rev Genet* 5: 213–222
12. Gruber TR (1993) Toward principles for the design of ontologies used for knowledge sharing. http://ksl-web.stanford.edu/KSL_Abstracts/KSL-93-04.html
13. OBO. Open Biomedical Ontologies. http://obo.sourceforge.net.
14. Mungall C (2004) OBOL: Integrating language and meaning in bio-ontologies. *Comp Funct Genomics* 6–7: 509–520
15. The Plant Ontology Consortium (2002) The Plant Ontology Consortium and Plant Ontologies. *Comp Funct Genomics* 3: 137–142
16. Ashburner M, Ball CA, Blake JA, Botstein D, Butler H, Cherry JM, Davis AP, Dolinski K, Dwight SS, Eppig JT et al. (2000) Gene ontology: tool for the unification of biology. The Gene Ontology Consortium. *Nat Genet* 25: 25–29
17. The Gene Ontology Consortium (2001) Creating the gene ontology resource: design and implementation. *Genome Res* 11: 1425–1433
18. Zimmermann P, Hirsch-Hoffmann M, Hennig L, Gruissem W (2004) GENEVESTIGATOR. *Arabidopsis* microarray database and analysis toolbox. *Plant Physiol* 136: 2621–2632
19. Khatri P, Draghici S (2005) Ontological analysis of gene expression data: current tools, limitations, and open problems. *Bioinformatics* 21: 3587–3595
20. Khatri P, Draghici S, Ostermeier GC, Krawetz SA (2002) Profiling gene expression using onto-express. *Genomics* 79: 266–270
21. Draghici S, Khatri P, Bhavsar P, Shah A, Krawetz SA, Tainsky MA (2003) Onto-Tools, the toolkit of the modern biologist: Onto-Express, Onto-Compare, Onto-Design and Onto-Translate. *Nucleic Acids Res* 31: 3775–3781

22. Zhang B, Schmoyer D, Kirov S, Snoddy J (2004) GOTree Machine (GOTM): a web-based platform for interpreting sets of interesting genes using Gene Ontology hierarchies. *BMC Bioinformatics* 5: 16
23. Lee HK, Braynen W, Keshav K, Pavlidis P. Ermine J (2005) Tool for functional analysis of gene expression data sets. *BMC Bioinformatics* 6: 269
24. Maere S, Heymans K, Kuiper M (2005) BiNGO: a Cytoscape plugin to assess overrepresentation of gene ontology categories in biological networks. *Bioinformatics* 21: 3448–3449
25. Wrobel G, Chalmel F, Primig M (2005) goCluster integrates statistical analysis and functional interpretation of microarray expression data. *Bioinformatics* 21: 3575–3577
26. Lottaz C, Spang R (2005) Molecular decomposition of complex clinical phenotypes using biologically structured analysis of microarray data. *Bioinformatics* 21: 1971–1978
27. Berardini TZ, Mundodi S, Reiser L, Huala E, Garcia-Hernandez M, Zhang P, Mueller LA, Yoon J, Doyle A, Lander G et al. (2004) Functional annotation of the *Arabidopsis* genome using controlled vocabularies. *Plant Physiol* 135: 745–755
28. Beckett P, Bancroft I (2005) M.T. Computational tools for *Brassica-Arabidopsis* comparative genomics. *Comp Funct Genomics* 6: 147–152
29. Gramene. www.gramene.org
30. Ware D, Jaiswal P, Ni J, Pan X, Chang K, Clark K, Teytelman L, Schmidt S, Zhao W, Cartinhour S et al. (2002) Gramene: a resource for comparative grass genomics. *Nucleic Acids Res* 30: 103–105
31. Ware DH, Jaiswal P, Ni J, Yap IV, Pan X, Clark KY, Teytelman L, Schmidt SC, Zhao W, Chang K et al. (2002) Gramene, a tool for grass genomics. *Plant Physiol* 130: 1606–1613
32. Soldatova LN, King RD (2005) Are the current ontologies in biology good ontologies? *Nat Biotechnol* 23: 1095–1098
33. Brazma A, Robinson A, Cameron G, Ashburner M (2000) One-stop shop for microarray data. *Nature* 403: 699–700
34. MIAME. www.mged.org/Workgroups/MIAME/miame_checklist.html
35. Zimmermann P, Schildknecht B, Craigon D, Garcia-Hernandez M, Gruissem W, May S, Mukherjee G, Parkinson H, Rhee S, Wagner U et al. (2006) MIAME/Plant – adding value to plant microarray experiments. *Plant Methods* 2: 1
36. MIAME-Tox. http://www.mged.org/MIAME1.1-DenverDraft.DOC)
37. Spellman PT, Miller M, Stewart J, Troup C, Sarkans U, Chervitz S, Bernhart D, Sherlock G, Ball C, Lepage M et al. (2002) Design and implementation of microarray gene expression markup language (MAGE-ML). *Genome Biol* 3:RESEARCH0046 Epub 2002 Aug 23
38. MAGE-ML.DTD. http://schema.omg.org/lsr/gene_expression/1.1/MAGE-ML.dtd
39. MGED Ontology draft. www.mged.org/Workgroups/MIAME/MIAMEv1.1-MAGE-OntologyDraft2v1.0.htm
40. Jenkins H, Hardy N, Beckmann M, Draper J, Smith AR, Taylor J, Fiehn O, Goodacre R, Bino RJ, Hall R et al. (2004) A proposed framework for the description of plant metabolomics experiments and their results. *Nat Biotechnol* 22: 1601–1606
41. Kaiser J (2002) Proteomics. Public-private group maps out initiatives. *Science* 296: 827
42. Orchard S, Hermjakob H, Apweiler R (2003) The proteomics standards initiative. *Proteomics* 3: 1374–1376
43. Orchard S, Taylor C, Hermjakob H, Zhu W, Julian R, Apweiler R (2004) Current status of proteomic standards development. *Exp Rev Proteomics* 1: 179–183
44. Jensen ON (2004) Modification-specific proteomics: characterization of post-translational modifications by mass spectrometry. *Curr Opin Chem Biol* 8: 33–41
45. Tyers M, Mann M (2003) From genomics to proteomics. *Nature* 422: 193–197

46. Anderson NL, Anderson NG (2002) The human plasma proteome: history, character, and diagnostic prospects. *Mol Cell Proteomics* 1: 845–867
47. de Lichtenberg U, Jensen LJ, Brunak S, Bork P (2005) Dynamic complex formation during the yeast cell cycle. *Science* 307: 724–727
48. Hermjakob H, Montecchi-Palazzi L, Bader G, Wojcik J, Salwinski L, Ceol A, Moore S, Orchard S, Sarkans U, von Mering C et al. (2004) The HUPO PSI's molecular interaction format – a community standard for the representation of protein interaction data. *Nat Biotechnol* 22: 177–183
49. DIP. http://dip.doe-mbi.ucla.edu
50. MINT. http://mint.bio.uniroma2.it/mint
51. MPact. http://mips.gsf.de/genre/proj/mpact
52. IntAct. www.ebi.ac.uk/intact
53. http://imex.sf.net
54. Pedrioli PG, Eng JK, Hubley R, Vogelzang M, Deutsch EW, Raught B, Pratt B, Nilsson E, Angeletti RH, Apweiler R et al. (2004) A common open representation of mass spectrometry data and its application to proteomics research. *Nat Biotechnol* 22: 1459–1466
55. Orchard S, Hermjakob H, Taylor C, Aebersold R, Apweiler R (2005) Human proteome organisation proteomics standards initiative pre-congress initiative. *Proteomics* 5: 4651–4652
56. Nesvizhskii AI, Aebersold R (2005) Interpretation of shotgun proteomic data: the protein inference problem. *Mol Cell Proteomics* 4: 1419–1440
57. Carr S, Aebersold R, Baldwin M, Burlingame A, Clauser K, Nesvizhskii A (2004) The need for guidelines in publication of peptide and protein identification data: Working Group on Publication Guidelines for Peptide and Protein Identification Data. *Mol Cell Proteomics* 3: 531–533
58. Keller A, Nesvizhskii AI, Kolker E, Aebersold R (2002) Empirical statistical model to estimate the accuracy of peptide identifications made by MS/MS and database search. *Anal Chem* 74: 5383–5392
59. Nesvizhskii AI, Keller A, Kolker E, Aebersold R (2003) A statistical model for identifying proteins by tandem mass spectrometry. *Anal Chem* 75: 4646–4658
60. Ahrens C, Jespersen H, Schandorff S (2005) Bioinformatics for Proteomics: Wiley, 249–272
61. Schwarz K, Schmitt I, Türker C, Höding M, Hildebrandt E, Balko S, Conrad S, Saake G (1999) Design Support for Database Federations. Springer-Verlag, 445–459
62. Hucka M, Finney A, Sauro HM, Bolouri H, Doyle JC, Kitano H, Arkin AP, Bornstein BJ, Bray D, Cornish-Bowden A et al. (2003) The systems biology markup language (SBML): a medium for representation and exchange of biochemical network models. *Bioinformatics* 19: 524–531
63. Adelberg A (1998) NoDoSE – A tool for semi-automatically extracting structured and semistructured data from text documents. In: Proceedings of the International Conference on Data Management, SIGMOD'98, ACM SIGMOD Record, 25
64. Sheth AP, Larson JA (1990) Federated database systems for managing distributed, heterogeneous, and autonomous databases. *ACM Computing Surveys* 22: 183–236
65. Batini C, Lenzerini M, Navathe SB (1986) A comparative analysis of methodologies for database schema integration. *ACM Computing Surveys* 18: 323–364
66. Sarkans U, Parkinson H, Lara GG, Oezcimen A, Sharma A, Abeygunawardena N, Contrino S, Holloway E, Rocca-Serra P, Mukherjee G et al. (2005) The ArrayExpress gene expression database: a software engineering and implementation perspective. *Bioinformatics* 21: 1495–1501

67. Parkinson H, Sarkans U, Shojatalab M, Abeygunawardena N, Contrino S, Coulson R, Farne A, Lara GG, Holloway E, Kapushesky M et al. (2005) ArrayExpress – a public repository for microarray gene expression data at the EBI. *Nucleic Acids Res* 33: D553–555
68. Dai M, Wang P, Boyd AD, Kostov G, Athey B, Jones EG, Bunney WE, Myers RM, Speed TP, Akil H et al. (2005) Evolving gene/transcript definitions significantly alter the interpretation of GeneChip data. *Nucleic Acids Res* 33: e175
69. Barrett T, Suzek TO, Troup DB, Wilhite SE, Ngau WC, Ledoux P, Rudnev D, Lash AE, Fujibuchi W, Edgar R (2005) NCBI GEO: mining millions of expression profiles – database and tools. *Nucleic Acids Res* 33: D562–566
70. Mukherjee G, Abeygunawardena N, Parkinson H, Contrino S, Durinck S, Farne A, Holloway E, Lilja P, Moreau Y, Oezcimen A et al. (2005) Plant-based microarray data at the European Bioinformatics Institute. Introducing AtMIAMExpress, a submission tool for *Arabidopsis* gene expression data to ArrayExpress. *Plant Physiol* 139: 632–636
71. Boyes DC, Zayed AM, Ascenzi R, McCaskill AJ, Hoffman NE, Davis KR, Gorlach J (2001) Growth stage-based phenotypic analysis of *Arabidopsis*: a model for high throughput functional genomics in plants. *Plant Cell* 13: 1499–1510
72. Craigon DJ, James N, Okyere J, Higgins J, Jotham J, May S (2004) NASCArrays: a repository for microarray data generated by NASC's transcriptomics service. *Nucleic Acids Res* 32: D575–577
73. Rhee SY, Beavis W, Berardini TZ, Chen G, Dixon D, Doyle A, Garcia-Hernandez M, Huala E, Lander G, Montoya M et al. (2003) The *Arabidopsis* Information Resource (TAIR): a model organism database providing a centralized, curated gateway to *Arabidopsis* biology, research materials and community. *Nucleic Acids Res* 31: 224–228
74. Yazaki J, Kishimoto N, Ishikawa M, Endo D, Kojima K (2002) The Rice Expression Database (RED): gateway to rice functional genomics. *Trends in Plant Sci* 7: 563–564
75. SGMD. http://psi081.ba.ars.usda.gov/SGMD/default.htm
76. Maizearray. www.maizearray.org
77. Shen L, Gong J, Caldo RA, Nettleton D, Cook D, Wise RP, Dickerson JA (2005) BarleyBase – an expression profiling database for plant genomics. *Nucleic Acids Res* 33: D614–618
78. Button DK, Gartland KM, Ball LD, Natanson L, Gartland JS, Lyon GD (2006) DRASTIC – INSIGHTS: querying information in a plant gene expression database. *Nucleic Acids Res* 34: D712–716
79. www.expasy.org/ch2d/2d-index.html
80. Desiere F, Deutsch EW, Nesvizhskii AI, Mallick P, King NL, Eng JK, Aderem A, Boyle R, Brunner E, Donohoe S et al. (2005) Integration with the human genome of peptide sequences obtained by high-throughput mass spectrometry. *Genome Biol* 6: R9
81. SBEAMS. www.sbeams.org/
82. Shannon P, Markiel A, Ozier O, Baliga NS, Wang JT, Ramage D, Amin N, Schwikowski B, Ideker T (2003) Cytoscape: a software environment for integrated models of biomolecular interaction networks. *Genome Res* 13: 2498–2504
83. Goffeau A, Barrell BG, Bussey H, Davis RW, Dujon B, Feldmann H, Galibert F, Hoheisel JD, Jacq C, Johnston M et al. (1996) Life with 6000 genes. *Science* 274: 546, 563–567
84. Adams MD, Celniker SE, Holt RA, Evans CA, Gocayne JD, Amanatides PG, Scherer SE, Li PW, Hoskins RA, Galle RF et al. (2000) The genome sequence of *Drosophila melanogaster*. *Science* 287: 2185–2195
85. *Arabidopsis* Genome Initiative (2000) Analysis of the genome sequence of the flowering plant *Arabidopsis thaliana*. *Nature* 408: 796–815
86. Bevan M, Walsh S (2005) The *Arabidopsis* genome: a foundation for plant research. *Genome Res* 15: 1632–1642

87. Yamada K, Lim J, Dale JM, Chen H, Shinn P, Palm CJ, Southwick AM, Wu HC, Kim C, Nguyen M et al. (2003) Empirical analysis of transcriptional activity in the Arabidopsis genome. *Science* 302: 842–846
88. DasGupta R, Kaykas A, Moon RT, Perrimon N (2005) Functional genomic analysis of the Wnt-wingless signaling pathway. *Science* 308: 826–833
89. Aebersold R, Mann M (2003) Mass spectrometry-based proteomics. *Nature* 422: 198–207
90. Kuster B, Schirle M, Mallick P, Aebersold R (2005) Scoring proteomes with proteotypic peptide probes. *Nat Rev Mol Cell Biol* 6: 577–583
91. Ideker T, Thorsson V, Ranish JA, Christmas R, Buhler J, Eng JK, Bumgarner R, Goodlett DR, Aebersold R, Hood L (2001) Integrated genomic and proteomic analyses of a systematically perturbed metabolic network. *Science* 292: 929–934
92. Chory J, Ecker JR, Briggs S, Caboche M, Coruzzi GM, Cook D, Dangl J, Grant S, Guerinot ML, Henikoff S et al. (2000) National Science Foundation-Sponsored Workshop Report: "The 2010 Project" functional genomics and the virtual plant. A blueprint for understanding how plants are built and how to improve them. *Plant Physiol* 123: 423–426
93. Cheng J, Sun S, Tracy A, Hubbell E, Morris J, Valmeekam V, Kimbrough A, Cline MS, Liu G, Shigeta R et al. (2004) NetAffx Gene Ontology Mining Tool: a visual approach for microarray data analysis. *Bioinformatics* 20: 1462–1463

Plant Systems Biology
Edited by Sacha Baginsky and Alisdair R. Fernie
© 2007 Birkhäuser Verlag/Switzerland

Integrated data analysis for genome-wide research

Matthias Steinfath[1], Dirk Repsilber[1], Matthias Scholz[2], Dirk Walther[2] and Joachim Selbig[1,2]

[1] *Institute for Biology and Biochemistry, University Potsdam, c/o MPI-MP, Am Mühlenberg 1, D-14476 Potsdam-Golm, Germany*
[2] *Max Planck Institute of Molecular Plant Physiology, Am Mühlenberg 1, D-14476 Potsdam-Golm, Germany*

Abstract

Integrated data analysis is introduced as the intermediate level of a systems biology approach to analyse different 'omics' datasets, i.e., genome-wide measurements of transcripts, protein levels or protein–protein interactions, and metabolite levels aiming at generating a coherent understanding of biological function. In this chapter we focus on different methods of correlation analyses ranging from simple pairwise correlation to kernel canonical correlation which were recently applied in molecular biology. Several examples are presented to illustrate their application. The input data for this analysis frequently originate from different experimental platforms. Therefore, preprocessing steps such as data normalisation and missing value estimation are inherent to this approach. The corresponding procedures, potential pitfalls and biases, and available software solutions are reviewed. The multiplicity of observations obtained in omics-profiling experiments necessitates the application of multiple testing correction techniques.

Introduction

Motivation and definition of integrated data analysis in the 'omics' era

Biological research aims at discovering knowledge about biological function of all components that make up a biological system. The term 'biological function', however, is used in different meanings by geneticists, cellular biologists, structural biologists, bioinformaticians, and biophysical chemists. Some use the word 'function' to refer to the general biochemical activity of the gene product, others refer to the cellular process in which the gene product is involved, while to others 'function' means an understanding of the details of the molecular mechanism of catalysis or recognition or in the genetic sense of a generalised phenotype. Of course, biological processes can only be projected onto just the level of current experimental investigation. For example, Somogyi and Sniegoski [1] proposed to project biological regulatory networks onto the transcriptome level. Other genome-wide studies are exclusively focussed on protein–protein interaction data. However, as for example transcript abundance is no reliable indicator of corresponding protein levels [2], this projection may conceal key parts of the process under study. Also, from a physiological point of view, the understanding of life at its different levels of organisation requires a comprehension of the functional interactions between the key components of cells, tissues, organs as well as how these interactions change in disease states or under environmental influences. As Noble [3] points out, the information to infer regulatory dependencies resides neither in the genome nor in the individual proteins that genes code for. This statement remains true in the current era of 'omic'-research, i.e., genome-wide measurements of transcript levels [4], protein levels and protein–protein interactions [5–7], metabolomics [8–10] and other genome-wide data collections. Here, any particular experimental technique will always deliver data restricted to the concrete level of biological organisation it is addressing. Therefore, it is necessary to integrate the analysis of different levels of biological organisation to identify functional biological themes revealing how biological molecules interact with each other [11].

It is this observation which constitutes the first motivation for what we call 'integrated data analysis', where experimental data from different levels of biological organisation are brought together to investigate functional interactions between the parts of the genome-wide parts lists established so far.

There is a second major motivation which relates to a data quality as well as the multiplicity problem inherent to most high-throughput studies. As a matter of fact, each experimental technique will not only measure the biological variation it was designed for, but also add technical variance and possibly a methodological bias to the results. Together with the high-throughput character of genome-wide experiments, this probably results in high rates of false positive discoveries. There are also technologically biased results, which among other effects are partly responsible for false positives. Hence, if data from two or more different platforms, different levels of gene expression or biological organisation point to the same candidates, for example, for differentially regulated genes, functional relevance of this result can be established with greater confidence. That is to say, it is more likely to have found a significant result when calling certain genes differentially regulated at the transcript level, if protein quantification results in similar findings, even if quantities of transcripts and proteins often do not correlate well [12]. Such integration can also be seen as a means of verification of isolated results, or as if gathering additional degrees of freedom to be able to test a given hypothesis together with the high level of signal errors typical for high-throughput experimental data [13]. Frequently, measured data are not independent across different experiments, but data obtained from distinct platforms are more likely to be independent observations.

Based on these two main motivations, Aitchison and Galitski [13] describe integrated data analysis as a way of going "from inventories to insights" and propose the following general approach: First, the discovery-based component, i.e., measuring and organising high-dimensional data from different gene expression or organisational levels, as well as storing and making it accessible in the form of query-based information retrieval from databases. This first level of integrated data analysis is also referred to as 'data integration', to distinguish it from the next components as outlined below. Data integration itself cannot explain the *dynamical* behaviour of the biological system and is not a replacement for a dynamical qualitative or quantitative model. However, data integration is needed to increase the information available for the individual unit, e.g., the gene of interest, by adding more measured values, thus making systems features well distinguished from an otherwise overwhelmingly noisy background. Current approaches to data integration of this type are not the focus of this chapter.

A second component of integrated data analysis of multiple and disparate data types typically takes a pattern finding and correlation analysis point of view. Here, the goal of analysis is to find common patterns in data of different origin. Reviewing possible approaches taking this view of analysis and introducing commonly used methods applicable in this context is the objective of this chapter.

Finally, as a third component of integrated data analysis, the formulation of quantitative as well as conceptual system models [14] (see also next chapter) is the ultimate goal.

In summary, we understand integrated data analysis as the combination of disparate experimental data (experiments, platforms, but in particular from heterogeneous biological omics-levels) for the validation of analysis results on the one hand, and to establish biological models for the functional interplay of the different levels of biological organisation aiming at a functional and holistic understanding [15].

This chapter will first review challenges and approaches to integrated data analysis before focussing in more detail on the correlation analysis approach. We will frequently refer to experimental design issues and the connection between modelling and analyses approaches. Additional approaches of integrated analysis are then discussed together with an overview of available software tools.

Challenges of integrated data analysis

At a basic level, experimental platforms vary considerably in precision and accuracy, dynamic ranges, linearity of response, and error sources, such that it is challenging enough to compare results between experiments at the same 'omics'-level, let alone between different 'omics'-technologies [15]. Furthermore, different technologies, even if measuring data within the same 'omics'-level may display distinct, or only partially overlapping views of the entities of interest. Consider, for example, protein–protein interaction data measured in yeast-2-hybrid assays and by co-immunoprecipitation. In the former case two proteins would be believed to interact if they exhibit some sort of direct physical binding, whereas in the latter case proteins that are part of the same complex would already be detected as interacting. These findings overlap only partly and are most different for large protein complexes. Another example is the comparability of different gene expression profiling platforms [16]. Often, specific probes used to measure specific mRNA abundance levels or relative amounts are not that specific and integrate over different subpopulations of the transcriptome, e.g., gene families, splice variants or other targets of cross hybridisation. These specificity problems are, of course, specific for different platforms. However, integrated data analysis aims not only at combining different experiments within a special design and level of biological organisation. Data are collected from a wide range of objects, e.g., the progression from molecules to cells to tissues to organisms to populations [15] remarks that difficulties regarding comparability tend to increase along this progression. Moreover, it is not at all straightforward to compare data originating from different approaches of experimental design. In this context, the distinction between data-driven and hypothesis-driven research is critical [17, 18]. This difference can probably best be demonstrated by surveying the different methods to cope with multiplicity (multiple testing correction). Results from a more observational type of research, often termed data-driven, have a screening type of character, where a high rate of false positives can be tolerated. Therefore, false positives are permitted applying only weak control of multiplicity. On the other hand, in a hypothesis testing setting, multiplicity is under strict control to guarantee a small false positive rate to permit statistically proven conclusions. However, on a finer scale, it is important to consider the underlying model and the associated biological question for a set of experiments before trying to integrate the 'functional data', as considered in the following paragraph.

To meet these challenges for integrated data analysis, normalisation procedures have to be applied. Consider two datasets chosen as candidates for integrated analysis. For each dataset an analysis starting from raw data, normalising for technical biases and filtering out outliers has already been conducted, applying some kind of high level analysis to test a hypothesis or for data mining purposes. In principle, the two datasets can now be combined taking either normalised data or results of the single analyses as basis for the integration. For example, combining two gene expression datasets from the same platform, it is possible to choose the normalised data level for integration. Normalisation in this context would 'align' the two raw datasets based on common gene identifiers, trying to correct for biases introduced by the individual microarrays or, for example, differences between spotting tips used for producing a spotted cDNA array. At an intermediate level of analysis, let us consider groups of genes as the basis of integration. As an example, consider the work of Ge et al. [19], where groups of co-expressed genes were found to include a significantly high proportion of genes coding for interacting proteins. For the highest possible level of integration consider the end-products of the single analyses – most general functional assignments to their objects of investigation. For example, co-expressed genes might be considered co-regulated via common transcription factors and playing a certain role in the physiology of certain cell types, organs or developmental states. Two such datasets can be compared at the level of functional assignments for groups of genes. To promote this kind of comparability of experimental results and to facilitate integrated analysis, ontologies describing biological function in a well-defined way have been introduced [20–22].

Missing values have to be mentioned as another important challenge to integrated data analysis. Here, we have to distinguish two cases. As an illustrative example consider two datasets measuring gene expression and protein levels across a defined common set of genes. Integrated data analysis thus has to deal with pairs of measurements as raw data of the integrated dataset. Missing values may arise either from different filtering techniques as applied to the two single datasets, or from the fact that for a certain subgroup of genes, it is impossible to measure either gene expression or protein level. In the first case, one would consider the missing value as a product of random noise, having thwarted just that value for the specific experimental outcome. Here, methods of missing value imputation might be applicable. If data quality allows the observation of possible non-linear behaviour in the measured data then the method of data imputation should enable to model this non-linearity. We will discuss this in greater detail in the section 'missing value estimation'.

Integrated data analysis and modelling

Specific biological questions and experimental designs are coupled via the choice of appropriate control groups, see for example [23] or [24]. This is most obvious for hypothesis driven reverse genetic studies where the differences in gene expression between mutants or genetically engineered strains are under investigation. Here, genotype replicates, i.e., individuals of a common genotype, constitute the groups for comparison. However, it is important to note that when studying environmental effects on gene expression, proteome or metabolism, or time series of developmental processes, the experimental comparisons also directly reflect the underlying question. Repsilber et al. [24] give examples for this general relationship for each of the principles of comparability. Here, we will now focus on a single example and then discuss the consequences for integrated data analysis.

Consider that many gene expression experiments are similar to a case-control study, where measurements of transcription levels are conducted for single time points only. On the other hand, others may focus on cyclic behaviour and gene expression rhythms or developmental phases of gene expression. For the latter type of studies, typically time series experiments are performed, keeping genotype as well as environmental influences as constant as possible. How can data from these two types of datasets be integrated and analysed together? Let us assume that the genotype used in the developmental series is part of a set of genotypes used for the genetic study. It becomes obvious that the underlying questions as well as the employed experimental designs are not matching. Therefore, it appears reasonable to consider only those data of time series that match the time of measurement for the genetic study. The remaining time series data cannot be used. This is of course a minimal scenario. The situation would be much more promising, if time series data were available for each of the genotypes under comparison. This would correspond to a cross-factorial design. Also, the underlying biological question would then be augmented from a genotype fixed time series study towards the question of time-resolved developmental differences between genotypes.

In summary, every approach of integrated analysis has to build upon a common model of biological function. For each dataset, the associated model limits the possible experimental designs. Hence, it is the intersection of the underlying models for which conclusions derived from integrated data are valid.

Examples of integrated data analysis at the correlation analysis level

Here, we will discuss a collection of examples for integrated data analysis such that the problems of experimental design, study type and data quality for the different levels of gene expression become evident. Only some of these examples come from plant biology, as system level investigations for microorganisms have been much more frequent in the past years. Multicellular organisms show an additional spatial level of organisation, and mostly also a diversity of cell types. Therefore, principal features and problems associated with integrated analysis are often better illustrated using microorganism studies.

Griffin et al. [12] studied transcriptome (mRNA levels) and proteome (protein levels) changes in yeast growing under two different carbon sources. They conclude that although the measurement of transcribed mRNA has proven to be very powerful for the discovery of molecular markers and the elucidation of functional mechanisms, analysed in isolation, it is not sufficient for the characterisation of biological systems as a whole. In their study they show that changes of mRNA levels and those of protein levels most likely account for different parts of regulatory responses to a change in growth medium. Here, the authors were primarily interested in those pairs of mRNA transcripts and corresponding proteins which showed highest discrepancies, because of potential post-translatory regulatory mechanisms. In other words, the underlying functional model suggested two basic classes of genes. One for which transcript levels would correlate well with protein levels, whereas for the other some additional posttranscriptional regulatory mechanism would result in weak correlations. This study illustrates how different levels of gene expression can be analysed together using a correlation analysis approach. This approach is different from correlating or clustering expression levels within transcriptome or proteome data, as shown in the next example.

Smith et al. [25] worked on comparative gene profiling looking for clusters of co-regulated mRNA levels to identify genes coding for novel peroxisomal proteins and proteins involved in peroxisome biosynthesis. For this type of analysis, the underlying functional model assumes that genes relevant for a particular biological process are co-regulated. In this example, proteins for peroxisome assembly or proteins responsible for the special peroxisome biochemical function are possibly co-regulated. Thus, proteins of unknown function but similar transcriptional profiles may have similar functions. Eisen et al. [26] were among the first to propose this type of analysis. Since then, many groups have followed their ideas. Tavazoie et al. [27], for example investigated common transcription factor binding sites in groups of genes which were clustered by common transcription profiles (for a review see [28]).

Ge et al. [19] studied the correlation of transcriptome (co-expression of genes) and proteome (interactions between proteins) and reported that they found a significantly larger amount of interacting proteins within the gene expression clusters than between proteins for genes from different gene expression clusters.

Ideker et al. [29] integrate three different datasets, protein–protein interaction data, protein-DNA, and transcript-levels.

While most integrated data analysis approaches have focused on transcript or protein levels and their mutual interactions, research by Kriete et al. [30] provides an example for correlating tissue data and gene expression levels, thereby including further levels of biological organisation for integrated analysis efforts.

Also metabolomics data are now more and more coming into view for systems biology approaches [31]. Urbanczyk-Wochniak et al. [32] used transcript and metabolite profiling approaches independently to discriminate between different developmental stages and transgenic lines of potato tuber preparations. Then, strong correlations between metabolite and transcript levels were evaluated. Many of the highly correlated pairs of transcript and metabolite data were supported by previous findings reported in the literature.

These examples may be complemented by the overview of a large collection of correlation analysis studies in Table 1 of Searls [15]. In the following section, we will focus on an example that demonstrates how integrated data analysis for a special dataset (see chapter by Steinhauser and Kopka) can be carried out taking a correlation analysis approach.

Integrating 'omics'-data using a correlation analysis approach

This type of analysis has multiple objectives. The first and more obvious one is to find out whether two different attributes of the same biological system contain information about each other. A more ambitious goal is to predict some properties by the knowledge of others. We will start with the simplest and continue with more sophisticated methods. This is motivated by the nature of biological data: While linear models can be easily applied, biological behaviour is often non-linear [33]. Furthermore, the original dimensions, e.g., transcript levels or metabolite concentrations, are often not those of the largest impact. In this case, a transformation of the co-ordinate system is preferable. Before an integrated data analysis can start, the data must be pre-processed at a lower level.

Data preparation and single level analysis

There are three main issues that need to be addressed prior to the actual data analysis: normalisation, missing value imputation, and dimensionality reduction. Raw data of gene expression experiments or metabolite experiments may come from different experiments rendering them different due to systematic differences in labelling, machine sensitivity, or loadings, etc. To make the results comparable and to avoid artificial correlations, normalisation is necessary. The second main problem is that for some attributes (e.g., concentration of substances) in some experiments no value may be available, either because no measurement was carried out or it was not reliable. Figure 1 illustrates this problem. Often, a gene by gene approach or metabolite by metabolite approach is applied to the analysis of gene expression or metabolite profile. However, higher order effects resulting from a combination of genes or metabolites cannot be discerned this way. Principal component analysis

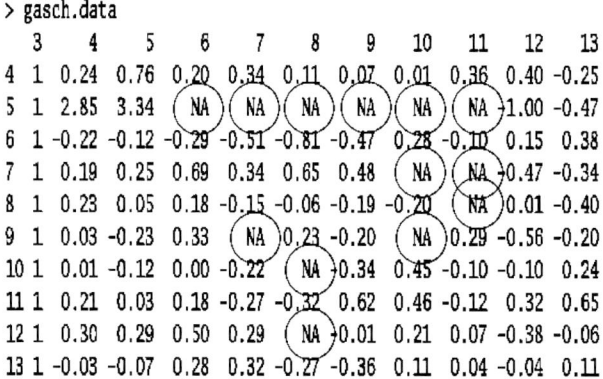

Figure 1. Missing data in a gene expression matrix. Gene expression data are from Gasch et al. [36]. Missing values are marked by a circle. It is obvious that reducing the dataset to samples (columns) or genes (rows) without missing values only 40% would remain of the entire dataset for further analyses. This motivates data imputation approaches as discussed in the text.

(PCA) and independent component analysis (ICA) are widely used to find the most significant of these combinations. Thus, the number of dimensions can be drastically reduced. We refer to this class of methods as dimensionality reduction methods.

Normalisation of metabolite profiles

The intensity of metabolite profiles may vary strongly due to machine sensitivity. This may result in great overall differences under various conditions. In this case, a normalisation of each metabolite intensity by the mean (or the median) of all metabolite intensities in one sample can be appropriate. However, to avoid systematic errors, the number of detected metabolites should be considerable, i.e., they should be a representative part of the total number of metabolites as this approach assumes that most metabolite levels do not change.

A similar approach was used by Scholz et al. [34]. The dataset contained metabolite profiles from different *Arabidopsis thaliana* lines and their crosses. Here, each component of a vector was divided by the l_2-vector norm. PCA was applied to the dataset which has been normalised in this way. Compared with other normalisation methods (like unit variance), this yielded the best visualisation and separation of biological states.

If internal controls are available, normalisation is much facilitated. In this case, all measured values could be divided by the median of the controls.

Missing value estimation

There are two simple methods for the estimation of missing values: Replacement by the mean or median of the metabolite intensities across different samples and replacement by the mean of the nearest neighbours. A new method was introduced by Scholz et al. [35]. This method is based on inverse non-linear PCA. It was applied to gene expression and metabolite data from *Arabidopsis thaliana* under cold stress. A proportion of known values were artificially removed from this dataset and estimated with this method so that the mean square error of this estimation could be calculated. Compared with linear methods or replacement by mean better results were obtained on non-linear datasets [35].

Principal component analysis and independent component analysis

Often, combinations of metabolites rather than individual ones have a larger impact on a biological question under investigation. Principal component analysis (PCA) is one of the classical methods to find such combinations. PCA is looking for a transformation of dimensions which yields the largest variance. The first principal component (PC) is then the linear combination which explains the largest amount of variation. The second PC explains the largest amount of the remaining variation. By omitting PCs with associated low variance, the number of dimensions can be reduced. PCs can be naturally ordered that way. PCA is an orthogonal transformation that results in uncorrelated components.

However, this does not mean that PCs are statistically independent. Independent component analysis (ICA) seeks dimensions which are statistically independent. However, a limitation of ICA is that it cannot be applied to high dimensional data spaces. To overcome this problem ICA can be applied to the most important PCs only. Scholz et al. [34] have demonstrated this method. Figure 2 illustrates both methods, demonstrating the advantages of ICA. The independent components (IC)

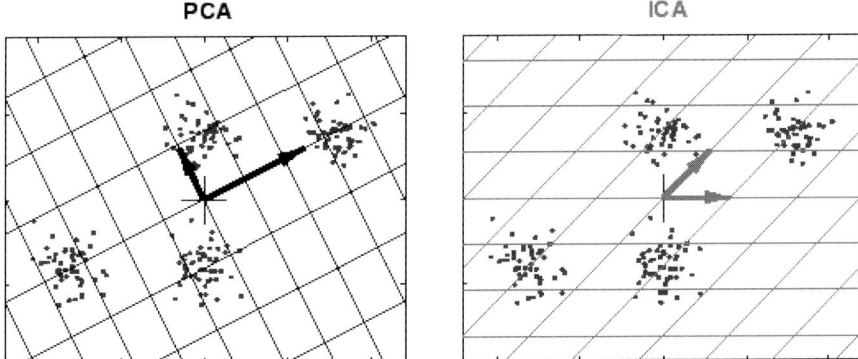

Figure 2. Comparing coordinate transformations for principal component analysis (PCA) and independent component analysis (ICA). PCA yields principal components capturing the largest variance in the dataset. The result is a transformation which can be illustrated as viewing the data in a new coordinate system the axis of which correspond to the first two principal components. By contrast, ICA optimises the coordinates for the transformed system to be as independent as possible. The resulting transformation yields a different view of the data, which is often more appropriate for separating factors that influence gene expression or metabolite concentrations.

can be ordered by their *kurtosis,* which is a measure of deviation from the normal distribution. Hence, ICA is advantageous when applied to non-normal distributions.

Pair-wise correlation analysis

Let us consider the example presented in the chapter by Steinhauser and Kopka. Gene expression levels and metabolite concentrations were measured for seven developmental stages. For 1,000 genes and 140 metabolites measurements were obtained from the same samples. The corresponding data are thus represented by a 7 x 1,000 matrix and a 7 x 140 matrix. The most basic level of integrated data analysis now is pair-wise correlation analysis. Here, gene-metabolite pairs are correlated under the assumption that they are independent of the effects from other metabolites or genes. Pearson correlation also only measures the linear dependence between genes and metabolites. One objective of this approach is to detect genes which are responsible for the modification of the metabolite content in biological systems [32]. However, there are different ways to correlate the data: first we can compare gene expression levels and metabolite concentrations at the same developmental stages. Alternatively, we can compare the gene expression levels at one stage and metabolite concentration at the next. In the latter case the obvious interpretation for a positive correlation between gene A and metabolite B would be that gene A synthesises an enzyme responsible for the production of B – so higher levels of A will be followed by an increase in the concentration of B. A synchronous correlation of A and B could indicate that there is the same effect as above but the time resolution is too low or both A and B are regulated by another gene C.

This method has the advantage that it can easily be implemented and interpreted, because of the simplicity of its underlying model. But often this model is too simple. Pearson correlation cannot measure non-linear dependencies and even a correlation coefficient of zero does not necessarily indicate statistical independence. The Spearman rank correlation, as an alternative, is robust against outliers, but it cannot detect non-monotonic dependencies. Since, in practice, very often non-linear and even non-monotonic dependence is observed, mutual information (MI), an entropy-based similarity measure, is widely used as an alternative. Butte and Kohane [37] were probably the first to applied MI to gene expression data. Steuer et al. [38] improved the method. Zero MI means statistical independence, maximal MI means that one variable is determined by the other.

Pair-wise correlation raises another problem: Are the calculated correlations significant? This is especially problematic as sample sizes are usually small. In our case we compare vectors of length 7. To answer the question of statistical significance we can use correlation tests [39]. But these tests are often dependent on certain assumptions about the distribution of the values and are only univariate tests. In our example, we have two (for the synchronous and the delayed case) x 1,000 (number of genes) x 140 (number of metabolites) correlations, thus we expected a considerable number of these correlations to be very high by chance. How to distinguish those from the true correlations is addressed by the theory of multiple testing. There will be a note on this theory later in this chapter.

Finally, it is possible that no pair-wise correlation is significant although there are strong linear relations between the two datasets. But instead of correlations between the columns of the matrices, there may be strong correlations between linear combinations of columns. This problem is addressed by canonical correlation analysis (CCA) and kernel canonical correlation analysis (KCCA), which we shall discuss now.

Canonical correlation analysis

Canonical correlation analysis (CCA) and the closely related kernel correlation analysis (KCCA) are well established methods for multivariate data analysis. CCA was originally introduced by Hotelling [40] and since then widely used in engineering and computer science. Hardoon et al. [41] have used KCCA for learning semantics of multimedia content by combining image and text data. Thereby, the retrieval of images given the texts was possible. Since the analogous problem of multiple datasets belonging to the same class of objects arise in systems biology, the method has recently been used in this area [42].

CCA is concerned with detecting and describing a linear relationship between sets of variables. As an example, let us imagine that we have two kinds of measurements for m objects. The vectors of these measurements for each object are denoted by $\mathbf{x}_i$ and $\mathbf{y}_i$ ($i=1,\ldots,m$) in the following. There may be n_x measurements for the first aspect and n_y for the second. The dataset can be represented by two matrices: $\mathbf{X}$ a ($m \times n_x$) matrix, and $\mathbf{Y}$ a ($m \times n_y$) matrix (see Fig. 3). In our example, $m = 7$ (number of developmental stages), $n_x = 1,000$ (number of genes), and $n_y = 140$ (number of metabolites).

The goal of CCA is to find new co-ordinate systems such that correlations between the two matrices are maximised. In the original co-ordinate systems the correlations were those between the columns of the matrices. In our example, we consider cor-

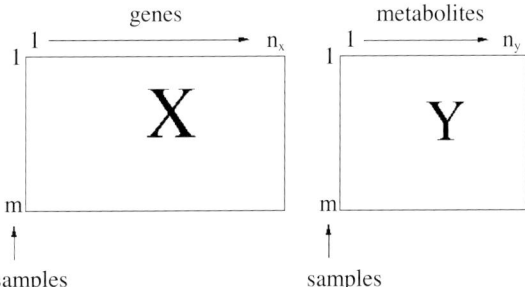

Figure 3. Data scheme as prepared for canonical correlation analysis (CCA). The gene expression data matrix, X, is arranged such that every row contains measurements for all n_x genes in sample i. The metabolite measurements, matrix Y, have been carried out for the same m samples. Typically there are less metabolites than genes, and hence $n_x < n_y$.

relations between gene expression levels and metabolite concentrations. The correlations which are maximised by CCA are those between linear combinations of the columns. Mathematically, this can be formulated as a constraint optimisation problem. The solution of this problem can be found by solving a combined eigenproblem. For details see [43] and [41].

The obtained eigenvectors pairs (**v, w**) are the basis vectors of a new co-ordinate system in which correlations are maximised. These vectors are called canonical vectors and the associated eigenvalues are equal to the correlation coefficients. As in the case of PCA, there is a natural order of the canonical vectors. The first canonical vector pair is the one that yields the highest correlation. The linear combinations (**Xv, Yw**) of columns yielding maximal correlation are calculated by multiplying of original data matrices (**X, Y**) with their corresponding canonical vectors (**v, w**). These combinations reveal higher order correlations between the datasets. In our example, **Xv** would represent a linear combination of gene expression values, **Yw** a combination of metabolite concentrations.

Another interesting property of CCA is that it solves the problem of finding linear combinations of columns of the respective other matrix which can be most accurately predicted by a least square regression [43].

There are some limitations to the CCA approach. First, there should not be many more dimensions (measurements) than objects, because in that case there will always be good correlations in the original co-ordinate system occurring by chance and the risk of overfitting is high. Another problem is that similar to the pair-wise correlation, there may be non-linear relations between the two datasets. Here, KCCA offers a solution that is widely applied.

Kernel canonical correlation analysis

This procedure is identical to a non-linear mapping of the original data into a usually high-dimensional feature spaces. Let φ_x and φ_y denote these mappings, then object vectors x_i and y_i ($I = 1,..., m$) are replaced by $\varphi_x(x_i)$ and $\varphi_y(y_i)$. The kernel functions $k_x(x_i, x_j)$, $k_y(y_i, y_j)$ are defined by the scalar products of these vectors. The calculation of this scalar product is substituted by a much simpler function. This substitution is referred to as the kernel trick in the literature [41]. Frequently used kernels are polynomial kernels and the Gaussian radial basis function (RBF). The kernels are represented by the matrices K_x and K_y with their components resulting from the scalar products $<\varphi_x(x_i), \varphi_y(y_i)>$. Analogously to CCA, we look for eigenvectors pairs (**α, β**), which solve the optimisation problems explained above. But instead of the original matrices **X** and **Y**, K_x and K_y are used. The vector with the largest eigenvalue corresponding to the highest correlation is the solution.

KCCA can also be used as a learning procedure. For this purpose, a dataset must be divided in two parts, a training set and a test set. In the training set all observations represented by the rows of the matrices are attributed to each other correctly. Therefore the KCCA is performed in the training set. The results can be used to retrieve the correct parings of objects in the test set, which were unknown. This can also be formulated as the prediction of an observation **y** by an observation **x**, e.g., of the metabolite profile by the gene expression profile.

Often, we are confronted with a large number of variables (e.g., genes), so that dimensionality reduction is necessary [41]. Also, kernel matrices have to be inverted which in some cases is not possible. Regularisation and dimensionality reduction methods are used to address these problems.

Yamanishi et al. [42] have extended and applied this approach to the problem of finding clusters of genes, which share similarities with respect to multiple biological attributes. More precisely, Yamanishi et al. [42] looked for operons in prokaryotes. Genes which form operons are close to one another on the chromosome, can code for enzymes belonging to the same pathways and have similar expression patterns. The authors used pathway information from KEGG [44], gene position information from BRITE, and data from microarray experiments as input. The data were derived from two organisms: Yeast and *E. coli*. Obviously it was necessary to adapt KCCA to more than two datasets. This was done in a straightforward manner: Instead of the correlation between a pair of matrices, the sum over correlations of all possible pairs was maximised. This method is referred to as multiple kernel canonical correlation analysis (MKCCA). Second, an integrated approach was developed (IKCCA). Here, the datasets are divided into two groups, the kernels are represented by the sum over the kernels of each group. In [42] gene position and expression together were compared to the pathway information. The authors investigated two questions: Are there strong correlations between the multiple attributes (pathways, genome, expression) and could this correlation be used to predict the operons correctly? Strong correlations were found, and 19 out of 26 genes were assigned to the correct operon by IKCCA. Ordinary KCCA and MKCCA had a somewhat lower rate of correct predictions.

These results show that KCCA or possible extensions should be tested in a wider range of questions in systems biology. For instance, the comparison of metabolite data and physiological data by KCCA may contribute to a prediction of physiological phenotype based on the metabolite profile in an early stage. A requirement for this is that a large number of samples (e.g., from different genotypes) will be measured in parallel.

Clustering methods

Genome-wide probing of the transcriptome, proteome or metabolome results in large datasets. Taking the physiological point of view, it is clear that many of the measured profiles should be strictly correlated, as it is known that frequently groups of genes and their products act together, for example in metabolic or signalling pathways. These groups of genes should be co-regulated, as it appears plausible that their action is required jointly or not at all. Again, different models of biological function will lead to different algorithms to find groups of co-regulated genes. Once such groups of genes are characterised for specific classes, say transcriptome and proteome, integrated data analysis can be applied to the condensed data.

Clustering of gene expression data is also often used to infer regulatory networks as discussed in the following chapter of this book.

Before discussing examples for integrated data analysis at the level of correlating gene groups, let us consider the process of clustering itself to understand the variety of available options. In general, unsupervised clustering methods are used to find groups of co-regulated genes. Firstly, a similarity measure of two given expression profiles has to be defined. Secondly, the algorithm for grouping together similar profiles has to be chosen. Both choices influence the overall results. In general, the underlying hypothesis should determine the algorithm to employ, but it is still not clear which clustering algorithm has to be chosen in which case. However, some general features seem already apparent [45]. In practice, similarity between profiles (e.g., protein or metabolite profiles) is often measured by employing a Euclidean distance measure, a Pearson correlation or a mutual information approach. Daub et al. [46] compared different algorithms to calculate mutual information as a measure to characterise dependencies between genes using their expression profiles, also allowing the detection of non-linear dependency structures. As far as the clustering algorithms are concerned, it is possible to choose between hierarchical and non-hierarchical approaches. Within the hierarchical approaches, agglomerative algorithms comprise hierarchical clustering methods together with their different linkage methods (single, average and maximum) [26, 47]. An example for using a divisive hierarchical algorithm can be found in the work by Alon et al. [48]. A popular example for a non-hierarchical divisive clustering algorithm is K-means clustering [27]. Furthermore, Self-Organising Maps [49] and Quality clustering [50] are being employed. Michaels et al. [51] demonstrated how different combinations of distance measures and clustering algorithms can result in very different gene groupings. Gibbons and Roth [45] suggest how to assess and compare the quality of different clustering approaches by assessing their ability to display groups of functionally related genes. Functional relation of genes was measured in terms of common attributes of gene ontology. Enrichment of clusters with functionally related genes was assessed using a mutual information measure. It is also conceivable to use a gene's functional information during the clustering.

Now let us consider two examples where clustering has been used serving an integrated data analysis approach. Ge et al. [19] proposed what they call a "transcriptome-interactome correlation mapping", where protein–protein interaction complexes are correlated to clusters of co-expression at the transcript level. In more detail, for every co-expression cluster the ratio of protein–protein interactions observed as related to the total number of possible interaction pairs within the cluster was scored. This score was represented in a colour coding scheme across the matrix of all pairs of co-expression clusters. Within clusters of gene expression profiles, protein interaction pairs and even triplets were significantly enriched which was not the case for the negative control of randomised interactions. The authors suggest that the observed correlations may help to identify expression clusters with relatively greater biological relevance. Also *vice versa*, they provide an example where the information from co-expression clusters was used to refine a hypothesis resulting from protein–protein interaction maps.

Data mining – Integration at the annotation level

In biomedical and molecular biology research the term 'data mining' is associated with several forms of integrated data analysis – at the highest possible level results from different experiments are integrated, mostly by searching databases for functional annotations for a given list of candidate genes from a genome-wide experiment. These searches range from manual browsing to an automated search strategy. Manually conducted searches are based on the associative memory of the individual researcher. However, for automated searches, standardised descriptions of gene identities and functions are an essential prerequisite. In this context, gene ontologies allow the integration of results from experiments covering different levels of biological organisation by providing a common descriptive framework and a controlled vocabulary [20, 21]. We also explicitly point to Table 2 in the review article by Searls [15] as an overview of integrated gene-centric data-mining resources. Methodologically, at this level of integrated analysis, we are confronted with a plethora of approaches, ranging from standard relational database queries, general-purpose visualisations, statistical learning approaches, to enhanced text retrieval within scientific literature.

The overall character of this level of integrated data analysis remains that of a data mining approach. The goal is to detect patterns in combinations of results. This should not be confused with a meta-analysis type of approach which aims at combining datasets of different nature capable to test complex hypotheses about biological function.

A note on multiplicity

As already pointed out before, experimental design issues are crucial for any integrated data analysis, especially when it comes to using disparate data from different experiments. Another issue that applies to all high-throughput data is the problem of multiple testing. In combining the significant results from multiple layers of biological organisation, sorting out the significant interactions has to be assessed with care. In principle, the same rationale applies here as for a single high-throughput experiment, such that available strategies for control of family-wise error rates should be applied. For the screening type of approach in high-throughput studies it has become apparent that possibly a more liberal criterion, the control of false discovery rates, is more suitable [52]. For a comprehensive review of these topics we refer to [53].

Software solutions for integrated data analysis

Two major platforms exist that provide a more general environment for statistical analysis, for the implementation of user-defined algorithms and as powerful platforms for data and pattern visualisation: The first we like to mention is the R-language [54, 55], including a collection of more specialised packages, for example the

Table 1. Contents a list of useful links, from where the tools mentioned above can be downloaded

R	www.r-project.org/
MATLAB	www.mathworks.com/products/matlab/
Bioconductor	www.bioconductor.org/
Cytoscape	www.cytoscape.org/
Osprey	biodata.mshri.on.ca/osprey/servlet/Index
Genevestigator	www.genevestigator.ethz.ch/
MAPMAN	gabi.rzpd.de/projects/MapMan/
MetaGenAlyse	metagenealyse.mpimp-golm.mpg.de/

Bioconductor package system [56]. R is a language and environment for statistical computing and graphics as a GNU project. R provides a wide variety of statistical functionalities such as linear and non-linear modelling, classical statistical tests, time-series analysis, classification, clustering, and graphical techniques, and is highly extensible. There are also functions for CCA and KCCA available for R. Secondly we refer to MATLAB [57], a commercial mathematical software system for solving a variety of mathematical and statistical problems and graphical representation of the results. Recently, MathWorks has launched the 'Bioinformatics Toolbox' as an extension to MATLAB for integrated analysis of sequence, protein and gene expression data. Another, more specialised system is the 'Gene Expression Dynamics Inspector' [58], a software enabling the analysis of gene expression profiles with Self-Organising Maps (SOM). Both R and MATLAB include many ready-to-use functions for expression data analysis, database access, multiple data format handling, normalisation and visualisation of data, and advanced statistical analysis. They also provide comprehensive, intuitive environments to rapidly extend and develop the tools needed for a particular analysis. For more specialised software tools in our context, we like to mention short selection of helpful applications.

Graphical representation of metabolite pathways together with the option to visually integrate gene expression information or other 'omics'-data is possible using software packages such as MAPMAN [59] or Genevestigator [60, 61]. Platforms for the visualisation of general interactions available include 'Osprey', a software for the visualisation and manipulation of complex interaction networks. Osprey builds data-rich graphical representations that are colour-coded for gene function and experimental interaction data [62]. An alternative and broadly used tool is 'Cytoscape' [63], an open source software project for integrating biomolecular interaction networks with high-throughput expression data and other molecular states into a unified conceptual framework. A software tool more focused on dimensionality reduction methods as well as correlation approaches is MetaGeneAlyse [64]. An overview over its functionality is given in Figure 4.

Integrated data analysis for genome-wide research 325

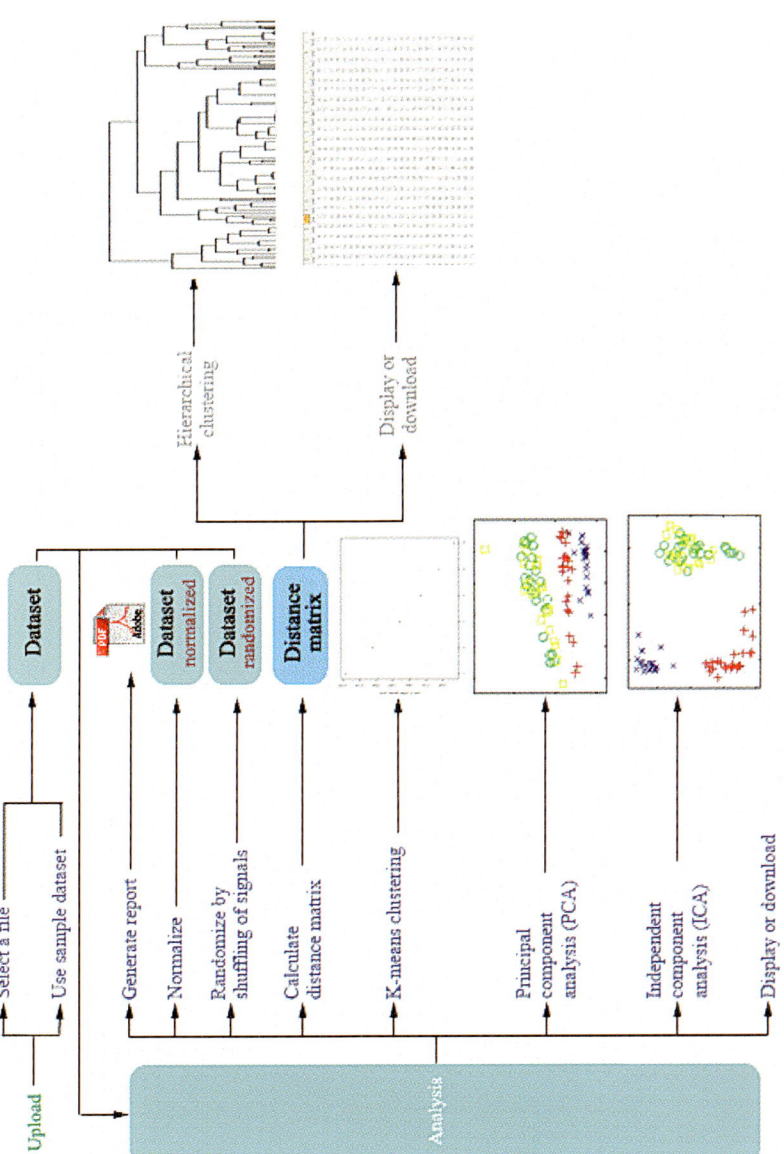

Figure 4. Data processing in MetaGenAlyse [64]. This software tool is accessible via the web that allows the analysis of integrated datasets that combine gene-expression data and metabolic data. After uploading, datasets can be normalised, clustered by various methods and results can be visualised. All calculations are carried out on a server, so even time- and memory-consuming analyses can be done independently of the performance of the client.

Summary

Integrated data analysis was introduced as the intermediate level of a systems biology approach to analyse different 'omics'-datasets, i.e., genome-wide measurements of transcripts, protein levels or protein–protein interactions, and metabolite levels aiming at assessing a coherent understanding of biological function.

Several examples demonstrated a variety of specific challenges inherent to this approach, such as the necessities for data normalisation and missing value estimation. We focussed on different methods of correlation analyses to highlight their relevance in integrated data analysis. At this level of data analysis, researchers typically are confronted with the fact that data from diverse experimental platforms are less correlated than assumed [11]. This can be due to technical as well as biological reasons. Technically, large amounts of noise in the data are often caused by multi-step experimental protocols, where each step may incorporate a substantial amount of variance or bias. Biologically, it seems plausible that interactions are often too complex to result in simple linear correlations. Non-linear correlations or such hidden in the original multidimensional representation of the data, however, need larger datasets for a reliable detection and more sophisticated methods.

After discussing important stumbling blocks one has to be aware of when combining genome-wide datasets, we also have introduced some of the available tools for such analyses. It is important to note that any sensible application of these tools has to build upon a sound understanding of how datasets have to be prepared for an integrated analysis, considering different biases and methods for their normalisation as well as multiplicity.

System-wide experiments require sound experimental designs such that data from disparate 'omics'-classes become usable in integrated analyses. Without such a design, the strict requirements for dataset combination frequently result in a very sparse intersection of testable hypotheses. In this respect, standardisation remains the most important task for future system-wide approaches.

References

1. Somogyi R, Sniegoski CA (1996) Modeling the complexity of genetic networks: understanding multigenic and pleiotropic regulation. *Complexity* 1(6): 45–63
2. Gygi S, Rochon Y, Franza B, Aebersold R (1999) Correlation between protein and mRNA abundance in yeast. *Mol Cell Biol* 19(3): 1720–1730
3. Noble D (2002) Modeling the heart-from genes to cells to the whole organ. *Science* 295(5560) 1678–1682
4. Grünenfelder B, Winzeler EA (2002) Treasures and traps in genome-wide datasets: case examples from yeast. *Nat Rev Genetics* 3: 653–661
5. Shevchenko A, Jensen O, Podtelejnikov A, Sagliocco F, Wilm M, Vorm O, Mortensen P, Shevchenko A, Boucherie H, Mann M (1996) Linking genome and proteome by mass spectrometry: large-scale identification of yeast proteins from two dimensional gels. *Proc Natl Acad Sci USA* 93(25): 14440–14445
6. Pandey A, Mann M (2000) Proteomics to study genes and genomes. *Nature* 405: 837–846

7. Walhout A, Vidal M (2001) Protein interaction maps for model organisms. *Nat Rev Mol Cell Biol* 2(1): 55–62
8. Fiehn O, Kopka J, Dormann P, Altmann T, Trethewey R, Willmitzer L (2000) Metabolite profiling for plant functional genomics. *Nat Biotechnol* 18(11): 1157–1161
9. Roessner U, Luedemann A, Brust D, Fiehn O, Linke T, Willmitzer L, Fernie A (2001) Metabolic profiling allows comprehensive phenotyping of genetically or environmentally modified plant systems. *Plant Cell* 13(1): 11–29
10. Fernie A, Trethewey R, Krotzky A, Willmitzer L (2004) Metabolite profiling: from diagnostics to systems biology. *Nat Rev Mol Cell Biol* 5(9): 763–769
11. Klipp E, Herwig R, Kowald A, Wierling C, Lehrach H (2005) *Systems biology in practice – concepts, implementation and application,* chapter 1.3, Wiley-VCH Verlag, Weinheim, Germany, 11–17
12. Griffin TJ, Gygi SP, Ideker T, Rist B, Eng J, Hood L, Aebersold R (2002) complementary profiling of gene expression at the transcriptome and proteome levels in *Saccharomyces cerevisiae*. *Mol Cell Proteomics* 1(4): 323–333
13. Aitchison JD, Galitski T (2003) Inventories to insights. *J Cell Biol* 161(3): 465–469
14. Wissel C (1992) Aims and limits of ecological modelling exemplified by island theory. *Ecol Model* 63: 1–12
15. Searls D (2005) Data integration: challenges for drug discovery. *Nat Rev Drug Discov* 4(1): 45–58
16. Park P, Cao Y, Lee S, Kim J, Chang M, Hart R, Choi S (2004) Current issues for DNA microarrays: platform comparison, double linear amplification, and universal RNA reference. *J Biotechnol* 112(3): 225–245
17. Aebersold R, Hood L, Watts J (2000) Equipping scientists for the new biology. *Nat Biotechnol* 18(4): 359
18. Weinstein JN (2002) 'Omic' and hypothesis-driven research in the molecular pharmacology of cancer. *Curr Opin Pharmacol* 2: 361–365
19. Ge H, Liu Z, Church GM, Vidal M (2001) Correlation between transcriptome and interactome mapping data from *Saccharomyces cerevisiae*. *Nature Genetics* 29: 482–486
20. Ashburner M, Ball C, Blake J, Botstein D, Butler H, Cherry J, Davis A, Dolinski K, Dwight S, Eppig J et al. (2000) Gene ontology: tool for the unification of biology. The Gene Ontology Consortium. *Nat Genet* 25(1): 25–29
21. The Plant Ontology Consortium (2002) The Plant Ontology Consortium and Plant Ontologies. *Comp Funct Genomics* 3: 137–142
22. Hazbun T, Malmstrom L, Anderson S, Graczyk B, Fox B, Riffle M, Sundin B, Aranda J, McDonald W, Chiu C et al. (2003) Assigning function to yeast proteins by integration of technologies. *Mol Cell* 12(6): 1353–1365
23. Wacholder S, McLaughlin JK, Silverman DT, Mandel JS (1992) Selection of controls in case-control studies. I. principles. *Am J Epidemiol* 135(9): 1019–1028
24. Repsilber D, Fink L, Jacobsen M, Bläsing O, Ziegler A (2005) Sample selection for microarray gene expression studies. *Meth Info Med* 44(3): 461–467
25. Smith JJ, Marelli M, Christmas RH, Vizeacoumar FJ, Dilworth DJ, Ideker T, Galitski T, Dimitrov K, Rachubinski RA, Aitchison JD (2002) Transcriptome profiling to identify genes involved in peroxisome assembly and function. *J Cell Biol* 158(2): 259–271
26. Eisen MB, Spellman PT, Brown PO, Botstein D (1998) Cluster analysis and display of genome-wide expression patterns. *Proc Natl Acad Sci USA* 95: 14863–14868
27. Tavazoie S, Hughes JD, Campbell MJ, Cho RJ, Church GM (1999) Systematic determination of genetic network architecture. *Nature Genetics* 22(3): 281–285

28. Qiu P (2003) Recent advances in computational promoter analysis in understanding the transcriptional regulatory network. *Biochem Biophys Res Commun* 309(3): 495–501
29. Ideker T, Ozier O, Schwikowski B, Siegel AF (2002) Discovering regulatory and signalling circuits in molecular interaction networks. *Bioinformatics* 18 (Suppl.1): S233–S240
30. Kriete A, Anderson MK, Love B, Freund J, Caffrey JJ, Young MB, Sendera TJ, Magnuson SR, Braughler JM (2003) Combined histomorphometric and gene-expression profiling applied to toxicology. *Genome Biol* 4: R32
31. Weckwerth W (2003) Metabolomics in systems biology. *Annu Rev Plant Biol* 54: 669–689
32. Urbanczyk-Wochniak E, Luedemann A, Kopka J, Selbig J, Roessner-Tunali U, Willmitzer L, Fernie A (2003) Parallel analysis of transcript and metabolic profiles: a new approach in systems biology. *EMBO Rep* 4(10): 989–993
33. Nilsson J, Fioetos T, Höglund M, Fontes M (2004) Approximate geodetic distances reveal biological relevant structure in microarray data. *Bioinformatics* 20(6): 874–880
34. Scholz M, Gatzek S, Sterling A, Fiehn O, Selbig J (2004) Metabolite fingerprinting: detection of biological features by independent component analysis. *Bioinformatics* 20: 2447–2454
35. Scholz M, Kaplan F, Guy CL, Kopka J, Selbig J (2005) Nonlinear PCA: a missing data approach. *Bioinformatics*, Advance Access published online 18 August 2005
36. Gasch AP, Spellmann PT, Kao CM, Carmel-Harel O, Eisen M, Storz, Botstein D, Brown PO (2000) Genomic expression programs in the response of yeast cells to environmental changes. *Mol Biol Cell* 11: 4241–4257
37. Butte A, Kohane IS (2000) Mutual information relevance networks: Functional genomic clustering using pair-wise entropy measurements. *Pac Symp Biocomput* 5: 415–426
38. Steuer R, Kurths J, Daub C, Weise J, Selbig J (2002) The mutual information: Detecting and evaluating dependencies between variables. *Bioinformatics* 18: S231–S240
39. Best DJ, Roberts DE (1975) Algorithm AS 89: The upper tail probabilities of spearman's rho. *Appl Stats* 24: 377–379
40. Hotelling H (1936) Relation between two sets of variates. *Biometrica* 28: 312–377
41. Hardoon D, Szedmak S, Shawe-Taylor J (2003) *Canonical correlation analysis; An overview with application to learning methods.* Technical Report CSD-TR-03-02. Department of Computer Science, University of London, UK
42. Yamanishi Y, Vert JP, Kanehisa M (2003) Extraction of correlated gene clusters from multiple genomic data by generalized kernel canonical correlation analysis. *Bioinformatics* 19: Suppl 1 i323–330
43. Kuss M, Graepel T (2003) *The geometry of kernel canonical analysis.* Technical Report No. 108, Max Planck Institute for Biological Cybernetics
44. Kanehisa M, Goto S, Kawashima S, Nakaya A (2002) The KEGG databases at GenomeNet. *Nucleic Acids Res* 30: 42–45
45. Gibbons F, Roth F (2002) Judging the quality of gene expression-based clustering methods using gene annotation. *Genome Res* 12(10): 1574–1581
46. Daub C, Steuer R, Selbig J, Kloska S (2004) Estimating mutual information using B-spline functions–an improved similarity measure for analysing gene expression data. *BMC Bioinformatics* 5: 118
47. Wen X, Fuhrman S, Michaels GS, Carr DB, Smith S, Barker JL, Somogyi R (1998) Large-scale temporal gene expression mapping of central nervous system development. *Proc Natl Acad Sci USA* 95: 334–339
48. Alon U, Barkai N, Notterman D, Gish K, Ybarra S, Mack D, Levine A (1999) Broad patterns of gene expression revealed by clustering analysis of tumor and normal colon tissues probed by oligonucleotide arrays. *Proc Natl Acad Sci USA* 96(12): 6745–6750

49. Tamayo P, Slonim D, Mesirov J, Zhu Q, Kitareewan S, Dmitrovsky E, Lander E, Golub T (1999) Interpreting patterns of gene expression with self-organising maps: methods and application to hematopoietic differentiation. *Proc Natl Acad Sci USA* 96(6): 2907–2912
50. Heyer L, Kruglyak S, Yooseph S (1999) Exploring expression data: identification and analysis of coexpressed genes. *Genome Res* 9(11): 1106–1115
51. Michaels GS, Carr DB, Askenazi M, Fuhrman S, Wen X, Somogyi R (1998) Cluster analysis and data visualization of large-scale gene expression data. *Pac Symp Biocomp* 3: 42–53
52. Storey JD, Tibshirani R (2003) Statistical significance for genomewide studies. *Proc Natl Acad Sci USA* 100(16): 9440–9445
53. Broberg P (2005) A comparative review of estimates of the proportion unchanged genes and the false discovery rate. *BMC Bioinformatics* 6: 199
54. Ihaka R, Gentleman R (1996) R: a language for data analysis and graphics. *J Comp Graph Stats* 5(3): 299–314
55. R Development Core Team (2005) *R: A language and environment for statistical computing*. R Foundation for Statistical Computing, Vienna, Austria
56. Gentleman RC, Carey VJ, Bates DM, Bolstad B, Dettling M, Dudoit S, Ellis B, Gautier L, Ge Y, Gentry J et al. (2004) Bioconductor: open software development for computational biology and bioinformatics. *Genome Biol* 5: R80
57. MathWorks IUC (2000) MATLAB
58. Eichler G, Huang S, Ingber D (2003) Gene Expression Dynamics Inspector (GEDI): for integrative analysis of expression profiles. *Bioinformatics* 19(17): 2321–2322
59. Thimm O, Bläsing O, Yves Gibon OB, Nagel A, Meyer S, Krüger P, Selbig J, Müller LA, Rhee SY, Stitt M (2004) MAPMAN: a user-driven tool to display genomics data sets onto diagrams of metabolic pathways and other biological processes. *Plant J* 37: 914–939
60. Zimmermann P, Hennig L, Gruissem W (2005) Gene-expression analysis and network discovery using Genevestigator. *Trends Plant Sci* 10(9): 407–409
61. Zimmermann P, Hirsch-Hoffmann M, Hennig L, Gruissem W (2004) GENEVESTIGATOR. *Arabidopsis* microarray database and analysis toolbox. *Plant Physiol* 136(1): 2621–2632
62. Breitkreutz B, Stark C, Tyers M (2003) Osprey: a network visualization system. *Genome Biol* 4(3): R22
63. Shannon P, Markiel A, Ozier O, Baliga N, Wang J, Ramage D, Amin N, Schwikowski B, Ideker T (2003) Cytoscape: a software environment for integrated models of biomolecular interaction networks. *Genome Res* 13(11): 2498–2504
64. Daub C, Kloska S, Selbig J (2003) MetaGeneAlyse: analysis of integrated transcriptional and metabolite data. *Bioinformatics* 19(17): 2332–2333

Plant Systems Biology
Edited by Sacha Baginsky and Alisdair R. Fernie
© 2007 Birkhäuser Verlag/Switzerland

Network analysis of systems elements

Daniel Schöner[1], Simon Barkow[2], Stefan Bleuler[2], Anja Wille[3], Philip Zimmermann[1], Peter Bühlmann[3], Wilhelm Gruissem[1] and Eckart Zitzler[2]

[1] *Plant Biotechnology, Institute of Plant Sciences, Rämistr. 2,*
[2] *Computer Engineering and Networks Laboratory, Gloriastr. 35,*
[3] *Seminar for Statistics, Leonhardstr. 27, Swiss Federal Institute of Technology (ETH), 8092 Zürich, Switzerland*
All authors are part of the Reverse Engineering Group at the Swiss Federal Institute of Technology (ETH), Zürich

Abstract

A central goal of postgenomic research is to assign a function to every predicted gene. Because genes often cooperate in order to establish and regulate cellular events the examination of a gene has also included the search for at least a few interacting genes. This requires a strong hypothesis about possible interaction partners, which has often been derived from what was known about the gene or protein beforehand. Many times, though, this prior knowledge has either been completely lacking, biased towards favored concepts, or only partial due to the theoretically vast interaction space. With the advent of high-throughput technology and robotics in biological research, it has become possible to study gene function on a global scale, monitoring entire genomes and proteomes at once. These systematic approaches aim at considering all possible dependencies between genes or their products, thereby exploring the interaction space at a systems scale. This chapter provides an introduction to network analysis and illustrates the corresponding concepts on the basis of gene expression data. First, an overview of existing methods for the identification of co-regulated genes is given. Second, the issue of topology inference is discussed and as an example a specific inference method is presented. And lastly, the application of these techniques is demonstrated for the *Arabidopsis thaliana* isoprenoid pathway.

Introduction

"No protein is an island entire of itself." [1]

A central goal of postgenomic research is to assign a function to every predicted gene. Molecular biology has so far focused on assaying genes and their products

---

[1] The original quote "No man is an island entire of himself" by the English poet John Donne (Donne 1624) was modified by A. Kumar and M. Snyder to express that proteins do not carry out their functions alone, but in pairs or complexes (Kumar 2002).

individually to gain insight into their molecular roles. Since genes often cooperate in order to establish and regulate cellular events the examination of a gene has also included the search for at least a few interacting genes. This requires a strong hypothesis about possible interactions partners, which has often been derived from what was known about the gene or protein beforehand. Many times, though, this prior knowledge has either been completely lacking, biased towards favored concepts, or only partial due to the theoretically vast interaction space.

With the advent of high-throughput technology and robotics in biological research, it has become possible to study gene function on a global scale, monitoring entire genomes and proteomes at once. These systematic approaches aim at considering all possible dependencies between genes or their products, thereby exploring the interaction space at a systems scale. The large amounts of data generated by these efforts eventually generate a whole network of interactions. As a consequence, within the scope of systems approaches the attention has shifted from the study of individual genes to the investigation of the intracellular network as a whole.

Current experimental standard, though, does not provide a complete view of one unified network. Depending on the experiments, only different aspects of this network come to the fore. As a result, distinct subnetworks can be considered that display different physiological functioning and consist of different units and connections; examples for specific subnetworks are:

- **The genetic regulatory network** describes the dependencies of genes on the transcription of other genes. The experimental source of this network is gene expression data derived from microarray or SAGE-experiments.
- **The genetic interaction network** depicts functional interactions between genes. These connections are inferred from phenotypic effects in cells where both genes have been knocked out or mutated.
- **The protein interaction network** consists of direct physical interactions between proteins. Yeast two-hybrid studies or copurification assays generate the data.
- **The metabolic network** comprises enzymes linked by compounds that serve as substrates or products of the relative biochemical reactions or *vice versa*.

For each of these networks the underlying experimental data have specific advantages and validities as well as typical drawbacks and biases. Microarrays, for instance, cover large fractions of the genome and advanced statistical models ensure good data quality. Yet the biological information is limited to regulation on the RNA level, and genetic dependencies have to be computationally inferred from their expression levels. Gene interaction screens (e.g., for synthetic lethality), in turn, have great power in directly identifying genes that are functionally related, but it is not clear how to interpret this relationship on a molecular level. In contrast, protein binding assays, such as yeast two-hybrid, indicate a common function through direct interaction of two proteins, more accurately reflecting the real molecular mechanism. Due to experimental limitations, though, they do not provide full genome coverage and the data is contaminated by a considerable amount of false positives. The interaction structure of the metabolic network is very rich in information, especially when reaction rates are also taken into account to study dynamic network behavior. However, knowledge about enzymatic reactions often comes from studies *in vitro*, it is often not clear whether the specific reaction would also occur in living cells.

Regardless of these issues, one can in general distinguish three levels of examination when analyzing intracellular networks on the basis of experimental data: (i) cluster identification, (ii) topology inference, and (iii) network inference.

Clustering aims at identifying groups of network entities. In the case of gene expression for instance, a popular approach is to group genes that exhibit similar expression profiles; thereby co-regulated or functionally related genes may be found. If focusing on protein interactions, one may cluster proteins according to the degree of overlap with regard to their interaction partners; here, the resulting groups may represent potential protein complexes. The main advantages of this approach are its general applicability and the availability of efficient algorithms. Hence, it is well suited for the global analysis of large data sets. The structural information, though, that clustering provides usually gives only a rough picture of the underlying network, and other methods are needed for inferring more specific network characteristics.

The goal of topology inference can be described as revealing the interactions between the network entities independently of the dynamics. Here, topology means structure comprising the components of the systems together with the connections. Considering the genetic regulatory level, the network topology defines which genes affect which other genes; however, it does not necessarily specify the kind of interaction, e.g., activation or inhibition. This type of analysis goes beyond clustering: it not only tries to detect groups of co-regulated genes, but also aims at identifying the transcription factors by which these genes are regulated. The algorithms used in this context can be computationally expensive and require a sufficient amount of experimental data in order to achieve reliable statements. Therefore, topology inference is only applicable to restricted scenarios in the order of some hundreds of genes.

Network inference represents the ultimate goal of network analysis, which can also be denoted as reverse engineering of the underlying system from experimental data. Here, the focus is on the functional relationships among the various entities in order to capture the network dynamics, which is essential for understanding and predicting the system's behavior. In the context of gene regulation for instance, a function relationship determines *how* a specific gene controls the transcription of another one; usually, one is interested in a quantitative description, e.g., using differential equations, that provides a basis for further mathematical analysis and simulation. Due to the complexity of living cells, though, the task of network inference is highly challenging and requires immense amounts of both experimental data and computing resources. Although some initial studies address this problem to date there does not exist a set of established methods as with, e.g., clustering.

This chapter provides an introduction to network analysis and illustrates the corresponding concepts on the basis of gene expression data. On the one hand, an overview of existing methods for the identification of coregulated genes is given in the following section. The issue of topology inference is discussed and as an example a specific inference method is presented. Moreover, the application of these techniques is demonstrated for the *Arabidopsis thaliana* isoprenoid pathway.

Clustering

Clustering algorithms are probably the most widely used tools in the analysis of gene expression data. Their goal is to find groups of genes that have similar expression patterns[2]. Two central questions need to be addressed in this respect: (i) when are the expression patterns of two genes similar and (ii) what does it mean biologically if two genes have similar expression patterns? A simple answer to the second question is that two genes with similar expression patterns are mechanistically related. This is the basic assumption behind all clustering approaches. Since there are many ways in which these two genes could be related (activation by the same transcription factor, one acting as transcription factor for the other, being involved in the same biological process and therefore regulated similarly by the cell, etc.) analysis of gene expression alone is generally not sufficient to reveal what kind of relation connects the genes. The nature of the interaction the genes exhibit may depend on the similarity measure, directly referring to the first question. A wide variety of similarity measures are used which capture different properties of the data. Similarity could mean similar trends over time, similar absolute values or similar ratios compared to a control experiment. The choice of similarity measure determines the kind of functional clusters that can be found. A direct consequence of this is that a cluster in itself is not a biological object. Some similarity measures are better suited to exhibit biological information.

One widely used similarity measure is the Pearson's correlation coefficient ρ. It measures the similarity between the expression patterns e_i, e_j of two genes i, j by describing how well the expression values of the first gene $e_{i1}, \ldots, e_{in}$ can be expressed as a linear function of the expression values of the second gene $e_{j1}, \ldots, e_{jn}$ where n is the number of chips,

$$\rho_{ij} = \frac{\sum_{k=1}^{n}(e_{ik} - \overline{e_i})(e_{jk} - \overline{e_j})}{\sqrt{\sum_{k=1}^{n}(e_{ik} - \overline{e_i})^2}\sqrt{\sum_{k=1}^{n}(e_{jk} - \overline{e_j})^2}} \quad (1)$$

where $\overline{e_i}$, $\overline{e_j}$ are the means of e_{ik}, e_{jk} respectively. The correlation coefficient lies in the interval $[-1, 1]$ and is close to 1 if the patterns are very similar. For example, if the two expression patterns are equal ρ equals 1 and if $e_{ik} = -e_{jk}$ for all k ρ equals -1. Each similarity measure has its own set of assumptions and requirements. For example the correlation coefficient is easily biased by outliers. A more detailed discussion of similarity measures and clustering methods can be found in most text books about microarray data analysis such as [3, 4].

A cluster analysis consists of three main steps: (i) choosing a mathematical representation reflecting the biological question, (ii) identifying an algorithm that solves the mathematical problem which in general means optimizing a score which de-

---

[2] The same methods can also be used to cluster the expression matrix columnwise, i.e., to group chips. So whenever we talk about groups of genes in the remainder of this section there is always an analog for the chips.

scribes the quality of a cluster or a clustering and (iii) analyzing the results using additional knowledge and data. Choosing a similarity measure is just one part of the first step. A complete mathematical problem formulation also contains a definition of the relation between clusters, e.g., whether clusters should be allowed to overlap. The following section groups existing approaches according to how they answer these questions.

General approaches

Looking at the development of clustering algorithms in the last few years one can clearly see a trend to include more and more biological considerations in the problem formulation. The following paragraph discusses two such aspects which had a large impact on the problem formulation and the clustering algorithms. Figure 1 presents the different categories of clustering algorithms.

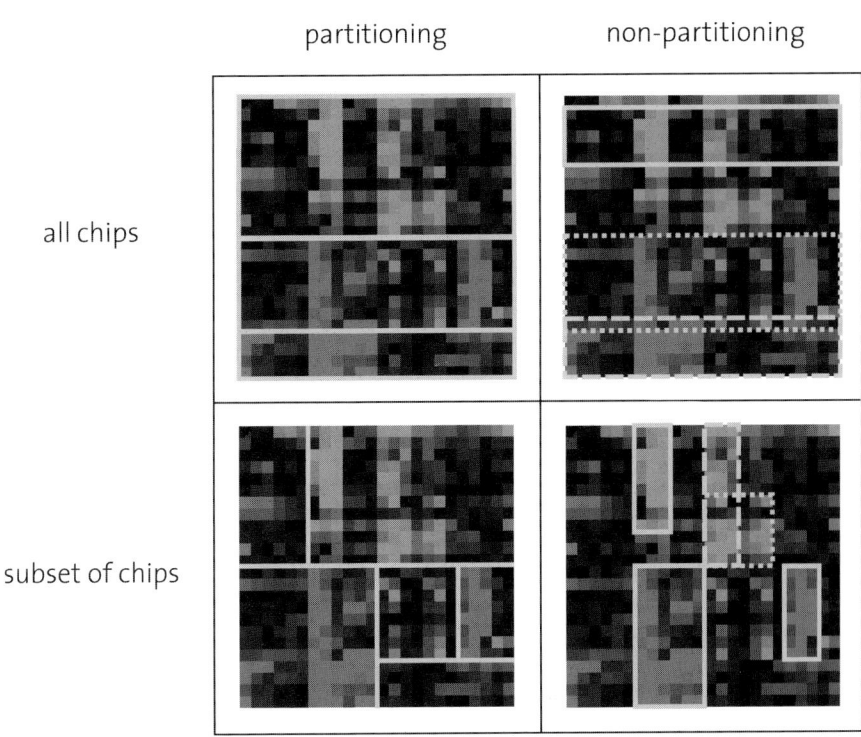

Figure 1. Categorization of clustering approaches. Schematic view of possible clustering results for each category. The approaches in the upper half consider all measurements when calculating the similarity between expression patterns. Those in the lower half look for genes that are similarly expressed over a subset of the conditions. Methods in the left half put each matrix element in exactly one cluster while those in the right half allow clusters to overlap and elements to be in no cluster.

Traditional clustering approaches such as k-means [5, 6], hierarchical clustering [7] and self organizing maps [8] put each gene in exactly one cluster. Methods in this category are most commonly used and have proven to be useful in many studies.

Frequently, clusters are interpreted as genes that are involved in the same biological process. Since some genes play a role in more than one distinct process it can make sense to include a gene in several clusters simultaneously, i.e., to allow clusters to overlap. Another issue concerns genes which do not fit well into any cluster; often, the goal of a cluster analysis is more to find significant groups of co-expressed genes than to determine for each gene in which cluster it best fits. In such a case some genes are best not assigned to any cluster. Several approaches have been proposed which follow one or both of these ideas and thus do not produce a partition of the matrix (shown in the upper right in Fig. 1), among them are fuzzy k-means [7] and CLICK [9].

Most clustering analyses are performed on a combination of datasets from different experiments. In such scenarios, it can be useful not only to look for groups of genes that have similar expression patterns over all measurements as traditional clustering algorithms do. Instead, one is interested in groups of genes that are co-expressed under certain conditions only. These groups potentially reflect genes that are responsible for a certain process which is not always active. The algorithm needs to select both a subset of genes and a subset of conditions. A few approaches follow this scheme while searching for a partition of the matrix [10, 11]. This is conceptionally similar to applying a traditional clustering algorithm in both dimensions. This idea is illustrated in the lower left of Figure 1.

A fourth category of methods combines both of the abovementioned refinements. In these approaches the focus is on finding strong local signals in the expression patterns. The goal is to find significant submatrices in the expression data containing similar patterns. These do not have to cover the whole matrix and in many approaches they can overlap. As a consequence the user does not have to set the number of clusters as it is necessary for classical clustering algorithm. The fourth category (sometimes including the third one) is referred to as biclustering methods [12–16]. An extensive review on biclustering methods was published in [17] and an evaluation of several biclustering methods can be found in [16].

Exploiting additional information

Continuing the trend to more biologically motivated problem formulations, many studies propose to include additional information in the clustering process.

Time courses
Time course gene expression experiments are a popular method for studying biological processes. In addition to the expression values, such measurements contain information about the time. Several methods try to exploit this information. One possibility is to focus on the changes between any two consecutive time points and discretize them into up, down and unchanged; genes which show the same changes

are then clustered together [18]. Many methods targeted to the analysis of time course experiments attempt to answer more specific questions than finding genes with similar expression patterns. Examples are the discovery of periodically expressed genes [19, 20] or the identification of time lags between the expression of different groups of genes [21]. Other methods concern the inference of causal relationships between expression of different genes [21]. (See [22] for a review of analysis methods for time series expression measurements.)

Multiple datasets
All methods discussed so far operate on one matrix of expression data. In contrast most biological studies analyze multiple datasets, e.g., multiple time course measurements. Most of the current clustering algorithms require to concatenate these datasets into one input matrix, thus losing the information about which measurements belonged to the same experiment and which did not. Correspondingly many studies which analyze clusters in gene expression data follow this path and mix different datasets into one [7, 23, 24]. In some cases the measured values from several experiments can be compared directly and the association with a specific experiment, e.g., a specific treatment, is not of interest in the analysis. In some cases, however, mixing measurements from different experiments is not possible or at least undesirable. In [25] an approach is presented which can find groups of genes that have similar expression patterns over multiple datasets without comparing the measurement values between the datasets directly.

Multiple datasets
Another direction of current research is to include additional information about the genes such as measurements of protein–protein interactions or membership in a specific pathway. The main idea is that these additional data complement the gene expression data and lead to a more specific statement about the functional relationships between the genes in a cluster. Several approaches combine distance on gene expression data with distance on a second type of data, e.g., distance in the metabolic network [26] or distance in the Gene Ontology classification [27].

A method which allows to integrate a variety of different data types was presented in [28]. The core idea is to represent all data as binary properties of a gene, e.g., for protein–protein interaction data a property represents an interaction to one specific gene. Each gene either has this interaction (value = 1) or it does not (value = 0). The resulting binary matrix is then analyzed using a biclustering algorithm.

Graphical models

In contrast to clustering approaches, network topology inference aims at exploring pairwise regulatory relationships between genes. Network inference can be based on different interactions models, such as logical (boolean) networks, probabilistic networks or kinetic networks.

Graphical models form a probabilistic network tool to analyze and visualize relationships between genes and can be used to provide first insights into the dependency structure of a genetic regulatory network. Genes are represented by vertices of a graph and dependencies between them are encoded by edges (see Fig. 2). An edge between two genes is drawn when the data (e.g., gene expression data) show strong statistical support for a direct relationship between the expression values of both genes.

Graphical modeling of genetic networks can be carried out with directed or undirected edges, with discretized or continuous data. Graphical models with directed edges are called Bayesian networks [29, 30]. Models with continuous data and undirected graphs, the so-called graphical Gaussian models, have the advantage that the dependency pattern can be completely described by the covariance (or correlation) matrix. For this reason, we present the concepts of graphical modeling based on this type of model. For a short introduction into the mathematical and statistical terminology of this section see Box 1.

When analyzing genetic regulatory associations from high-throughput biological data, such as gene expression data, the activity of thousands of genes is monitored over relatively few samples. This implies that the statistical techniques applied to estimate the dependency structure between genes must provide a high estimation accuracy even for few available samples.

The simplest method to model the dependence structure would be in a so-called reference or covariance graph [31, 32] where edges represent the marginal dependence (correlation) between genes. The covariance structure can be accurately estimated and easily interpreted even with a large number of variables (genes) and a small sample size. However, the covariance graph contains only limited information since the effect of the remaining genes on the relationship between two genes is ignored.

For example, assume four genes $g_1, ..., g_4$ where g_1 regulates the remaining three genes such that up- and downregulation of g_1 leads to down- and upregulation of the g_2, g_3 and g_4. Assume further that the log ratios of gene expression values follow a normal distribution with covariance matrix Σ as given in Figure 2. Although there is no direct dependence between g_2, g_3 and g_4, all correlation coefficients are different from 0. Therefore, the reference graph would be complete. However, by the inverse of the covariance matrix (Σ^{-1}) it can be shown that, for example, the relationship between g_2 and g_4 is caused by the dependencies between g_1 and g_2 on the one hand and g_1 and g_4 on the other ($\Sigma^{-1}_{24} = 0$).

To find direct relationships between genes, one has to look at conditional dependencies, i.e., the dependence between two genes conditional on all the remaining genes. This means that the effect of the remaining genes is "taken out" before the relationship between the two genes under consideration is explored. As mentioned above, conditional dependence patterns can be studied by the inverse Σ^{-1} of the covariance (or correlation matrix). Whenever an element is 0, there is no conditional dependence (or direct relationship) between two genes and no edge between the corresponding nodes in the graph. Consequently, the inverse of the covariance matrix rather than the covariance matrix itself should be used to infer regulatory relationships between genes. The structure of the conditional relationships within a group of genes can be exhaustively explored with help of the so-called Markov

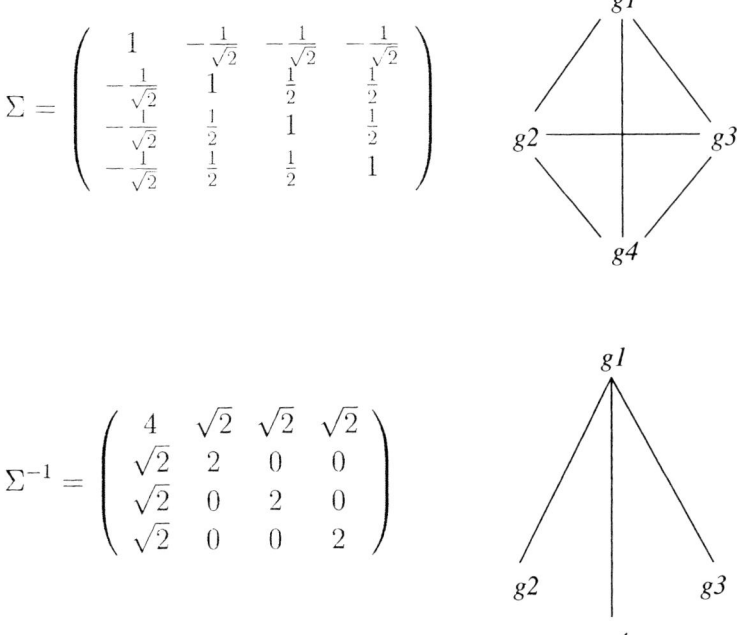

Figure 2. Covariance matrix and graph (upper panel) and the corresponding matrix and graph of a graphical Gaussian model (lower panel).

properties which imply that two unconnected vertices are independently conditional on their common neighbors [33, 34]. The graphs are also called conditional independence graphs.

Graphical models have become increasingly popular for inferring genetic regulatory networks based on the conditional dependence structure of gene expression levels [29, 30, 35, 36]. However, since one does not know Σ^{-1} in advance, this matrix has to be estimated from the data. For large number of genes, this estimation of Σ^{-1} is highly inaccurate as errors do accumulate when inverting the estimated covariance matrix Σ. Therefore, the standard graphical modeling approach should be adjusted to accommodate the small sample size of the gene expression data.

For this purpose, models that exploit the sparsity of regulatory networks have recently gained attention [37–40]). One possibility is to restrict the number of edges per gene in the graph [37]. Another possibility is to assume that indirect dependencies between genes can be only mediated by a single third gene at a time. Therefore, one does not have to condition on all genes at a time, i.e., one does not have to invert the complete covariance matrix Σ. Instead, one applies graphical modeling only to small subnetworks with three genes to explore the dependence between two of the genes conditional on the third one. These subnetworks are then combined for inference on the complete network. This simplified approach makes it possible to include many genes in the network while studying dependence patterns in a more complex and exhaustive way than with only pairwise relationships. One main benefit of the presented approach over graphical models is that one can easily test on a large scale how well additional genes can be integrated in the network. This allows to select additional genes with similar expression patterns in a fast and efficient way.

Box 1: Statistical terminology

Matrix – A matrix is a rectangular array of numbers. The numbers are called elements. The elements of a matrix are accessed by their row and their column indices. For example, in Fig. 2, two matrices Σ and Σ^{-1} are displayed and two exemplary elements are $\Sigma_{23} = \frac{1}{2}$ and $\Sigma^{-1}_{23} = 0$ referring to the relationship between the second and the third gene.

Inverse of a matrix – Matrices can be added up, subtracted from each other and multiplied. The inverse of a matrix is the opposite or the reciprocal value of a matrix with respect to multiplication, such as 2 and 0.5 are reciprocal scalars.

Σ – Covariance matrix whose elements Σ_{ij} are computed as

$$\Sigma_{ij} = \sum_{k=1}^{n} (e_{ik} - \overline{e_i})(e_{jk} - \overline{e_j}), \qquad (2)$$

compare Equation (1).

ρ – Correlation matrix whose elements ρ_{ij} are the Pearson's correlation coefficients,

$$\rho_{ij} = \frac{\sum_{k=1}^{n}(e_{ik} - \overline{e_i})(e_{jk} - \overline{e_j})}{\sqrt{\sum_{k=1}^{n}(e_{ik} - \overline{e_i})^2}\sqrt{\sum_{k=1}^{n}(e_{jk} - \overline{e_j})^2}}, \qquad (3)$$

compare Equation (1). The correlation matrix is the normalized covariance matrix, all elements are between –1 and 1. Elements ρ_{ij} that are close to –1 or 1 indicate a strong relationship between the corresponding genes, elements close to 0 indicate independence.

$\Omega = \Sigma^{-1}$ – Inverse covariance matrix which represents the reciprocal value of the covariance matrix. Elements close to 0 indicate that two genes have no direct relationship after the effect of all other genes is taken out.

Some more theory on graphical models

Let q be the number of genes in the network, n be the number of observations for each gene. The vector of log scaled gene expression values is assumed to follow a multivariate normal distribution with covariance matrix Σ.

In the covariance graph, an edge between vertex i and j ($i \neq j$) is drawn if and only if the correlation coefficient is different from 0,

$$\rho_{ij} \neq 0, \quad \rho_{ij} = \frac{\Sigma_{ij}}{\sqrt{\Sigma_{ii}\Sigma_{jj}}}. \tag{4}$$

The covariance graph as a representation of the marginal dependence structure between variables is simple to interpret and can be accurately estimated even if q is very large in comparison to sample size n. However, as mentioned before this graph is often not sufficient to capture complex conditional dependence patterns.

In the graphical Gaussian model (conditional independence graph), an edge between vertex i and j is drawn when the *partial* correlation coefficients ω_{ij} is different from 0. The partial correlation coefficients ω_{ij} which measure the correlation between genes i and j conditional on all other genes in the model are calculated as

$$\omega_{ij} = \frac{-\Omega_{ij}}{\sqrt{\Omega_{ii}\Omega_{jj}}},$$

where Ω_{ij}, $i,j = 1, \ldots, q$ are the elements of the inverse covariance matrix $\Omega = \Sigma^{-1}$.

To learn the conditional independence structure of the graph, it is necessary to determine which partial correlation coefficients ω_{ij} are 0. This can be carried out via likelihood methods where each ω_{ij} is estimated and tested against the null hypothesis $\omega_{ij} = 0$ [33]. An edge between genes i and j is drawn if the null hypothesis is rejected. Since the estimation of the partial correlation coefficients involves matrix inversion, estimates are only reliable for a large number of observations when many genes are involved. Modeling of the graph is commonly carried out in a stepwise backward manner starting from the saturated model where all edges are included in the graph. From this model edges are removed consecutively to find a good model with as few edges as possible.

Graphical modeling based on the sparsity assumption of the network combine statistical features from the covariance and the conditional independence graph. In this respect, they can be viewed as striking a balance between the covariance and the conditional independence graph. Let i,j be a pair of genes. The correlation coefficient ρ_{ij} is the commonly used measure for co-regulation. For examining possible effects of other genes k on ρ_{ij}, we consider triples of genes i, j, k with $k = \{1, \ldots, q\} \setminus \{i,j\}$. For each k, the first-order partial correlation coefficient $\omega_{ij|k}$,

$$\omega_{ij|k} = \frac{\rho_{ij} - \rho_{ik}\rho_{jk}}{\sqrt{(1-\rho_{ik}^2)(1-\rho_{jk}^2)}}$$

is computed and compared to ρ_{ij}. If the expression levels of k are independent of i and j, the first-order partial correlation coefficient would not differ from ρ_{ij}. If on the other hand, k is co-regulating both genes, i and j, one would expect $\omega_{ij|k}$ to be close to 0. Here, we use the terminology, that k 'explains' the correlation between i and j.

Therefore, one draws an edge between two genes i and j when there is no single gene k that explains the correlation between i and j. In other words, for each k the null hypothesis $\omega_{ij|k} = 0$ is tested. An edge between i and j is drawn when this hypothesis is rejected for all k. These edges represent direct relationships that cannot be explained by any other gene and can also be used to analyze the topology of genetic regulatory networks.

The next section illustrates how graphical Gaussian models and biclustering can be applied to study genetic regulatory networks. The methods are applied on pathways controlling isoprenoid synthesis in *Arabidopsis thaliana*.

Case study

Introduction: The isoprenoid biosynthesis pathways

Isoprenoids comprehend the most diverse class of natural products and have been identified in many different organisms including viruses, bacteria, fungi, yeasts, plants, and mammals. In plants, isoprenoids play important roles in a variety of processes such as photosynthesis, respiration, and regulation of growth and development, in protecting plants against herbivores and pathogens, attracting pollinators and as allelochemicals. Two distinct pathways are responsible for the biosynthesis of isoprenoid precursors: the mevalonate dependant (MVA) and the plastidic methy-D-erythritol 4-phosphate (MEP) pathway. Whereas the mevalonate pathway is responsible for the synthesis of sterols, sesquiterpenes, and the side chain of ubiqinone the mevalonate independent pathway is employed for the synthesis of isoprenes, carotenoids, and the side chains of chlorophyll and plastoquinone. Although both pathways operate independently under normal conditions interaction between them has been repeatedly reported. A cross-talk regulation between both pathways coordinates the expression of MVA and MEP genes in response to internal and external factors. Reduced flux through the MVA pathway after treatment with lovastatin can be compensated by the MEP pathway. However, inhibition of the non-mevalonate pathway leads to reduced levels in carotenoids and chlorophylls indicating a unidirectional transport from isoprenoid intermediates from the chloroplast to the cytosol [41, 42].

Application of biclustering to the study of isoprenoid biosynthesis

In order to identify groups of genes that coordinately respond to individual factors, a data matrix containing 90 genes from the MEP, MVA and downstream pathways, as well as 140 response conditions [43, 44] was biclustered using the BiMax algorithm implemented in BicAT (see Box 2). The data was discretized at a two-fold change level using the complementary pattern option. This allows for grouping both up- and downregulated genes into the same biclusters. More than 100 biclusters were generated. From these, one illustrative example is shown (see Fig. in Box 2). This bicluster contains 12 genes that respond together to *A. tumefaciens* and senescence/cell death, but are not necessarily coordinately expressed under other condi-

Box 2: Biclustering analysis toolbox

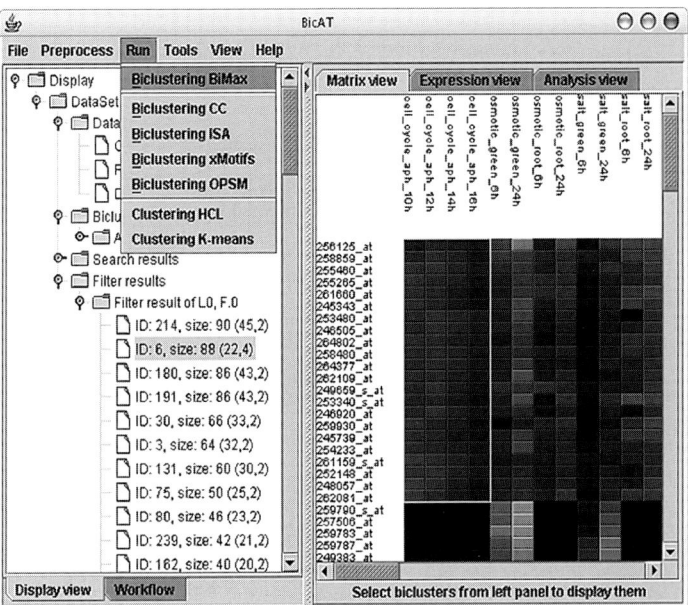

For cluster analysis of biological data, several freely available tools exist. As an example, we present BicAT ([45]), a software that provides a graphical user interface for various traditional clustering as well as biclustering techniques. Since every algorithm is subject to a specific mathematical problem formulation, it can be useful in practice, to try out different algorithmic approaches. Therefore, BicAT provides five biclustering and two standard clustering procedures. To our best knowledge, BicAT is the only tool available that offers more than one biclustering algorithm. Like other tools, it offers facilities for data preparation, inspection, and postprocessing. To find biclusters containing genes under consideration, the tool offers filtering and searching utilities which produce lists of biclusters. The selected bicluster appears in the upper left corner of the heatmap (see Fig.). Interesting biclusters can be exported to a file or can be further investigated by a gene pair analysis, which summarizes the outcome of a biclustering run as a whole. In the gene pair analysis, it is calculated how often each pair of genes occurs together in the same bicluster, which in turn can give an indication of a possible functional relationship of the two genes. The resulting gene–gene matrix with the according counts can be exported for further visualization and derivation of gene interconnection graphs with external tools, e.g., BioLayout [46].

tions. Two groups of genes that respond oppositely in these two conditions can be distinguished. The first group is upregulated in response to *A. tumefaciens* and downregulated under senescence/cell death. In these two treatments, cell elongation and division are activated and respressed, respectively. All genes from this group encode cytosolic proteins, and four of six are involved in sterol biosynthesis, of which one is known as cell elongation protein DWARF1 and is a key enzyme in this pathway. The pattern observed in this bicluster is biologically interesting, because sterols are known to be involved in membrane fluidity and stability. Since cell elongation and division require an extension of lipid membranes, an activation of genes encoding proteins from this pathway is a prerequisite for normal cell development. The second group shows the opposite responses and involves mainly genes encoding plastid-targeted proteins, but also two cytosolic proteins (AACT1 and SQP1), which possibly gives an indication about cross-talk between the MEP and MVA pathways.

Application of graphical modeling

We applied graphical modeling to elucidate the regulatory network of the two isoprenoid biosynthesis pathways in *Arabidopsis* (as reviewed in [47]). In order to gain

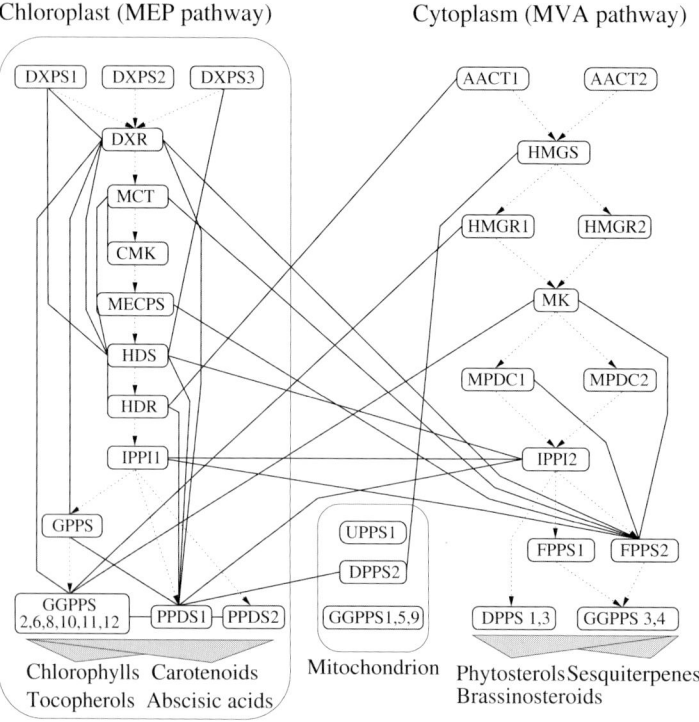

Figure 3. Conventional graphical Gaussian modeling of the isoprenoid pathways. The dashed arrows mark the metabolic network and are not part of the graphical model.

better insight into the crosstalk between both pathways on the transcriptional level, gene expression patterns were monitored under various experimental conditions using 118 GeneChip (Affymetrix) microarrays [38]. For the construction of the genetic regulatory network, we focused on 40 genes, 16 of which were assigned to the cytosolic pathway, 19 to the plastidial pathway and five are located in the mitochondrion. The genetic interaction network among these genes was first constructed employing graphical Gaussian modeling with backward selection. This was carried out with the MIM 3.1 program [33]. A separate regulatory module was found in the MEP but not in the MVA modules. However, a high level of co-expression between the genes AACT2, MK, MPDC1, FPPS2 suggests a separate regulatory module in the MVA pathway. The genes in the MVA pathway did not form a separate regulatory structure, even when more edges were included in the model.

Figure 4 shows the network model obtained when using the graphical model under the sparsity assumption. Since we find a module with strongly interconnected genes in each of the two pathways, the graph is split up into two subgraphs each displaying the subnetwork of one module and its neighbors.

In the MEP pathway, the genes DXR, MCT, CMK, and MECPS are nearly fully connected (left panel of Fig. 4). From this group of genes, there are a few edges to genes in the MVA pathway. Similarly, the genes AACT2, HMGS, HMGR2, MK, MPDC1, FPPS1 and FPPS2 share many edges in the MVA pathway (right panel of Fig. 4). The subgroup AACT2, MK, MPDC1, FPPS2 is completely interconnected. From these genes, we find edges to IPPI1 and GGPPS12 in the MEP pathway. In contrast to the graphical model without the sparsity assumption, the method could now identify connections between AACT2 and the three genes MK, MPDC1 and FPPS2. The detection of the additional gene module in the MVA pathway is in good agreement with earlier findings that within a pathway, potentially many consecutive or closely positioned genes are jointly regulated [48].

Comparison of biclustering and graphical modeling results

The results from this bicluster analysis were compared to those obtained using graphical modeling. Although the dataset used for the graphical modeling analysis was completely independent of the one selected for bicluster analysis, similar relationships were obtained. For example, AACT1 was associated with plastidic genes in both analyses. Similarly, FPS2 and MPDC2 were closely associated in the graphical modeling approach and were found together in the first group of the presented bicluster. However, due to the different problem formulations of each approach, one should not expect to always find similar relationships between genes. The results from a biclustering analysis will tend to shed light to smaller subsets of biological processes or responses, while graphical modeling and several other methods reveal gene relationships based on all conditions. For example, two genes that show the same response in a subset of conditions may be only weakly correlated throughout most conditions that occur in nature.

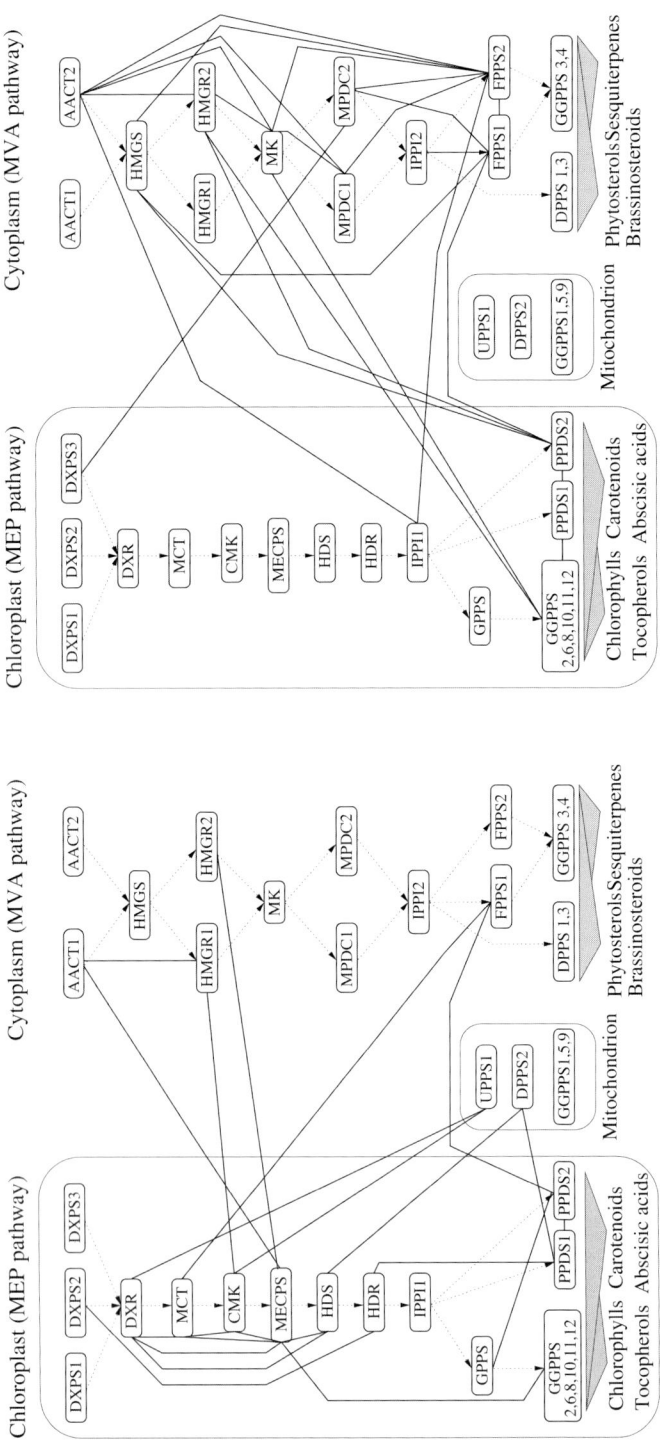

Figure 4. Graphical model of the isoprenoid pathways based on the sparsity assumption. Left panel: subgraph of the gene module in the MEP pathway, right panel: subgraph of the gene module in the MVA pathway.

Box 3: Clustering yeast interaction data

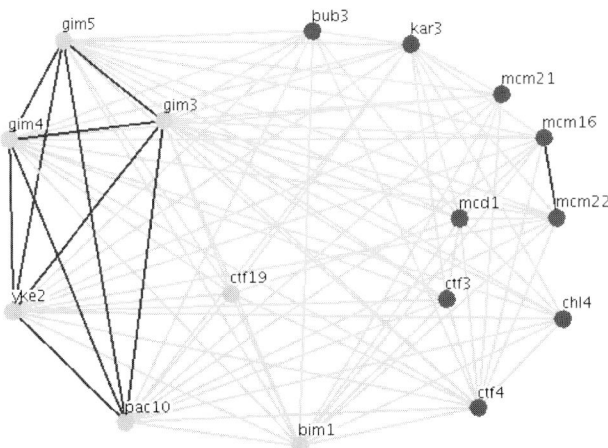

In principle clustering methods are not only applicable to gene expression data, but can be used with various biological measurements (protein or genetic interactions, etc.). In this example data from several synthetic lethal screens in *Saccharomyces cerevisiae* [49] were biclustered using the BiMax-algorithm (see Box 2, [45]). A bicluster is presented that exclusively contains genes/proteins involved in cell division. In the picture, genetic interactions (bright edges) are combined with protein interactions (dark edges). The shading of the nodes represent the function annotations for each gene: Bright: Cell organization; Dark: Cell cycle/cell growth. The bicluster contains five members of the prefoldin complex (gim3, gim4, gim5, yke2, pac10), genes involved in chromosome packing and -segregation (mcm16, mcm21, mcm22, mcd1, ctf3, ctf4, chl4), and genes that control microtubuli formation (bub3, kar3). All of these processes are necessary for proper cell division. Thus, clustering synthetic lethal or other binary interaction data can help to identify functional modules. Further validation of the bicluster using other data types indeed confirms part of the bicluster to be a protein complex. Although the genes of the prefoldin-complex do not show interaction in the synthetic lethal screen they all share the same genetic interaction partners. Therefore, they can be put in a shared cellular context, which is further supported by common function annotation and most strongly by the fact that their respective proteins also directly bind to each other as retrieved from MIPS-database [50]. This illustrates how experimental data can be combined to increase biological significance of clustering results and how clustering can be used for function annotation.

Picture produced with osprey [51] function annotations obtained from GRID-database [52].

Summary

This chapter has provided a general introduction to network analysis and illustrated basic concepts on the basis of the isoprenoid pathways in *Arabidopsis thaliana*. Although the discussion has focused on the genetic regulatory level and mRNA profiling data, similar considerations hold for other network levels as demonstrated in Box 3. Furthermore, the presented computational methods only cover a fraction of all currently available methods for the analysis of large scale biological data. New techniques inspired by novel mathematical models keep emerging, which makes network analysis an ever-changing field of systems biology.

Since the amount and the diversity of biological data continues to grow, a crucial part of network analysis is also the combination and integration of different experimental results. Various difficulties arise in this context, e.g., it has been shown that the outcome of different large scale screens overlap only in parts ([53]). On the one hand, this raises issues about the reliability of the data. Genes involved in the same biological scenario should ideally show a high correlation in many different experiments. That this is rarely the case for most genes is likely due to the problem that each experiment has its specific bias so that systematic errors occur in the measurements. For instance, estimates of the false positive rate of genome wide protein-interaction data based on gene expression profiles and interactions between homologous genes in other organisms mount to 50% ([54]). On the other hand, the small overlap of independent large scale studies reflects the fact that different experimental approaches yield outcomes that complement each other, since each experiment reveals only parts of the intracellular network. It is a major challenge of systems biology to bring these data together in an appropriate way to increase data reliability and to exploit the plentitude of information in order to construct network models that describe the biological system as completely as possible. In this regard, the significance of merely static networks depicting which gene has a relationship with which other genes for biological research is restricted. Intracellular biological networks are not static formations but highly dynamic structures. They perpetually change not only during the life cycle of the cell and developmental stages of multicellular organisms but they also have to respond to environmental cues constantly receiving input from the extracellular space. Indeed, there is much that has to be learned about the dynamics of cellular networks. Our understanding of the cell's metabolism for example and of complex diseases that can be triggered by many factors, such as cancer, is clearly limited with regard to its dynamical properties. Therefore, studying the dynamic features of intracellular networks will be an essential part of future research.

References

1. Donne J (1624) *Meditation XVII: Devotions Upon Emergent Occasions.* McGill-Queens's Univ. Press, Montreal
2. Kumar A, Snyder M (2002) Proteomics: Protein complexes take the bait. *Nature* 415: 123–124

3. Kohane IS, Kho AT, Butte AJ (2003) *Microarrays for an Integrative Genomics.* MIT Press
4. Wit E, McClure J (2004) *Statistics for Microarrays,* Wiley
5. Tayazoie S, Hughes JD, Campbell MJ, Cho RJ, Church GM (1999) Systematic determination of genetic network architecture. *Nat Genet* 22: 281–285
6. Soukas A, Cohen P, Socci ND, Friedman JM (2000) Leptin-specific patterns of gene expression in white adipose tissue. *Genes Dev* 14: 963–980
7. Eisen MB, Spellman PT, Brown PO, Botsteinl D (1998) Cluster analysis and display of genome-wide expression patterns. *PNAS* 95: 14863–14868
8. Tamayo P, Slonin P, Mesirov J, Zho Q, Kitareewan S, Danitrovsky E, Lander ES, Golob TR (1999) Interpreting patterns of gene expression with self-organizing maps: Methods and application to hematopoietic differentiation. *PNAS* 96: 2907–2912
9. Sharan R, Maron-Katz A, Shamir R (2003) Click and expander: A system for clustering and visualizing gene expression data. *Bioinformatics* 19: 1787–1799
10. Hartigan JA (1972) Direct clustering of a data matrix. *J Am Statistical Assoc* 67: 123–129
11. Kluger Y, Basri R, Chang JT, Gerstein M (2003) Spectral biclustering of microarray cancer data: Co-clustering genes and conditions. *Genome Research* 13,703–16. http://bioinfo.mbb.yale.edu/e-print/biclusters/all.pdf.
12. Cheng Y, Church GM (2000) *Biclustering of Gene Expression Data.* pp 93–103. http://cheng.ecescs.uc.edu/biclustering.
13. Tanay A, Sharan R, Shamir R (2002) Discovering statistically significant biclusters in gene expression data. *Bioinforrnatics* 18: S136–S144
14. Murali TM, Kasif S (2003) *Extracting Conserved Gene Expression Motifs from Gene Expression Data.* Vol. 8, pp 77–88
15. Bergmann S, Ihmels J, Barkai N (2003) Iterative signature algorithm for the analysis of large-scale gene expression data. *Phys Rev E Stat Nonlin Soft Matter Phys* 67: 031902
16. Prelić A, Bleuler S, Zimmermann P, Wille A, Bühlmann P, Gruissem W, Hennig L, Thiele L, Zitzler E (2005) A systematic comparison and evaluation of biclustering methods for gene expression data. *Bioinformatics* 22: 1122–1129
17. Madeira SC, Oliveira AL (2004) Biclustering algorithms for biological data analysis: A survey. *IEEE Transactions on Computational Biology and Bioinformatics* 1: 24–45
18. Ernst J, Nau GJ, Bar-Joseph Z (2005) Clustering short time series gene expression data. *Bioinformatics 21 Suppl 1:* i159–i168
19. Luan Y, Li H (2004) Model-based methods for identifying periodically expressed genes based on time course microarray gene expression data. *Bioinformatics* 20: 332–339
20. Wichert S, Fokianos K, Strimmer K (2004) Identifying periodically expressed transcripts in microarray time series data. *Bioinformatics* 20: 5–20
21. Qian J, Dolled-Filhart M, Lin J, Yu H, Gerstein M (2001) Beyond synexpression relationships: Local clustering of time-shifted and inverted gene expression profiles identifies new, biologically relevant interactions. *J Mol Biol* 314: 1053–1066
22. Bar-Joseph Z (2004) Analizing time series gene expression data. *Bioinformatics* 20: 2493–2503
23. Gasch AP, Spellman PT, Kao CM, Carmel-Harel O, Eisen MB, Storz G, Botstein D, Brown PO (2000) Genomic expression programs in the response of yeast cells to environmental changes. *Mol Biol Cell* 11: 4241–4257
24. Segal E, Shapira M, Regev A, Pe'er D, Botstein D, Koller D, Friedman N (2003) Module networks: Identifying regulatory modules and their condition-specific regulators from gene expression data. *Nature Genetics* 34: 166–176
25. Bleuler S, Zitzler E (2005) *Order Preserving Clustering over Multiple Time Course Experiments,* LNCS. (Springer), No 3449, pp 33–43

26. Hanisch D, Zien A, Zimmer R, Lengauer T (2002) Coclustering of biological networks and gene expression data. *Bioinformatics* 18: S145–S154
27. Speer N, Spieth C, Zell A (2004) *A Memetic Co-Clustering Algorithm for Gene Expression Profiles and Biological Annotation.* (IEEE), Vol 2, pp 1631–1638
28. Tanay A, Sharan R, Kupiec M, Shamir R (2004) Revealing modularity and organization in the yeast molecular network by integrated analysis of highly heterogeneous genomewide data. *Proc Natl Acad Sci USA* 101: 2981–2986
29. Friedman N, Linial M, Nachman I, Pe'er D (2000) Using bayesian networks to analyze expression data. *J Comput Biol* 7: 601–620
30. Hartemink A, Gifford D, Jaakkola T, Young R (2001) *Using graphical models and genomic expression data to statistically validate models of genetic regulatory networks.* PSB01. pp 422–433
31. Cox D, Wermuth N (1993) Linear dependencies represented by chain graphs (with discussion). *Statist Sci* 8: 204–218
32. Cox D, Wermuth N (1996) *Multivariate dependencies: Models analysis and interpretation.* Chapman & Hall, London
33. Edwards D (2000) *Introduction to Graphical Modelling.* Springer Verlag; 2nd edition
34. Lauritzen S (1996) *Graphical Models.* Oxford University Press
35. Toh H, Horimoto K (2002) Inference of a genetic network by a combined approach of cluster analysis and graphical gaussian modeling. *Bioinformatics* 18: 287–297
36. Wang J, Myklebost O, Hovig E (2003) Mgraph: Graphical models far microarray data analysis. *Bioinformatics* 19: 2210–2211
37. Friedman N, Nachman I, Pe'er D (1999) *Learning Bayesian network structure from massive datasets: The "Sparse Candidate" algorithm,* UAI. pp 206–215
38. Wille A, Zimmermann P, Vranova E, Furholz A, Laule O, Bleuler S, Hennig L, Prelic A, von Rohr P, Thiele L et al. (2004) Sparse graphical gaussian modeling of the isoprenoid gene network in arabidopsis thaliana. *Genome Biol* 5: R92
39. Magwene P, Kim J (2004) Estimating genomic coexpression networks using first-order conditional independence. *Genome Biol* 5: R100
40. de la Fuente A, Bing N, Hoeschele I, Mendes P (2004) Discovery of meaningful associations in genomic data using partial correlation coefficients. *Bioinformatics* 20: 3565–3574
41. Laule O, Fürholz A, Chang H, Zhu T, Wang X, Heifetz P, Gruissem W, Lange M (2003) Crosstalk between cytosolic and plastidial pathways of isoprenoid biosynthesis in *Arabidopsis thaliana*. *Proc Natl Acad Sci USA* 100: 6866–6871
42. Bick J, Lange B (2003) Metabolic cross talk between cytosolic and plastidial pathways of isoprenoid biosynthesis: Unidirectional transport of intermediates across the chloroplast envelope membrane. *Arch Biochem Biophys* 415: 146–154
43. Zimmermann P, Hennig L, Gruissem W (2005) Geneexpression analysis and network discovery using genevestigator. *Trends Plant Sci* 10: 1360–1385
44. Zimmermann P, Hirsch-Hoffmann M, Hennig L, Gruissem W (2004) Genevestigator. arabidopsis microarray database and analysis toolbox. *Plant Physiol* 136: 2621–2632
45. Barkow S, Bleuler S, Prelic A, Zimmermann P, Zitzler E (2005) Bicat: A biclustering analysis toolbox. *unpublished*
46. Enright A, Ouzounis C (2001) Biolayout – an automatic graph layout algorithm for similarity visualization. *Bioinformatics* 17: 853–854
47. Rodriguez-Concepcion M, Boronat A (2002) Elucidation of the methylerythritol phosphate pathway for isoprenoid biosynthesis in bacteria and plastids. a metaboc milestone achieved through genomics. *Plant Physiol* 130: 1079–1089

48. Ihmels J, Levy R, Barkai N (2004) Principles of transcriptional control in the metabolic network of *Saccharomyces cerevisiae*. *Nat Biotechnol* 22: 86–92
49. Tong AH, Lesage G, Bader GD, Ding H, Xu H, Xin X, Young J, Berriz GF, Brost RL, Chang M et al. (2004) Global mapping of the yeast genetic interaction network. *Science* 303: 808–813
50. Mewes HW, Amid C, Arnold R, Frishman D, Goldener U, Mannhaupt G, Musterkotter M, Pagel P, Strack N, Stumpflen V et al. (2004) MIPS: Analysis and annotation of proteins from whole genomes. *Nucleic Acids Research* 32: D41–44
51. Breitkreutz B-J, Stark C, Tyers M (2003) Osprey: A network visualization system. *Genome Biology* 4: R22
52. Breitkreutz B-J, Stark C, Tyers M (2003) The grid: The general repository for interaction datasets. *Genome Biology* 4: R23
53. von Mehring C, Krause R, Snel B, Cornell M, Oliver S, Fields S, Bork P (2002) Comparative assessment of large-scale data sets of protein-protein interactions. *Nature* 417: 399–403
54. Deane C, Salwinski L, Xenarios I, Eisenberg D (2002) Protein interactions: Two methods for the assessment of the reliability of high throughput ovservations. *Mol Cell Proteomics* 1: 349–356

Index

abi4 74
abiotic stress 56
abscisic acid (ABA) 56
accurate mass measurement 204
acetophenone 196
Affymetrix 59
– ATH1-22K array 74
– chip 70
AG-8K array 75
ageing 55
aleurone 75
alfalfa trichome 206
alkaloids 196
allele and quantitative characters 23
amines 196
analyte 180
analytical HPLC 200
anthocyanidin 196
anthranoid 196
anthraquinone 196
apoplast 74
Arabidopsis 81
Argonaute proteins 100
aristolochic acid 196
artemisinin 196
ascorbate oxidase 74
ascorbic acid 55
ATH1-22K array 77, 78
At-MIAMEexpress 299
A-type cyclin 79
aurone 196
avidin affinity purification 127
azoxylglycoside 196

BEL1 75
benzenoid 196
beta-oxidation of fatty acids 228

biclustering 343
bottom-up systems biology 1, 8

calibration, mass 176
–, retention time 176
canonical correlation analysis 319
capillary/nano-HPLC-QtofMS 200
cell cycle 79, 82, 92
– plate 81
central metabolic intermediates 237
– metabolism 216
centrality 257, 258
chemical diversity 195
chromatin immunoprecipitation (ChIP) 88, 92
chromatographic resolution 203
chromatography, ion exchange 126
–, strong cation exchange (SCX) 128
clustering 322, 334
– coefficient 258
coefficient of variation 128
computational strategies, in the identification of miRNAs 104
condensed tannin 196
connectivity 256
– theorem 7
controlled vocabularies 280
correlation coefficient 341
coumarin 196
covariance matrix 338, 340
coverage inventory 185
Cy3 57
Cy5 57
cyanogenic glycoside 196
cycloheximide (CHX) 90
cytokinesis 81

data heterogeneity   279
– integration   291
– standards   278
– warehousing   292
database, ArrayExpress   285, 295
–, BarleyBase   299
–, Database Resource for the Analysis of Signal Transduction In Cells (DRASTIC)   299
–, federated databases   292
–, Gene Expression Omnibus (GEO)   285, 297
–, Maize Olignucleotide Array Project   299
–, PeptideAtlas   300
–, Rice Expression Database (RED)   299
–, soybean genomics and microarrays database   299
–, Swiss-Prot   279
de novo sequencing, mass spectrometric protein identification method   155
description logic   281
dexamethasone (-DEX)   88, 89, 91
dibenzofuran   196
Dicer   100
Directed Acyclic Graph (DAG)   281
diterpene   233
DMAPP   231
DNA microarray   55
double stranded RNA, sources of   100
D-type cyclin   79
dye-swap experiment   59, 60

electrophoresis, 2-D fluorescence difference gel electrophoresis (DIGE)   118
–, SDS-PAGE   116
–, two-dimensional gel electrophoresis (2-DE)   116
electrospray ionization (ESI), mass spectrometric ionization   145
emergent system property   2
emerging technology   197
endoplasmic reticulum, assembly   81
enrichment, of precursor   215
epistasis   24
ethylmethanesulfonate (EMS)   37
exact quantification   178
experimental design   173, 174
expression microarray   57

FERTILIZATION INDEPENDENT SEED (FIS) PcG proteins   88
findMiRNA algorithm   106
flavonoid   196
– metabolism   94
fluorescent cell sorting   205
Fourier transform ion cyclotron mass resonance spectrometry (FTMS)   204
– transform ion cyclotron resonance (FT-ICR), mass analyzer   148
freezing, enhanced sensitivity   56
functional genomics   197

G1/early S-phase   79
GA 20 oxidase   78
GC/MS   205
GC-EI-MS, response   179
GC-MS profiling   185
GC-MS profiling, limitation   190
GCxGC   203
GDP-D-mannose   55
GDP-L-galactose   55
gene ontology (GO)   92
general proteomics standards (GPS)   288
GeneSpring   63
genetic interaction network   332
– regulatory network   332
Genevestigator   299
genotype, QTL mapping   24
germplasm   40
gibberellic acid (GA)   75
glucocorticoid receptor   89
glucosinolate   196
glutamate   225
glycolysis   219
GO categories, overrepresented   284
Golm Metabolome Database (GMD)   185
graphical model   337

$H_2^{18}O$ 131
histone   80
homoduplex   41
Human Proteome Organization (HUPO)   288
hydroxybenzoic acid   196
hydroxycinnamic acid   196

Index 355

independent component analysis (ICA)   316, 317
INference of cell SIGnaling HypoTheseS (INSIGHT)   299
interactome   248
internal standards (IS)   176
interval mapping   26
isoelectric focusing (IEF)   116
isoflavonoid   196
isopentenyl diphosphate (IPP)   231
isoprenoid biosynthesis   342
isothiocyanate   196
isotopic tracer   214

kernel canonical correlation analysis   320
kinesin   80, 81
kinetic model   8
– parameter   5
KNOX   75

labeling, $^{14}N/^{15}N$ 121
–, ^{18}O 129
–, cleavable ICAT   124
–, ICAT   123
–, iTRAQ   124, 127
–, Leu-d3   122
–, metabolic   121
–, SILAC   122
–, VICAT   124
laser microdissection   205
LC/MS   198
LCxLC   203
LEAFY (LFY)   90
L-galactose   55, 56
L-gulono-1,4-lactone oxidase   56
lignan   196
lignin   196
Lowess smoothing   65

malic enzyme   227
mannose   55
marker loci   24
mass analyzer, different types of   145
– fragment, selective   180
– fragment, specific   180
– isotopomer   178
– resolution   203
– spectral tag (MST)   180
– spectral tag, identification   181

matrix assisted laser desorption ionization (MALDI), mass spectrometric ionization   143
MEDEA (MEA)   89
Medicago   199, 201, 202, 207
metabolic control analysis (MCA)   6
– flux   11, 213
– network   332
– profiling   195
metabolite identification   180
– identifier   185
– instability   191
metabolome   249
– studies   174
metabolomics   172, 197, 204, 310
methyl jasmonate   199
mevalonate pathway   231
mevalonate-independent pathway   231
microarray   81
MIRFINDER   104
MiRscan   104
miRseeker   104
mis-match (MM) fragment   70
missing value estimation   317
mitogen activated protein kinase (MAPK) cascade   79
model plant   35
– solving   230
– species   198
modeling   217
molecular evidence, in confirming miRNA existence   107
– marker   23
– network   4
monoterpene   233
M-phase   79
MPSS (massively parallel signature sequencing)   103
MS signal intensity, ion intensity   133
– signal intensity, protein abundance indices (PAI)   133, 134
multidimensional chromatographic method   203
multiple testing   323
mutagenesis   35
mutant collection   35
mutation detection   40
mzXML   280, 289

nano-HPLC, peptide separation for protein identification   158
natural product   195
noise, analysis of   13
non-protein amino acid   196
normalisation   64, 316
NPR1   81
numerical transformation   179

^{18}O enriched water (H$_2$^{18}O)   131
ontology, Functional Genomics Ontology (FuGO)   280, 289
–, Gene Ontology (GO) Consortium   282
–, Mass Spectrometry Standards (PSI-MS)   288
–, MGED ontology   287
–, Molecular Interaction Standards (PSI-MI)   288
–, Open Biomedical Ontologies (OBO)   282
–, Plant Ontology (PO)   282
–, Proteomics Standards Initiatives (PSI)   288
Orbitrap mass analyzer   147, 204
oxidative pentose phosphate pathway   219
2-oxoacid-dependant dioxygenase (2ODD)   75

parallel targeted profiling   197
patellin   81
pathogenesis resistance (PR) protein   74
peak capacity   200
Pearson correlation   318
PEPC/PK flux   226
peptidase   129
peptide mass fingerprinting (PMF), mass spectrometric protein identification method   149
perfect match (PM) oligonucleotide   70
periodic expression   93
phenanthrene   196
phenocopies, metabolic   175
phenol   196
phenolics   196
phenotype, metabolic   175
–, silent   175
phenylpropanoid   196
PHERES1 (PHE1)   90

phosphoenolpyruvate carboxylase (PEPC)   226
phosphorylation, of proteins   159
photosynthesis   56
phylogenic shadowing   104
polyacetylene   196
polycomb group (PcG)   88
polyine   196
polyketide   196
polymorphism   32
polysaccharide   56
posttranslational modification (PTM), of proteins   159
principal component analysis (PCA)   316, 317
processing, data   178
profiling, constraints   172
–, potential   172
–, technology   172
programmed cell death (PCD)   55
protein interaction network   332
protein-protein interaction   310

quadrupole ion trap mass spectrometer (LC-QITMS)   200
Quantitative Trait Loci (QTL)   22, 26

5' RACE, in validation of miRNA targets   108
redox regulation   81
RefSeq initiative   279
Resource Description Framework (RDF)   193
respiration   56
retention time index (RI)   176
–, ion current   176
–, quantity   176
retrobiosynthetic analysis   217, 230
RITS   102
RNA dependant RNA polymerase   102
RNAi   96
RNA-induced silencing complex (RISC)   100
Robust Multichip Average (RMA)   71

saponin biosynthesis   199
secondary metabolite   195
senescence   55

separation efficiency 200
sequence of metabolites 215
454 sequencing 103
serine protease, trypsin 130
short-term labeling 214, 226
silicon-cell model 2
small RNAs 100, 103, 108, 109
– secreted protein 95
software, ASAPRatio 301
–, EXPRESS 125
–, MSQuant 132
–, Peptide Prophet 290
–, Protein Prophet 290
–, XPRESS 301
Spearman rank correlation 318, 319
species-specific miRNAs, identification of 106
srnaloop 104
stable isotope 178
standardization of data 280
Stanford Universities cDNA microarray 76
steady-state labeling 216, 226
steroidal and triterepenoid saponin 196
stilbenes 196
substrate of respiration 224
sucrose 217
sugar sensing mutant 56
synthetic peptides, AQUA 133, 134
Systems Biology Experiment Analysis Management System (SBEAMS) 300
– Biology Markup Language (SBML) 292

Targeting Induced Local Lesions IN Genomes (TILLING) 35, 96
taxol 196
terpenoid 196, 230
thiosulfinate 196
time-of-flight 203
top-down systems biology 1, 7
trans-acting siRNAs (ta-siRNAs) 103
transcript level 310
transcription factor 75
transcriptome 57, 249
transitive RNA interference 102
tricarboxylic acid cycle (TCA cycle) 223
trichome 205
triterpene saponin 205
tubulin 80

ultra-performance liquid chromatography mass spectrometry (UPLC-MS) 200
UPLC-QtoFM base-peak ion chromatogram 202
UV-B irradiation 56

variation filter 92
Virtual Plant project 302
vitamin C 55
vtc1 mutant 56

wheat 62
wrapper 293

xanthone 196

zeaxanthin 56
zoom IPG strip 117

The EXS-Series
Experientia Supplementa

Experientia Supplementa (EXS) is a multidisciplinary book series originally created as supplement to the journal *Experientia* which appears now under the cover of *Cellular and Molecular Life Sciences*. The multi-authored volumes focus on selected topics of biological or biomedical research, discussing current methodologies, technological innovations, novel tools and applications, new developments and recent findings.

The series is a valuable source of information not only for scientists and graduate students in medical, pharmacological and biological research, but also for physicians as well as practitioners in industry.

Published volumes:

Proteomics in Functional Genomics, EXS 88, P. Jollès, H. Jörnvall (Editors), 2000
New Approaches to Drug Development, EXS 89, P. Jollès (Editor), 2000
The Carbonic Anhydrases, EXS 90, Y.H. Edwards, W.R. Chegwidden, N.D. Carter (Editors), 2000
Genes and Mechanisms in Vertebrate Sex Determination, EXS 91, G. Scherer, M. Schmid (Editors), 2001
Molecular Systematics and Evolution: Theory and Practice, EXS 92, R. DeSalle, G. Giribet, W. Wheeler (Editors), 2002
Modern Methods of Drug Discovery, EXS 93, A. Hillisch, R. Hilgenfeld (Editors), 2003
Mechanisms of Angiogenesis, EXS 94, M. Clauss, G. Breier (Editors), 2004
NPY Family of Peptides in Neurobiology, Cardiovascular and Metabolic Disorders: from Genes to Therapeutics, EXS 95, Z. Zukowska, G.Z. Feuerstein (Editors), 2006
Cancer: Cell Structures, Carcinogens and Genomic Instability, EXS 96, L.P. Bignold (Editor), 2006